T0141898

Transactions on Computer Systems and Networks

Series Editor

Amlan Chakrabarti, Director and Professor, A. K. Choudhury School of Information Technology, Kolkata, West Bengal, India

Transactions on Computer Systems and Networks is a unique series that aims to capture advances in evolution of computer hardware and software systems and progress in computer networks. Computing Systems in present world span from miniature IoT nodes and embedded computing systems to large-scale cloud infrastructures, which necessitates developing systems architecture, storage infrastructure and process management to work at various scales. Present day networking technologies provide pervasive global coverage on a scale and enable multitude of transformative technologies. The new landscape of computing comprises of self-aware autonomous systems, which are built upon a software-hardware collaborative framework. These systems are designed to execute critical and non-critical tasks involving a variety of processing resources like multi-core CPUs, reconfigurable hardware, GPUs and TPUs which are managed through virtualisation, real-time process management and fault-tolerance. While AI, Machine Learning and Deep Learning tasks are predominantly increasing in the application space the computing system research aim towards efficient means of data processing, memory management, real-time task scheduling, scalable, secured and energy aware computing. The paradigm of computer networks also extends it support to this evolving application scenario through various advanced protocols, architectures and services. This series aims to present leading works on advances in theory, design, behaviour and applications in computing systems and networks. The Series accepts research monographs, introductory and advanced textbooks, professional books, reference works, and select conference proceedings.

Saurabh Mani Tripathi ·
Francisco M. Gonzalez-Longatt
Editors

Real-Time Simulation and Hardware-in-the-Loop Testing Using Typhoon HIL

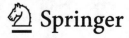 Springer

Editors
Saurabh Mani Tripathi
Department of Electrical Engineering
Kamla Nehru Institute of Technology
Sultanpur, India

Francisco M. Gonzalez-Longatt
Department of Engineering
Centre for Smart Grid
University of Exeter
Exeter, UK

Department of Electrical Engineering,
Information Technology and Cybernetics
University of South-Eastern Norway
Notodden, Norway

ISSN 2730-7484   ISSN 2730-7492   (electronic)
Transactions on Computer Systems and Networks
ISBN 978-981-99-0226-2   ISBN 978-981-99-0224-8   (eBook)
https://doi.org/10.1007/978-981-99-0224-8

This Springer imprint is published by the registered company Springer Nature Singapore Pte Ltd.
The registered company address is: 152 Beach Road, #21-01/04 Gateway East, Singapore 189721,
Singapore

Foreword

The story of Typhoon HIL started in 2008, when motivated by a real-life challenge to completely change the way power electronics control systems are designed and tested, Typhoon HIL's founding team embarked on a quest to bring *Hardware in the Loop* (HIL) testing for power electronics and motor drives to reality. A couple of years later, we succeeded in demonstrating the first of kind 1 μs time step real-time HIL simulation for power electronics that set the record for the smallest time step HIL simulation by two orders of magnitude. More importantly, we proved that ultra-high fidelity HIL for power electronics is possible, and for the first time, there was a clear path towards a fully integrated *model-based design workflow* where design, testing, and system integration can be done using the same models, from start to finish, all in one unified workflow.

At the time there were many good off-line simulation software for the design and simulation of power converters. Also, there were good solutions for converter testing and validation. But the gap was HIL testing to connect model-based design with model-based testing. Indeed, it was HIL that unlocked wider acceptance of integrated and unified *Model-Based Engineering* (MBE) as the key process for the design, testing, and lifecycle maintenance of power electronics control systems. And as with any new innovation that enables radical departure from a widely accepted practice of the time, it took more than a decade until HIL became adopted as a cornerstone of control design and testing thus replacing high-power lab testing of controllers.

From day one, our vision was to develop integrated and easy-to-use MBE tools to empower control engineers developing power electronics controls, to elevate the design processes, empower teams to collaborate, thus unleashing creative energy, and radically accelerating the pace of innovation. On the one hand, we understood the importance of power electronics, and on the other hand, we were unsatisfied with the status quo in terms of design and testing tools. This nexus was further amplified by the imminent need to accelerate the transformation of our civilization's energy systems toward a 100% clean and sustainable energy future and the need to have better testing, design and integration tools. Indeed, in the past 10+ years, power electronics converters and their applications have been driving renewable integration, energy storage proliferation, and electrification of transportation.

This first issue of the *Real-time Simulation and Hardware in the Loop Testing* serves as an impressive demonstration of both the breadth and depth of power electronics applications with particular emphasis on control design, testing, and system-level integration challenges. We believe that this book, with in-depth treatment of grid forming inverter control design, advanced protection concepts, interoperability challenges between protection and DERs, and microgrid and distribution system controls, will empower and motivate the community to push the boundaries of control design and optimization both in terms of performance, robustness, and interoperability. Furthermore, it paves the way for further exploration of machine learning and AI applications to the problem of control and coordination of the future truly cyber-physical grid.

We want to express our deepest gratitude and appreciation to all the authors for their deep, technically rigorous, and original contributions. We have been deeply inspired and genuinely amazed by the authors' creative ways of using Hardware in the Loop and real-time simulation. This has energized us to continue improving our software and hardware tools and to continue co-creating with authors and the research and development community at large.

This book will not only help the users and power electronics and power systems communities, but it will also motivate all the developers of *Model-Based Engineering* (MBE) tools for power to continue improving their solutions and processes that will further accelerate our civilization's transition to a clean energy future.

September 2022

Ivan Celanovic
Co-founder and President
Typhoon HIL, Inc
Newton, MA, USA

Preface

Real-time simulation is a concept that refers to the execution of a computer-based model of a physical system by matching simulation time with actual "wall clock" time to replicate the fundamental behaviour of a physical system. Nowadays, it is increasingly used by industry and academia. Real-time simulation tools have become the most important tool to validate conventional/proven and unconventional/unproven design approaches for complex energy systems. For the purpose of de-risking equipment in complex electrical systems and to provide evidence of proper functionality under a wide range of realistic dynamic conditions—safely, repeatedly and economically; the newer design approaches need to be tested stringently through in-depth simulation before deployment at large scale.

Hardware-in-the-loop (HIL) testing framework is an approach to combining real-time simulation and hardware experimentation using signal interfaces between hardware devices and real-time computational systems, allowing rigorous testing of new design approaches for electrical systems at the required complexity. In the HIL test framework, the actual controllers, as devices under test, are usually connected in a closed loop with the simulated power stage running on the real-time simulator.

The purpose of real-time simulation/HIL testing is to provide unbeatable evidence of acceptable electrical system performance (during normal, abnormal and degraded conditions) in accordance with the given functional requirements. Typhoon HIL has a successful trajectory by developing a powerful framework in the rapidly growing field of ultra-high-fidelity controller-hardware-in-the-loop (C-HIL) simulations for power electronics, micro-grids and distribution networks.

This book is an edited collection that explores the fundamental concepts of real-time simulation/hardware-in-the-loop testing using "Typhoon HIL" for complex electrical systems. This book integrates the coverage of underlying theory and acclaimed methodological approaches as well as high-value applications of real-time simulation and HIL testing—all from the perspectives of eminent researchers around the globe utilising Typhoon HIL.

Chapter 1 highlights the critical technical aspects behind the Typhoon HIL toolchain. In addition, an example of a simple C-HIL test setup featuring an actual controller has been demonstrated to illustrate how a simple test environment can be quickly built and parameterised. Chapter 2 throws light upon the importance of power electronic converters, their modelling and control functionalities in grid-connected systems. A case study for a single-phase grid-connected PV inverter simulation using the Typhoon HIL 402 device is also presented.

Chapter 3 presents grid forming control techniques for power electronic converters as a solution to low rotational inertia systems; the controllers have been implemented and tested using the modelling and real-time simulation framework of Typhoon HIL. Chapter 4 presents various model predictive control (MPC) strategies for grid-connected converters and describes their implementation, testing and validation using the Typhoon HIL platform.

An optimal programmed pulse width modulation (PWM) strategy coordinated with the virtual synchronous machine (VSM) concept for grid-connected multilevel converters in accordance with the current harmonic content limits of the IEEE 1547 standard has been proposed in Chap. 5. In addition, the real-time operation of a grid-connected three-phase neutral point clamped converter was carried out in Typhoon HIL 402 to demonstrate the performance of the proposed approach. A non-linear predictive current control scheme for a single-phase shunt active power filter (SAPF) using selective harmonic compensation has been presented in Chap. 6. The proposed non-linear predictive current control scheme has been implemented and tested using the Typhoon HIL 402 device.

A practical experimental setup for the development and controller testing of xEV applications has been proposed in Chap. 7 by demonstrating a C-HIL setup for field-oriented control (FOC) of a permanent magnet synchronous motor (PMSM) using the Typhoon HIL 602+ simulator and AURIX^{TM} microcontroller. Chapter 8 provides a theoretical and conceptual introduction to electric vehicle digital twins that can be used as a platform for research and development of electric vehicle pertinent technologies. In Chap. 9, the authors derive large and small signal models for primary controllers and demonstrate the effect of primary controller parameters on steady-state and transient behaviour by showing the performance of time domain simulations on the HIL. In Chap. 10, the authors have used Typhoon HIL real-time simulation platform for modelling a reconfigured IEEE-33 bus distribution system to assess the effect of diverse harmonic order frequency on network parameters and its subsequent impact on the hosting capacity of the network.

The authors of Chap. 11 systematically introduce modelling and simulation for testing purposes of non-directional over-current protection relays in Virtual HIL (VHIL) to help power engineers evaluate protective relay settings under more realistic conditions. In Chap. 12, the 8-bus transmission system is implemented under a soft real-time simulation platform that allows the determination of the sequence of operation, fault current detection capability and operating time of directional over-current relays under solid three-phase to ground fault conditions. In Chap. 13, a VHIL platform for line protection with distance protection relay has been developed and tested using a real-time HIL simulation validated by theory-based calculation and DIgSILENT PowerFactory software with three-phase and single-line-to-ground short-circuit fault cases. Finally, Chap. 14 proposes and demonstrates a cyber-physical co-simulation framework between Typhoon HIL and OpenDSS to solve the problem of modelling complex distribution networks.

The editors hope this book will cater to understanding and familiarity with the real-time simulation of complex electrical systems, specifically focusing on HIL modelling, simulation and testing. It will also make the readers conversant for real-time validation of the unconventional/unproven design approaches for power systems and power electronics applications using "Typhoon HIL".

Sultanpur, India   Saurabh Mani Tripathi
Exeter, UK/Notodden, Norway   Francisco M. Gonzalez-Longatt

Acknowledgments The editors are grateful to all the authors for their valuable contribution to this edited book. The editors also thank all the reviewers who have generously given their time to review the chapter manuscripts. The editors thank the CEO and co-founder of Typhoon HIL, GmbH, for granting the necessary permission, and the staff of Springer Nature for their continued support throughout the press production process of this edited book.

Contents

Editors and Contributors

About the Editors

Saurabh Mani Tripathi is currently an Associate Professor in electrical engineering and Founder & Coordinator of the Power & Energy Research Centre (Centre of Excellence) at Kamla Nehru Institute of Technology, Sultanpur, India. He is a recipient of the prestigious "IEI Young Engineers Award 2018–19" presented by the Institution of Engineers (India). He is also a recipient of the "Best Teachers Award 2020" presented by Dr. A.P.J. Abdul Kalam Technical University, Lucknow, India. He authored and edited several books and has published numerous research papers in various journals and conferences as well. He is a member of several professional societies such as the Institution of Engineers (India) and the International Association of Engineers. He is a Life Member of the Indian Society for Technical Education. His areas of current interest include renewable energy systems, electrical drives, real-time simulation, and hardware-in-the-loop testing.

 **Francisco M. Gonzalez-Longatt** is currently a Professor in Electrical Energy Systems at the Centre for Smart Grid, University of Exeter, Founder & Leader of the DIgEnSys-Lab (Digital Energy Systems Laboratory) at the University of South-Eastern Norway, Norway, where he is an honorary full professor in Electrical Power Systems. He has been involved in several industrial research projects and consultancy worldwide. In addition, he is the author/editor of numerous books (Spanish and English) and a member of the editorial board of several leading journals. His primary research interest is digital technologies as an enabler of net-zero energy systems and smart grids.

Contributors

Mohammad Amin Department of Electric Power Engineering, NTNU, Trondheim, Norway

Fernanda Carnielutti Federal University of Santa Maria - UFSM, Santa Maria, Brazil

Sergio Costa Typhoon HIL, Novi Sad, Serbia

Paulo Manuel De Oliveira De Jesús Department of Electrical and Electronical Engineering, University of the Andes, South America, Colombia

Marko Gecic Infineon Technologies AG, Neubiberg, Germany

Adrien Genic Typhoon HIL, Novi Sad, Serbia

Debomita Ghosh Birla Institute of Technology, Mesra, Jharkhand, India

Francisco Gonzalez-Longatt University of South-Eastern Norway, Porsgrunn, Norway;
Centre of Smart Grid, University of Exeter, Exeter, UK

Felipe B. Grigoletto Federal University of Pampa - UNIPAMPA, Alegrete, RS, Brazil

Rajesh Gupta Electrical Engineering Department, Motilal Nehru National Institute of Technology Allahabad, Prayagraj, Uttar Pradesh, India

Ahteshamul Haque Advance Power Electronics Research Lab, Department of Electrical Engineering, Jamia Millia Islamia (A Central University), New Delhi, India

Ivana Isakov Faculty of Technical Sciences, University of Novi Sad, Novi Sad, Serbia

Swati Kumari Birla Institute of Technology, Mesra, Jharkhand, India

Daniel M. Lima Federal University of Santa Catarina - UFSC, Blumenau, Brazil

Luiz A. Maccari Jr. Federal University of Santa Catarina - UFSC, Blumenau, Brazil

Henrique Magnago UFSM, Santa Maria, Brazil

Azra Malik Advance Power Electronics Research Lab, Department of Electrical Engineering, Jamia Millia Islamia (A Central University), New Delhi, India

Jorge R. Massing UFSM, Santa Maria, Brazil

Hugo Mendonça Escuela Técnica Superior de Ingenieros Industriales, Universidad Politécnica de Madrid, Madrid, Spain

Vinícius F. Montagner Federal University of Santa Maria - UFSM, Santa Maria, Brazil

Martha Nohemi Acosta Montalvo University of South-Eastern Norway, Porsgrunn, Norway

Sandeep Ojha Electrical Engineering Department, Motilal Nehru National Institute of Technology Allahabad, Prayagraj, Uttar Pradesh, India

Felipe Antonio Gómez Olaya Department of Electrical and Electronical Engineering, University of the Andes, South America, Colombia

Caio R. D. Osório Typhoon HIL, Novi Sad, Serbia

Alexandre T. Pereira UFSM, Av. Roraima, Santa Maria, Brazil

Le Nam Hai Pham University of South-Eastern Norway, Porsgrunn, Norway

Humberto Pinheiro Federal University of Santa Maria - UFSM, Santa Maria, Brazil

Jose Miguel Riquelme-Dominguez Escuela Técnica Superior de Ingenieros Industriales, Universidad Politécnica de Madrid, Madrid, Spain

Sourav Kumar Sahu Birla Institute of Technology, Mesra, Jharkhand, India

Juan David Hernández Santafé Department of Electrical and Electronical Engineering, University of the Andes, South America, Colombia

Dimas A. Schuetz Federal University of Santa Maria - UFSM, Santa Maria, Brazil

Pawan Sharma Department of Electrical Engineering, UiT The Arctic University of Norway, Narvik, Norway

Márcio Stefanello UNIPAMPA, Av. Tiaraju, Alegrete, Brazil

Jonas R. Tibola Federal University of Santa Maria - UFSM, Santa Maria, Brazil

Ivan Todorović Faculty of Technical Sciences, University of Novi Sad, Novi Sad, Serbia

Raju Wagle Department of Electrical Engineering, UiT The Arctic University of Norway, Narvik, Norway

Abbreviations

ANN	Artificial neural network
ANSI	American national standards institute
APF	Active power filter
API	Application programming interface
ATOM	ARU-connected timer output module
BA	Bee algorithms
BMS	Battery management systems
CB-PWM	Carrier-based pulse-width modulation
CC	Control code
CCS-MPC	Continuous control set model predictive control
C-HIL	Controller hardware-in-the-loop
CMIL	Controller-model-in-the-loop
CPU	Central processing unit
CT	Current transformers
DAB	Dual-active bridge
DE	Differential evolution
DER	Distributed energy resource
DFT	Discrete Fourier transform
DI	Digital input
DLF	Distribution load flow
DOCP	Directional overcurrent protections
DOCR	Directional overcurrent relays
DSADC	Delta-sigma analog-to-digital converter
DSG	Digital signal controller
DSO	Distribution system operator
DSP	Digital signal processor
DUT	Device under test
dVOC	Dispatchable virtual oscillator control
ECU	Electronic control unit
EES	Electrical energy storage
EG	Embedded generation

EV	Electric vehicle
FCS-MPC	Finite control set model predictive control
FFT	Fast Fourier transform
FOC	Field-oriented control
FPGA	Field programmable gate array
FRT	Fault ride-through
GA	Genetic algorithm
GCC	Grid-connected converter
GDS	Gate drive signals
GTM	General timer module
GUI	Graphical user interface
HC	Hosting capacity
HEV	Hybrid electric vehicles
HIL	Hardware-in-the-loop
HLF	Harmonic load flow
HSM	Hardware security module
ICE	Internal combustion engine
IDE	Integrated development environment
IEEE	Institute of electrical and electronics engineers
iLLD	Infineon low level drivers
IM	Induction motors
IOs	Inputs/outputs
LCT	Low-carbon technologies
LUT	Look up table
LVRT	Low voltage ride through
MBSE	Model-based systems engineering
MGCS	Microgrid control system
MIL	Model-in-the-loop
MPC	Model predictive control
MTU	Memory test unit
NPC	Neutral-point clamped
NR	Newton-Raphson
OBC	On-board chargers
OC	Operational Condition
OpenDSS	Open distribution system simulator
OP-PWM	Optimal programmed pulse-width modulation
PC	Personal computer
PCC	Point of common coupling
PCC	Predictive current control
PEC	Power electronic converter
PFC	Power factor correction
P-HIL	Power hardware-in-the-loop
PIL	Processor-in-the-loop
PLL	Phase-locked loop
PMSM	Permanent magnet synchronous motor

PoC	Point of connection
PP-PWM	Pre-programmed pulse-width modulation
PR	Proportional resonant
PS	Power stage
PSO	Particle swarm optimization
PV	Photovoltaic
PVDG	Photovoltaic distributed generation
PWM	Pulse-width modulation
R&D	Research and development
RCA	Relay characteristic angle
RMS	Root-mean square
ROCOF	Rate of change of frequency
RTDS	Real-time digital simulator
RTS	Real-time simulator
SAPFs	Shunt active power filters
SCADA	Supervisory control and data acquisition
SCH	Selective current harmonic
SCR	Short circuit ratio
SG	Synchronous generator
SHE	Selective harmonic elimination
SHE-PWM	Selective harmonic elimination pulse-width modulation
SIL	Software-in-the-loop
SISO	Single-input-single-output
SMU	Safety management unit
SPC	Standard processing core
SPE	Sensor pattern evaluation
SPV	Solar photovoltaic
SRFPI	Synchronous reference frame proportional-integral
SS	State space
SVM	Space-vector modulation
SVM^2PC	Space-vector modulated model predictive control
SV-PWM	Space vector pulse-width modulation
SynC	Synchronverter
TDD	Test-driven design
THD	Total harmonic distortion
TIM	Timer input module
TLM	Transmission line model
TOM	Timer output module
TSO	Transmission system operator
VADC	Versatile analog-to-digital converter
VHDL	VHSIC hardware description language
VHIL	Virtual hardware-in-the-Loop
VHSIC	Very high speed integrated circuit
VOC	Virtual oscillator control
VSC	Voltage source converter

VSG	Virtual synchronous generator
VSM	Virtual synchronous machine
xEV	Any kind of vehicle that utilizes electric motor traction
ZOH	Zero-order hold

Chapter 1
Introduction to Typhoon HIL: Technology, Functionalities, and Applications

Caio R. D. Osório, Adrien Genic, and Sergio Costa

Abstract This first chapter provides an introduction to the hardware-in-the-loop (HIL) approach and Typhoon HIL, in particular, including a brief overview of its history, achievements, and vision. Real-time simulation challenges are introduced. Throughout the chapter, key technological aspects and functionalities behind the Typhoon HIL toolchain are discussed, highlighting how this seamlessly integrated solution enables the creation of high-fidelity models for hardware-in-the-loop-based real-time simulations and performs automated tests for dynamic and complex systems that go from single high switching frequency power electronics converters to larger microgrid systems.

Keywords Control validation · Hardware-in-the-loop · High-fidelity · Model-based testing · Real-time simulation · Typhoon HIL

1.1 Introduction

Power-electronics-based technologies are in continuous and accelerated development, leading to a significant cost reduction and increased reliability in different components and devices in the past decades. Motivated by the need to digitize, decarbonize, and decentralize electric energy systems, these advancements enabled global transformations in the energy and electrical power industries. For instance, modern power systems have evolved from a centralized generation framework with unidirectional power flow to dynamic and complex smart grids, characterized by a high penetration of distributed, intermittent renewable energy sources, energy storage systems, smart relays, and the possibility of consumers also acting as producers (e.g., prosumers). Paradigm shifts are also present in other power electronics applications, such as the growing market share of electric vehicles in the automotive industry; the burgeoning interest in more electric and environmentally friendly shipboard

C. R. D. Osório (✉) · A. Genic · S. Costa
Typhoon HIL, Novi Sad, Serbia
e-mail: caio.osorio@typhoon-hil.com

© The Author(s), under exclusive license to Springer Nature Singapore Pte Ltd. 2023
S. M. Tripathi and F. M. Gonzalez-Longatt (eds.), *Real-Time Simulation and Hardware-in-the-Loop Testing Using Typhoon HIL*, Transactions on Computer Systems and Networks, https://doi.org/10.1007/978-981-99-0224-8_1

1

and airplane power systems in the marine and aerospace industries; and the advancement in high-efficiency and low-cost electric motors, electric drives, and powertrains (Liserre et al. 2010; Osório et al. 2021; Chemali et al. 2016; Rommel 2019; Xu et al. 2021).

As a common point in these applications, one can look to the presence of highly dynamical switching converters that, besides its own complexity, include additional features such as intricate digital control, protection capabilities, and advanced communication systems. As a consequence, a major engineering challenge has been to be able to design, implement, and validate, in a timely manner, high-quality and economically viable solutions that comply with multiple development and operational requirements, such as electric vehicle integration standards and grid codes (Osório 2020; Knezović et al. 2017).

In this direction, as the intricacy of controlling power electronics, microgrids, and power systems rise, the ability to reduce development time and costs is a key trait. It is not efficient to wait until advanced stages of a project to carry out tests or to wait for prototypes to be built in order to manually verify the integration of different hardware and software, as well as to assess the performance of the overall system. If this strategy is adopted, it can significantly prolong development time and cost, in addition to limiting testing flexibility due to hardware constraints and safety precautions (Dinavahi et al. 2001; Vekić et al. 2012; Khan et al. 2017).

To overcome that, and to increase the efficiency of engineering processes, hardware-in-the-loop (HIL) simulations have been increasingly used by industry and academia. In this testing framework, the device under test can be directly connected to a real-time simulation, enabling efficient closed-loop, model-based automated testing. HIL proved to be reliable and comprehensive in accelerating the development cycle by allowing testing to start early in the development process, all while improving flexibility, coverage, and security in the verification and validation process. As a testament to that, HIL tools have been used by the automotive and aerospace industries for decades, and have proven effective for testing and pre-commissioning of microgrids, shipboard power systems, validation of energy storage systems, motor drives, and other applications (Genic et al. 2017; Salcedo et al. 2019; Jonke et al. 2016; Zelic et al. 2020; Abdelrahman et al. 2018; Amin et al. 2019; Badini and Verma 2019).

Since its foundation in 2008, Typhoon HIL has supported industry and academia by providing high-fidelity hardware-in-the-loop real-time emulators for electrical systems, with continuous development driven by extensive user feedback. By means of vertically integrated test solutions, Typhoon HIL enables model-based development, test-driven design and the development of digital twin models to assess the technical feasibility of complex systems from the early stages of development all the way to pre-certification, including verification and validation of controls, protection, the communication layer, system integration, and interoperability testing. Some Typhoon HIL devices and features of the toolchain are illustrated in Fig. 1.1.

For a better understanding of how Typhoon HIL toolchain has been recognized as a powerful solution for real-time hardware-in-the-loop simulation in different applications, throughout this chapter, technical details about the technology, methodology,

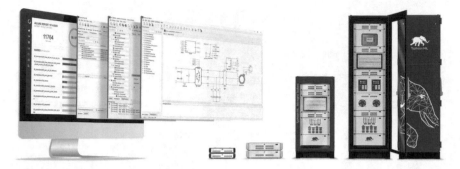

Fig. 1.1 Typhoon HIL testing solution

and functionalities are presented. It is also worth mentioning that several tutorials, videos, and knowledge-based articles are available online, detailing the features presented in this chapter (Typhoon HIL 2023, a, b).

1.2 Model-Based System Engineering and HIL Testing

For a long time, control system testing was done manually, relying on small-scale or large full-scale power hardware. In traditional development and validation cycles, such as those following the V-model, these tests would often occur only in the verification and commissioning stage after a physical prototype has been developed. In order to meet cost, time, and quality requirements, model-based systems engineering (MBSE) has emerged as a powerful methodology. In this framework, physical systems and prototypes can be replaced by virtual models, which enable the execution of exhaustive simulations in a safe and flexible environment, saving time, and reducing costs from the specification to the commissioning and maintenance phase. A graphical representation of how MBSE can be applied to support different steps of the development cycle is shown in Fig. 1.2.

Depending on the specifications, level of abstraction, application, and device under test, different testing setups can be considered, as illustrated in Fig. 1.3. These approaches include model-in-the-loop (MIL), software-in-the-loop (SIL), controller hardware-in-the-loop (C-HIL), and power hardware-in-the-loop (P-HIL).

In the MIL and SIL approaches, both control and power stages (i.e., controller and plant) are simulated in a virtual environment (V-HIL), generally not requiring real-time execution. In the MIL approach, the controller is modeled together with the power layer, while in the SIL approach, the actual control software is considered in the simulation.

Testing setups that feature a mix of physical systems and virtual models are collectively referred to as hardware-in-the-loop (HIL). This means that some physical

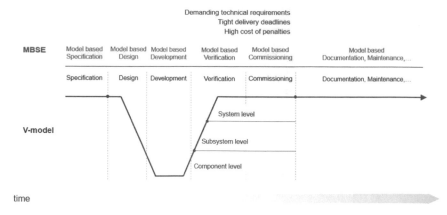

Fig. 1.2 V-model: graphical representation of MBSE applied in different steps of the development cycle

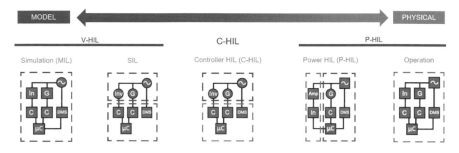

Fig. 1.3 Common methodologies for model-based systems engineering

part of the system is connected to the real-time simulation, which could be part of the power hardware (P-HIL) or part of the controller hardware (C-HIL).

In the P-HIL approach, the focus is on testing power hardware. Therefore, power amplifiers can be used to link the real-time simulator to the actual power hardware device under test via analog input/output signals or communication protocols. For instance, current and voltage references can be sent from the real-time simulation to the amplifier using analog outputs, while the feedback signals of the device under test are sensed by the power amplifier, and then sent to the simulation using the real-time simulator analog inputs. As a drawback, although testing software in the presence of actual power provides very accurate results, it usually involves higher cost, lower flexibility, and the need for additional safety precautions.

On the other hand, the C-HIL approach stands out as an effective solution for testing controllers, combining high fidelity, reduced cost, high testing flexibility, and a safe environment. That is possible thanks to the advancements in computing technologies, such as discussed in Sect. 1.4, which enable the development of real-time simulation devices capable of emulating a device's power stage (physical layer)

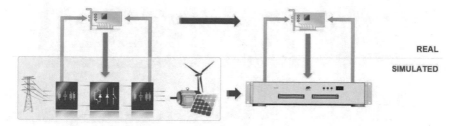

Fig. 1.4 Concept of controller hardware-in-the-loop

with high precision, even considering demanding high switching power electronics applications.

With the C-HIL approach, the actual controller under test can be directly connected to the modeled plant in real-time simulations, allowing for closed-loop evaluation even before a prototype of the plant is available, as illustrated in Fig. 1.4. This allows for verification and validation of control hardware, software, and firmware in a realistic environment, which provides more flexibility and security than fully physical prototypes, as well as higher fidelity when compared to fully simulated environments. C-HIL also enables engineering teams to automate test cases and to perform several evaluations effortlessly, allowing discovery of performance and integration issues as soon as they arise, iteratively improving the performance of the system being developed. In a similar manner, once experimental prototypes or the actual plant are running, digital twins can be built and the C-HIL approach can be used to validate controller continuous improvements; software lifecycle maintenance; quality assurance processes; and to perform tests that can be hard to replicate, dangerous, or potentially destructive to lab equipment.

The Typhoon HIL toolchain supports all aforementioned testing scenarios, with a targeted focus on the C-HIL approach.

1.3 About Typhoon HIL

Typhoon HIL Inc. was founded in 2008 as a startup, thanks in part to the investment provided by Ray Stata, Founder of Analog Devices. Typhoon HIL today is a multinational corporation that is the current technology leader in the rapidly growing field of ultra-high-fidelity controller Hardware-in-the-Loop (C-HIL) technology. The company mission is to "Engineer and promote environmentally sustainable power technologies that scale," with the aim of laying the groundwork for building a sustainable future.

Typhoon HIL serves its customers with custom solutions comprised of fully vertically integrated software and hardware for model-based testing and development of power electronics, e-mobility, microgrids, and distribution networks. Typhoon HIL solutions aim to support its users through the entire span of their product's lifecycle,

Fig. 1.5 Typhoon HIL coverage of different market and application segments

starting from design and development, throughout validation and verification stages driven by automated testing, all the way to integration and maintenance. Engineering services provided by the company help in technology adoption, system bring-up, and scaling, to speed up project progress and success. Since its establishment, the company successfully brought to market a number of HIL products, installing over 1000 HIL systems worldwide in both industry and academia.

The company's primary R&D center in Serbia features a multidisciplinary team of experts in the fields of power electronics, signal electronics, real-time and application-specific software, computer architectures, electricity distribution, protection and control, industrial power system management, integration of distributed energy sources, and communication protocols. As a result, the Typhoon HIL environment has competences to cover multiple applications in fields such as microgrids, drives, e-mobility, battery energy storage systems, marine power systems, and so on (see Fig. 1.5 for illustration).

In addition to the corporate headquarters in Boston, MA and the main R & D center in Serbia, the company has offices in Switzerland, Brazil, Canada, France, and soon Germany. The company also works together with over 20 value-added resellers, distributors, and engineering centers worldwide which facilitate both development and production, as well as successful communication to serve the global market.

1.4   Typhoon HIL Technology

When performing HIL testing, it is imperative that the simulation runs in real time; the elapsed time when running the digital model of a physical system must match exactly with the real-world time, also known as the wall-clock time. In this context, Typhoon HIL devices are high-performance computers designed and built for real-time simulation of power-electronics-based systems. This makes a HIL device an important tool for several applications where the behavior of a device should be tested before prototyping, including model-based control development, test-driven design, pre-commissioning, virtual system integration, and interoperability testing of modern power-electronics-enabled technologies. Simulations run on these devices have also proven useful when acting as a high-fidelity replica, or "digital twin," of a power electronic device or power system, such as for replicating faults encountered by a real device in customer support applications or by creating a sandbox environment for SCADA operator control training in microgrids. But what are the challenges in performing real-time simulations?

Real-Time Simulation Challenges

Real-time simulation of power-electronics-based systems (e.g., microgrids, EV drivetrains, shipboard power systems) is challenging since these applications comprise switching converters that operate at ever-increasing frequencies, especially considering the advancements on the semiconductor devices. Therefore, to be able to simulate switching effects with high accuracy, very short simulation time steps are required, as well as high-resolution sampling of the switch gate drive signals (GDS), advanced processing capability, and ultralow latency. As a consequence, power electronics applications comprise complex and highly dynamic systems that are highly demanding to simulate in real time with high fidelity (Osório et al. 2021; Majstorovic et al. 2011; Pallo et al. 2017).

Real-time simulation devices run in discrete time and typically employ linear solvers with fixed time steps. To encompass the switching dynamics in efficient simulations, a piece-wise linear approach can be used. In this context, power converters can be modeled based on ideal switches, and for every switch permutation, a time-invariant linear state space model, called mode, is defined. A single mode is applied over each simulation time step, and the simulation dynamically changes among modes throughout execution. As an advantage, modeling switches as ideal do not introduce non-physical behavior, as may be the case in simulation approaches where the switches are replaced with simplified equivalents. Moreover, it is possible to pre-compute the system matrices and to store them in the solver memory, during compilation. On the other hand, since theoretically each and every semiconductor can be either conducting or open, the number of modes increases exponentially with the number of switches, thus increasing exponentially the memory capacity required (Osório et al. 2021; Majstorovic et al. 2011).

Another important challenge for real-time simulation of power electronics applications is related to the effective time resolution of the digital inputs used to drive the converters, which are usually pulse-width-modulated (PWM) signals. When an actual controller hardware is being used to generate the GDS, its clock (and therefore the time instant where its outputs are updated) is not synchronized with the simulation clock. In this context, if the sampling period of the PWM signals is equal to the simulation time step, the transitions between on and off states can only be detected at the subsequent sampling, as illustrated in Fig. 1.6. This inaccuracy in identifying the exact instant at which the transitions occur may lead to significant sampling errors, causing imprecise duty cycle detection and, therefore, inaccurate simulation results. When offline simulations are performed, this drawback can be mitigated by using variable step solvers or by reducing the simulation time step as much as necessary to make the sampling errors become negligible. However, this happens at the price of longer execution times, which is not a viable solution for real-time simulations where the model response calculation must be finished within the predefined simulation step (Lian and Lehn 2005).

The challenges described so far focused primarily on the real-time simulation of power converters, where time constants in the order of nanoseconds are required in order to precisely reproduce switching effects and obtain accurate simulation results. On the other hand, when testing, for instance, the secondary or tertiary control layers of power systems such as microgrids, models tend to be large and simulation run times may reach days or weeks, with time constants in the order of minutes or hours. In this sense, it is possible to see that different applications present different requirements, such as high time resolution and long-term stability, which may demand different modeling approaches and processor capabilities, posing a significant challenge. A chart illustrating the wide range of time scales of interest within a microgrid application is illustrated in Fig. 1.7.

In addition to that, as mentioned before, real-time simulations are essential when real elements are present in the loop. As a consequence, the hardware-in-the-loop simulation devices must be robust and present suitable interfaces, allowing easy access to multiple inputs, outputs, and connection with a wide range of possible devices under test, including supporting the specific communication protocols those devices may use. At the same time, real-time simulations and the hardware-in-the-loop testing framework aim to reduce time and costs in the development cycle, and thus must not overwhelm engineers with additional concerns. Therefore, it is important to provide a solution that, although technically advanced, is user-friendly and easy to get used to. In this context, the HIL solution should suit different application-

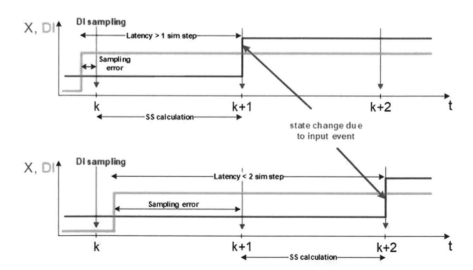

Fig. 1.6 Illustration of state space (SS) calculation and the respective state (X) change due to a digital input (DI) event with sampling period equal to the simulation time step. Sampling error and latency depend on when the DI changes with respect to the simulation time step (Osório et al. 2021; Typhoon HIL 2023c)

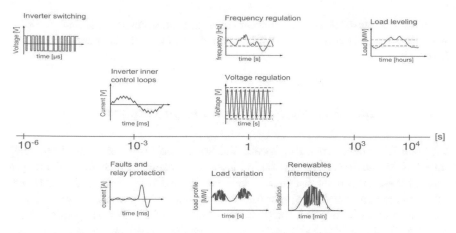

Fig. 1.7 Graph illustrating how the time scale of interest varies in a microgrid application according to the phenomenon to be observed and the objectives of the simulation (Typhoon HIL 2017)

specific systems with easily deployable preset configurations, while still providing flexibility for more experienced HIL users to develop bespoke solutions for custom applications.

Challenges for real-time simulation solutions include the following:

- Achieving very short simulation time steps and low latency to represent highly dynamic power electronics systems with accuracy.
- Reducing memory capacity requirements.
- Improving the effective PWM time resolution.
- Coping with large models and different application-specific requirements.
- Hardware with suitable interfaces and support for industry standard communication protocols.
- User friendliness, flexibility, and easy to get used to.

Typhoon HIL Testing Solutions

Aiming to overcome the challenges mentioned in this section, Typhoon HIL provides a vertically integrated solution, comprising of real-time simulator hardware and a dedicated software toolchain (Typhoon HIL Control Center). The technology stack is seamlessly integrated from Typhoon HIL's application-specific processors and robust numerical solver all the way to the model building interface, supervisory system, and testing automation solution, in a single easy-to-use and affordable toolchain.

In the next subsections, the Typhoon HIL real-time simulator hardware and software technology is presented, as well as how they address the challenges described here.

1.4.1 Typhoon HIL Real-Time Simulation Platform

Typhoon HIL simulators are hardware platforms specialized for high-fidelity real-time HIL simulations of power-electronics-based systems, which are enabled by a state-of-the-art processor design seamlessly integrated with a fully embedded compiler. As mentioned before, proper real-time simulation of power-electronics-based systems requires high-speed, low-latency, scalable, and flexible computation technologies. Typhoon HIL devices achieve that by using a programmable, application-specific, hybrid architecture that combines CPU (central processing unit) and FPGA (field programmable gate array) technologies, seamlessly integrated with the software toolchain.

The current line-up includes two generations of devices. The third-generation devices (HIL402, HIL602+, and HIL604) support simulation steps down to 500 ns, while oversampling digital inputs with 6.5 ns resolution. These devices have proved themselves in numerous industrial applications, even in some cases with switching frequencies exceeding 100 kHz. To further improve simulation fidelity for high switching frequency applications, such as high-speed drives and DC-DC resonant converters, fourth-generation devices (HIL404 and HIL606) support even lower simulation steps, down to 200 ns, with digital input sampling resolution of 3.5 ns. More details about the current device line-up, including the number of processing cores, model capacity, time resolution, number of analog and digital inputs/outputs (IOs), and connectivity support with industry standard protocols can be seen in Fig. 1.8. In addition, it is worth noting that thanks to the modular design, multiple device units can be stacked together and paralleled, behaving as a single larger simulator.

All Typhoon HIL devices share a common multi-processor architecture, which contains a proprietary multi-core FPGA solver, system CPUs, and user CPUs, as illustrated in Fig. 1.9. A summary of their functions is given as follows:

- **Typhoon FPGA Solver:** The multi-core FPGA solver is used to simulate the electrical layer of the model, optimized for time-exact simulation.
- **System CPUs:** General-purpose processors indirectly controlled by the user, typically used to simulate low dynamics phenomena of certain electrical domain components or to handle communication protocol stacks.
- **User CPUs:** General-purpose processors that are under direct user control, responsible for the simulation of model components that don't belong to the electrical domain, such as mechanical, thermal, and signal processing components. User CPUs can also be used for the development of controller algorithms within the model, using MIL and SIL approaches or rapid control prototyping.

HIL Simulators	HIL402	HIL404	HIL602+	HIL604	HIL606
Processing cores (number of cores) Model capacity	Up to 4	Up to 4	Up to 6	Up to 8	Up to 8
Detailed converter models (1ph/3ph)	8 / 4	8 / 4	12 / 6	16 / 8	16 / 8
Average converter models (3ph)	8	12	10	10	24
Distribution network simulation	✔	✔	✔	✔	✔
Time resolution					
Minimum simulation step	500 ns	200 ns	500 ns	500 ns	200 ns
DI sampling resolution	6.2 ns	3.5 ns	6.2 ns	6.2 ns	3.5 ns
I/O					
Analog I/O (per unit)	16/16	16/16	16/32	32/64	32/64
Digital I/O (per unit)	32/32	32/32	32/32	64/64	64/64
Connectivity					
USB	✔	✔	✔	✔	✔
Ethernet	✔	✔	✔	✔	✔
CAN		✔	✔	✔	✔
RS232		✔	✔	✔	✔
EtherCAT					✔
SFP		✔			✔
Time synchronization (PPS and IRIG-B)				✔	✔
Paralleling		Up to 4 units	Up to 4 units	Up to 16 u	Up to 16 u

Fig. 1.8 Current Typhoon HIL devices line-up

The HIL606 device is illustrated in Fig. 1.10, where (a) illustrates the front view, highlighting the analog and digital IOs, and (b) illustrates the back view, highlighting the expanded connectivity options, which increases flexibility for multiple protocols.

1.4.1.1 Multi-Core FPGA

As mentioned in the previous section, real-time emulation imposes a rigid time limitation for the calculation of model responses which is not present in offline simulations. Therefore, to achieve very short simulation time steps, computation of complex models has to be parallelized.

In this context, the Typhoon HIL multi-core FPGA solver is the key technology that enables high-fidelity real-time emulation of power-electronics-based systems. The FPGA is optimized for time-exact simulation, and thanks to its paralleled architecture and low latency connection between the processing elements, it is capable of reducing the memory requirements and of running complex models with time steps down to 200 ns, as shown in Fig. 1.8. Moreover, its processing blocks are tightly integrated with the input/output (IO) stage, ensuring very low and fully predictable loopback latency, which allows seamless interfacing with external controllers.

The main FPGA solver computation elements are depicted within the red box in Fig. 1.9. Their descriptions are given below:

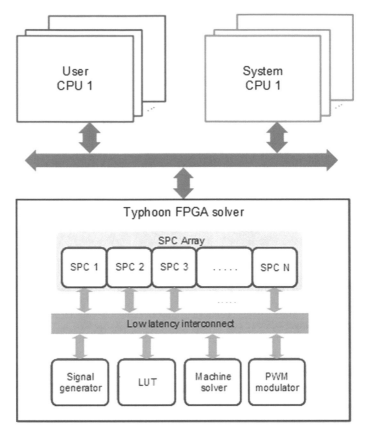

Fig. 1.9 Typhoon HIL device architecture, encompassing: user CPU, System CPU, and Typhoon FPGA solver with multiple Standard Processing Cores (SPCs)

- **Standard Processing Core (SPC):** Responsible for the simulation of electrical circuits consisting of linear passive elements (both constant and time-varying), converter blocks (built with ideal switches), and contactors based on ideal or non-ideal switches. Dedicated communication lines interconnect the different SPC blocks, allowing variable exchange with a single simulation step delay.
- **Signal Generator:** Built-in, runtime tunable block responsible for generating independent voltage and current sources, as well as other arbitrary waveforms synchronously and at the full simulation rate. Linear interpolation is applied if the waveform sample rate is lower than the simulation rate.
- **Look-Up Table (LUT):** Block used to simulate nonlinear elements such as PV panels, batteries, nonlinear passive components, and saturable transformers.
- **Machine Solver:** Each unit can emulate a single electrical machine model including its electromagnetic part, mechanical part, and speed measurement devices such as encoder and resolver.

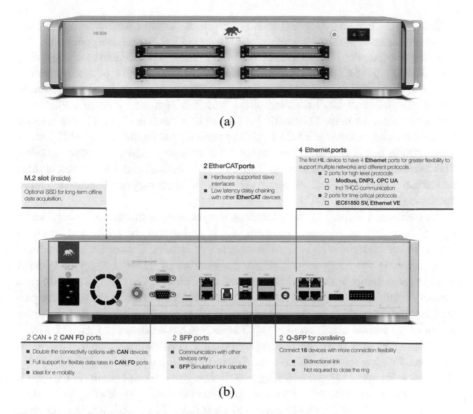

Fig. 1.10 HIL606 device: **a** front view; **b** back view

- **PWM Modulator:** It consists of a multi-channel triangular PWM modulator that can be used both internally, to drive converter models built in the model, and externally, using the digital outputs. The modulator runs on the FPGA internal clock with a built-in dead time generator.

Notice that the resources available on the Typhoon HIL FPGA solver enable it to support various elements used in different electrical applications. This architecture is scalable and is used on all Typhoon HIL devices. Nevertheless, every device has a number of different configurations, which differ in number and size of the computational elements available. The user can choose the configuration in order to optimize FPGA resources for the specific application under test. For instance, time-varying elements are not supported for all configurations, while the number of signal generators or machine solvers depends on the device configuration.

The FPGA solver also provides the means for emulating switch turn-on and turn-off delays in addition to real-time calculation of semiconductor losses, enabling detailed power converter modeling even with converter blocks that consist of ideal switches.

In addition, oversampling methods are available to meet the rigorous accuracy requirements for real-time simulation of high switching frequency applications. By enabling oversampling in Typhoon HIL simulations, it is possible to reach sampling periods down to 3.5 ns (fourth-generation devices) for all digital inputs, significantly improving the PWM resolution for higher switching frequencies. One of these methods is called Global GDS Oversampling, and it is enabled by default when you create a new model in the Schematic Editor. As illustrated in Fig. 1.11, this method is based on high-resolution PWM sampling (i.e., digital inputs are sampled several times within one simulation step), event time stamping, and error compensation. The oversampling methods used in Typhoon HIL devices and how enhancements on sampling resolution improve real-time simulation accuracy are discussed in more details in (Osório et al., 2021).

As mentioned before, another challenge is to be able to simulate with high accuracy a wide range of applications, going from unit models of power converters to larger power systems with a significant level of details. Converters are modeled using a piece-wise linear model, where the number of linear state space systems grows exponentially with the number of switches, thus exponentially increasing the memory required. The Typhoon HIL solution addresses this challenge using the parallel computing enabled by the multi-core FPGA solver. To illustrate that, consider a simulation which contains $n + m = 12$ switches, as illustrated in Fig. 1.12.

With 12 switches, one converter has 2^{12} possible permutations, meaning that 4096 state space matrix representations would have to be stored in memory. On the other hand, by using paralleling computing and dividing the circuit equally in two SPCs ($n = m = 6$), each core would have 2^6 possible permutations, leading to 64 state space matrix representations per core, and therefore 128 in total, which represents a significant reduction in memory capacity requirements. Moreover, given that the time needed to simulate the model is a function of the size and density of state

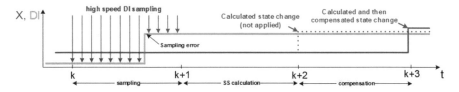

Fig. 1.11 Illustrative representation of the GDS Global Oversampling method. DI represents the digital input (GDS or PWM signal) and X illustrates the state change due to the input event (Osório et al. 2021; Typhoon HIL 2023c)

Fig. 1.12 Example of how circuit partitioning allows for reducing memory capacity requirements and achieving faster simulation rates

space matrix, splitting the model in different cores also allows for achieving faster simulation rates. It is worth mentioning that circuit partitioning can be done both between SPCs of a single HIL device and between multiple HIL devices. Moreover, enabling parallel computing of complex power electronics or power system models is done in a very easy and straightforward way, by using dedicated components in the Schematic Editor of the Typhoon HIL Control Center tool that will be detailed later.

1.4.1.2 Embedded Compiler

The Typhoon HIL compiler is fully embedded, allowing compilation of high-fidelity models optimized for real-time execution without third-party tools and without requiring expertise in low-level programming. The compilation is fully automated and accessible through one click, converting the graphical representations built in the Typhoon HIL Schematic Editor to sets of instructions for both FPGA and CPU processors.

Throughout the compilation process, the Typhoon HIL compiler provides a detailed report, which warns about defects and possible instabilities in the circuit. It also shows how the model is distributed in the HIL devices, its processing cores, what are the resources being utilized, time utilization within the simulation step, and also sub-optimal model characteristics, providing guidance for further model optimization. An excerpt of a compiler log is shown in Fig. 1.13. In this example, it is possible to verify that the model was divided into three subcircuits, as well as the partial list of components in each SPC and the hardware utilization analysis.

1.4.1.3 Hardware Interface and Accessories

Real-time simulation platforms must be flexible to enable hardware-in-the-loop simulations for various applications, which require suitable interfaces between the simulator and the large variety of possible devices under test. For that purpose, Typhoon HIL real-time simulators comprise several digital and analog inputs and outputs, as illustrated in Fig. 1.10a. Moreover, a number of dedicated interface systems are offered, which can be chosen according to key factors such as the number of signals and the signal conditioning requirements (based on the voltage and current levels of the device under test). The IO voltage levels and sample rates on Typhoon HIL devices can be easily found in the documentation (Typhoon HIL 2023e, f).

As the simplest interface possible, wires can be used to directly interface the real-time simulator with external controllers. For that purpose, Typhoon HIL offers a HIL Breakout Board, which simplifies the wiring between the control hardware and the HIL system, as shown in Fig. 1.14a. Nevertheless, note that this kind of interface requires matching of the voltage levels of the devices. As an alternative, dedicated interface boards can be used, where printed circuit boards are responsible for the signal conditioning, comprising connectors that are compatible with both the

```
Compiling model for device with id 0

PWM Modulators scheduling completed.

Circuit is divided into 3 subcircuits.

Partial list of components in subcircuit (SPC) 0:
     Grid Side Converter1
     Rotor Meter
     Chopper

Partial list of components in subcircuit (SPC) 1:
     Grid Meter
     S1
     R1

Partial list of components in subcircuit (SPC) 2:
     Tr2
     Vrms_grid
     Vs

Communication lines scheduling completed.

Running Device specific hw utilization analysis:
     Standard processing core utilization:      3 out of 6    50.0%
     Signal generator utilization:              3 out of 12   25.0%
     Look up tables utilization:                0 out of 8    0.0%
     Machine solver utilization:                1 out of 2    50.0%
     Parallel DTV Conv. Detectors utilization:  0 out of 3    0.0%
     PWM channels utilization:                  6 out of 12   50.0%
```

Fig. 1.13 Example compiler log

(a)   (b)

Fig. 1.14 HIL interfaces: **a** HIL breakout board; **b** HIL TI launchpad interface

real-time simulator and the device under test. As an example, Typhoon HIL offers off-the-shelf plug-and-play interface boards for Texas Instruments controllers and Launchpads, as illustrated in Fig. 1.14b.

Multiple conditioning systems packaged in a dedicated enclosure, called HIL Connect, are also offered, as illustrated in Fig. 1.15a. This approach provides great flexibility once it supports all major types of connectors, allowing the user to connect the device under test to the emulator with the exact same cables that would be used in the real system. HIL Connect systems can be customized according to particular requirements and specifications.

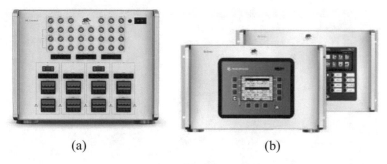

(a)   (b)

Fig. 1.15 HIL interfaces: **a** HIL connect; **b** Packaged interface with HIL compatible controllers

In addition to these, a selection of pre-packaged third-party device controllers with standardized, reproducible interfaces are also available. Known as HIL Compatible interfaces, these solutions are C-HIL ready, and can be connected to real-time simulators using only a set of standard cables, as illustrated in Fig. 1.15.

1.4.2 Typhoon HIL Control Center

Typhoon HIL Control Center is a fully integrated toolchain that enables users to build models, parametrize components, run HIL-based real-time simulations, and perform automated tests. This means any user can access the full potential of the developed hardware technology in an easy and straightforward way, without requiring experience in low-level programming. In addition, by means of a Virtual HIL device, the toolchain can also be used to verify real-time ready models even without controller hardware and before having an actual HIL device available, further facilitating the test-driven development process.

The initial window of the software can be seen in Fig. 1.16. The main resources include the modeling tool, the real-time graphical interface, and test development tools, as described in the following sections. If you are interested to raise your skill and knowledge of the Typhoon HIL toolchain, a HIL Fundamentals course is available on the HIL Academy platform which provides a detailed explanation and interactive demonstration of the tools described here (Typhoon HIL 2023b).

Schematic Editor

Schematic Editor is a software environment where real-time ready models are built and compiled using a user-friendly and intuitive interface, as illustrated in Fig. 1.17. The models can be developed from scratch, by dragging and dropping any number of the hundreds of pre-built components easily accessible using the Library Explorer tool shown in the left side of Fig. 1.17. The library includes pre-packaged converters, transformers, renewable sources, electrical machines, passive components, and oth-

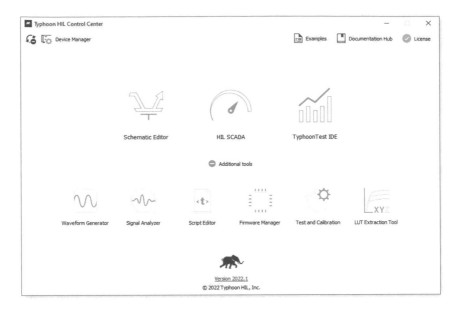

Fig. 1.16 Typhoon HIL control center

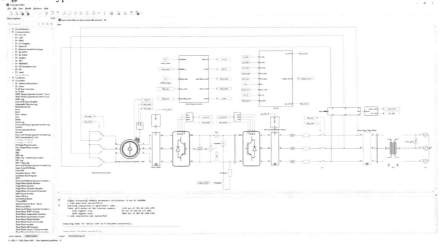

Fig. 1.17 Schematic editor interface displaying a wind turbine with doubly-fed induction generator model

ers, which are optimized for fast compilation and real-time executions, in addition to being easily parametrized for different domain-specific applications.

As an example, Fig. 1.18 shows a three-phase inverter component, which is part of the converters library. This component can be used in conjunction with passive and other components to build tailor-made models for custom applications. Notice

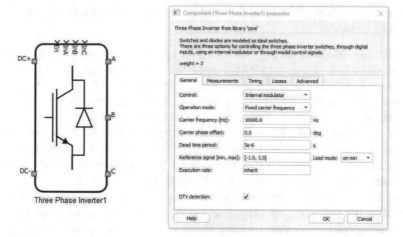

Fig. 1.18 Example of a pre-packaged converter (three-phase inverter) and its general properties window

that instead of having to use individual switches and diodes, converters like this one are available as pre-packaged components optimized for real-time execution, with a specialized runtime logic that allows for reducing the number of modes, thereby reducing memory capacity requirements.

Figure 1.18 also shows the general properties of the three-phase inverter component when the internal modulator control type is selected, highlighting the ease of configuring different parameters. In this case, the controller can be modeled within Schematic Editor using signal processing components, providing the control signals (InA, InB, and InC) for the converter. If a HIL device is available, the modulation is done with high resolution by the dedicated PWM modulator unit in the FPGA. Alternatively, if an external controller is available and it is properly interfaced with the HIL device, the converter can be directly controlled by HIL digital inputs. Additional tabs on the properties window also allow enabling the emulation of turn-on and turn-off switching delays (Timing tab) as well as semiconductor losses (Losses tab).

As mentioned in the challenges section, it is also important to provide a solution that suits users with different levels of expertise and different application-specific systems with easily parameterizable models. With that in mind, besides the default libraries, Typhoon HIL provides domain-specific toolboxes with component-level building blocks optimized for different model depths and requirements, making the task of building complex models even easier. One example of this is the Microgrid Toolbox, which contains distributed energy resources such as diesel generators, PV power plants, wind power plants, and energy storage systems that can be built using different component types. The choice of which type of component to use when building a model depends on the device under test, the purpose of the simulations, the testing requirements, and also on the hardware resources that the user has available, as described below:

- **Switching components:** Recommended for system-level testing of real converter controllers that require detailed power electronics models and accuracy in emulating the switching behavior in order to interface the PWM outputs. These components include pre-implemented control subsystems that can be freely modified by the user, as well as extensive control gains parametrization.
- **Average components:** Behavioral twins of the switching component models in terms of parametrization and dynamics, but consumes significantly less computation resources, making them the better choice for situations where the switching dynamics can be neglected and a PWM interface is not needed.
- **Generic components:** Based on average models, recommended for microgrid applications and energy management systems, where the simulations focus on testing top-level controllers responsible for steady-state regulation of voltage and frequency as well as load/energy management. Dedicated communication user interfaces are available for communication testing and troubleshooting interoperability issues. These components also include useful built-in functionalities such as voltage and frequency droop, ramping, low-voltage ridethrough (LVRT), and voltage and current protection, as well as self-tuning and grid support features.

Another toolbox of note is Typhoon HIL's Communication Toolbox, which incorporates many standard protocols from various industry and research applications. Most modern engineering system employs some sort of critical, digital communication protocol. Testing these communication protocols is important to verify the proper functioning of the device under test in an integrated system, including interoperability and pre-certification testing, communication fault testing, and cybersecurity testing, among others.

Applications that require communication testing extend to several industry fields, such as automation, energy generation transmission and distribution, automotive, aerospace, and marine. In academic research, communication protocols are also used to implement co-simulation interfaces and integrate different laboratory equipment. Table 1.1 shows the protocols available in Typhoon HIL Control Center, organized by the application where they are most commonly used in HIL tests. To understand and choose which protocol is suitable for an application, different requirements must be considered. Common requirements include flexibility, criticality, determinism, number of devices, standards, robustness, data types, security level, remote or local access, speed, and hardware setup.

Typhoon HIL Control Center also includes several examples for various applications, which can be used as starting point to build different models. The examples library is organized by application area and includes descriptions of the models often coupled with application notes, which makes it easy to navigate. Figure 1.19 shows the schematic of the terrestrial microgrid example using generic components, available in the Examples Explorer.

Table 1.1 Communication protocols supported by Typhoon HIL Control Center as of the 2022.4 software release (future releases will include support for additional protocols)

General/Industry	Energy/Microgrids	Automotive
Modbus Server (slave)	DNP3 Outstation	CAN
Modbus Client (master)	IEC 61850 GOOSE	J1939
Modbus SunSpec	IEC 61850 Sampled Values	CAN-FD
Ethernet Variable Exchange (TCP/UDP)	IEC 61850 MMS Server	CANOpen Slave
OPC UA	IEEE C37.118 PMU Server	ISO 15118-2 EVCC
EtherCAT Slave	IEEE C37.118 PMU Client	ISO 15118-2 SECC
Precision Timing	Modbus Server (Slave)	
SFP Aurora	Modbus Client (Master)	
PROFINET IO	IEC 60870 Server	
Serial/UART		
SPI (slave)		

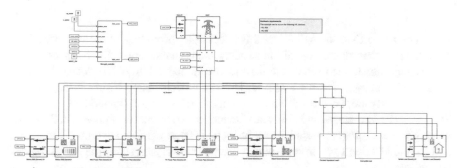

Fig. 1.19 Example of model built in the schematic editor for a terrestrial microgrid using generic components

HIL SCADA

Once the model is compiled, it can be loaded to HIL SCADA, a real-time graphical user interface that enables operating, controlling, and monitoring of the simulation. The time elapsed while the model is being simulated is called simulation runtime and for real-time simulations this will exactly match the wall-clock time.

During simulation, HIL SCADA can be used to modify simulation inputs such as signal generator variables and power plant inputs, as well as to observe signals in real time or capture them for further analysis. That can be done by easily dragging and dropping action and monitoring widgets available in the widgets library. Custom libraries can also be created, according to the user's needs. In addition, Typhoon HIL API (application programming interface) functions and Python code can be used to achieve more flexibility when programming widgets. Users don't need to be experts

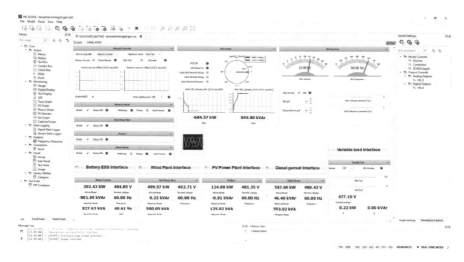

Fig. 1.20 SCADA panel built for the model shown in Fig. 1.19

in Python to do that; an API Wizard is accessible, providing all control variables from the model settings, in addition to capture commands and other features.

It is important to mention that models can be loaded to HIL SCADA even if a real-time simulator device is not connected. This can be done using Virtual HIL (V-HIL), which is a software module within the Typhoon HIL toolchain capable of emulating Typhoon HIL four-series and six-series devices on a personal computer. On the other hand, given the software-based nature of Virtual HIL, there is no external IO support and the models don't run in real time, with runtime varying according to the model complexity.

The SCADA panel illustrated in Fig. 1.20 is built for the model shown in Fig. 1.19 and is accessible from the same Example Explorer model. It is worth noting the presence of different widgets, configured to display the most relevant information about the microgrid. Users can also access and control every DER by accessing its respective interface sub-panels.

Typhoon Test IDE

Testing is a crucial aspect in the development of new power electronics devices, microgrid controllers, distribution management systems, and other applications. At the same time, due to an increasing number of required tests and ever-changing standards, manually testing all necessary conditions is often unfeasible. Therefore, using HIL and test automation together means it is possible to increase test coverage as well as to continuously improve the development cycle and the final product.

To address that, the Typhoon Test IDE (integrated development environment) is a testing tool specialized for power electronics and power systems, allowing for the creation and running of automatic tests for different applications with interactive

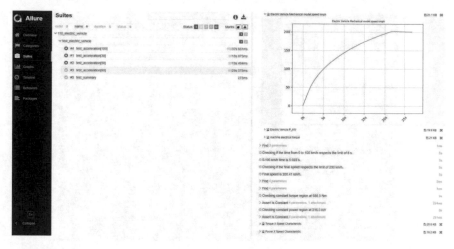

Fig. 1.21 Example of Allure report automatically generated when running tests in Typhoon Test IDE for the electric vehicle example available in the Example Explorer

automatic reports. This is possible since Typhoon HIL software is Python based, and the tests can be done using the Typhoon HIL API, which is a set of Python functions that allows the user to control the simulation environment, parametrize components, load, run, and interact with the models. The TyphoonTest IDE automation tool runs using pytest, providing automatic report generation with Allure. An example of an automatically generated report is shown in Fig. 1.21, considering the test of an electric vehicle drivetrain. In this report, the user can easily verify which conditions are passed or not in the tests, in addition to checking more detailed information.

Using HIL with test automation can bring benefits for several applications. For instance, in industry, test automation can be used in the development cycle, including a very large number of tests, operating conditions, and parameters; in device pre-certification, by running certification-like tests as part of the development process; and in the commissioning process. In Academia, use cases include testing of new methodologies for a wider range of conditions; benchmarking different methodologies, by using the same test procedure and metrics to compare them; and automatically generating results, which allow for quick updates to reports and papers, if needed. For laboratories and certification bodies, automatized pre-certification tests can improve certification turnaround.

Fig. 1.22 Example of C-HIL setup with a Typhoon HIL402 real-time simulator, TI LaunchPAD LAUNCHXL-F28379D, and a dedicated interface board

Additional Tools

As shown in Fig. 1.16, Typhoon HIL Control Center also includes the following additional tools:

- **Waveform Generator:** It is used to generate *current* × *voltage* and *power* × *voltage* curves for photovoltaic generation, as well as custom waveforms that can be later imported to the simulation.
- **Signal Analyzer:** It is used to visualize and analyze data exported from Typhoon HIL simulations or dynamically import data from HIL SCADA.
- **Firmware Manager:** It is used to configure the firmware of the HIL device, as well as to change between different device configurations.
- **Test and Calibration:** It is used to calibrate the HIL device using the HIL Calibration Card.
- **LUT Extraction Tool:** It is used to automatically convert images from datasheet charts and graphs into useful data file formats.

1.5   C-HIL Setup Example

An example of a C-HIL testing setup is illustrated in Fig. 1.22, encompassing a Typhoon HIL402 real-time simulator, a HIL TI Launchpad Interface, and a Texas Instruments Launchpad LAUNCHXL-F28379D.

To build a testing setup such as this one and to be able to execute C-HIL real-time simulations, it is first necessary to build an appropriate model of the power stage

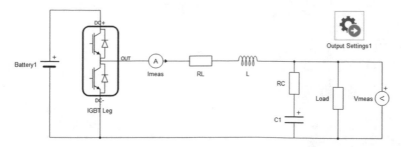

Fig. 1.23 Model built in Typhoon HIL Schematic Editor for real-time simulation of a buck synchronous converter

Table 1.2 C-HIL interface specification table for example shown in Fig. 1.23

Signal type[a]	Simulation range	Controller range	Description
Digital input 1	0/1	0/3.3 V	Top switch PWM signal
Digital input 2	0/1	0/3.3 V	Bottom switch PWM signal
Analog output 2	[0–50] V	[0–3.3] V	Voltage measurement
Analog output 1	[0–30] A	[0–3.3] V	Current measurement

[a] Signal types are defined here from the HIL device. HIL digital inputs come from the controller's digital outputs, while HIL analog outputs are measured through controller's analog inputs

using Schematic Editor. In this model, the user must identify all interface points between the controller and the power stage model, including inputs, outputs, and all necessary voltage and current measurements. When building the schematic, an easy way to keep track of all relevant signals is to create a table with their names, description, and voltage levels.

To demonstrate this, Fig. 1.23 shows the power stage model of a synchronous buck converter, built for real-time simulation. Table 1.2 lists the inputs and outputs of the model, as well as the voltage range specified in the controller.

Configuration of the digital inputs is set directly in the IGBT leg properties, as shown in Fig. 1.24a, and should be done considering the proper mapping of the controller digital outputs to the HIL digital inputs through the interface. The configuration of the analog outputs can be done using the Output Settings component, as shown in Fig. 1.24b, taking into account the appropriate signal mapping.

In the output settings, the Scaling and Offset parameters must be defined from the values obtained in simulation in order to ensure proper conditioning of the signals from the real-time simulator (HIL analog outputs) to the controller (digital signal processor inputs). These values can be calculated as follows:

$$\text{Scaling} = \frac{V_{\text{ph}}^{\max} \, K_{\text{int}}}{V_{\text{ctrl}}^{\max} - V_{\text{ctrl}}^{\text{zero}}}, \quad \text{offset} = \frac{V_{\text{ctrl}}^{\text{zero}}}{K_{\text{int}}}, \tag{1.1}$$

             (a)   (b)

Fig. 1.24 Interface settings: **a** digital input settings; **b** analog output settings

where K_{int} is the gain of the interface board; V_{ctrl}^{zero} is the voltage in the controller inputs that represents 0 V in the physical system; V_{ctrl}^{max} is the maximum rated voltage at the controller inputs; and V_{ph}^{max} is the physical values represented by V_{ctrl}^{max}.

Once the controller is properly connected to the real-time simulator and the model is built, validated, and fully parametrized, then the C-HIL simulation is ready to run.

It is worth mentioning that the real-time ready model shown in Fig. 1.23 can be validated offline in a preliminary stage without the controller, and the results can be compared with offline simulations or mathematically calculated responses. The offline simulation can be performed using the Virtual HIL (V-HIL) environment which, as mentioned in Sect. 1.4.2, enables simulation of real-time ready models without a HIL device.

1.6   Conclusions

This introduction highlights some of the key challenges to performing real-time simulation testing, and how the specific hardware and software solutions in the Typhoon HIL real-time simulation platform work to overcome them. An example of a simple C-HIL testing setup featuring a real controller is shown to demonstrate how such a testing environment can be easily built and parameterized.

HIL testing plays a growing and increasingly critical role in the development and improvement of new power electronics and power systems technologies. Controller hardware-in-the-loop (C-HIL) testing solutions, such as those provided by Typhoon HIL, stand out as effective solutions for testing, validating, and troubleshooting real controllers in a safe, realistic environment.

In the following chapters, real examples of HIL testing solutions using the Typhoon HIL toolchain are presented in detail.

References

Abdelrahman AS, Algarny KS, Youssef MZ (2018) A novel platform for powertrain modeling of electric cars with experimental validation using real-time hardware in the loop (HIL): a case study of GM second generation chevrolet volt. IEEE Trans Power Electron 33(11):9762. https://doi.org/10.1109/TPEL.2018.2793818

Amin M, Abdel Aziz GA (2019) A hardware-in-the-loop realization of a robust discrete-time current control of pma-synrm for aerospace vehicle applications. IEEE J Emerg Select Top Power Electron 7(2):936. https://doi.org/10.1109/JESTPE.2018.2890592

Badini SS, Verma V (2019) A novel mras based speed sensorless vector controlled pmsm drive. In: 2019 54th international universities power engineering conference (UPEC), pp 1–6. https://doi.org/10.1109/UPEC.2019.8893607

Celanovic NF (2017) Typhoon HIL blog – digitalization of microgrids and electrical distrubution networks. www.info.typhoon-hil.com/blog/microgrid-digitalization

Chemali E, Preindl M, Malysz P, Emadi A (2016) Electrochemical and electrostatic energy storage and management systems for electric drive vehicles: state-of-the-art review and future trends. IEEE J Emerg Select Top Power Electron 4(3):1117. https://doi.org/10.1109/JESTPE.2016.2566583

Dinavahi V, Reza Iravani M, Boncrt R (2001) Real-time digital simulation of power electronic apparatus interfaced with digital controllers. IEEE Trans Power Delivery 16(4):775. https://doi.org/10.1109/61.956769

Genic A, Gartner P, Almeida M, Zuber D (2017) Hardware in the loop testing of shipboard power system's management, control and protection. In: 2017 IEEE vehicle power and propulsion conference (VPPC), pp 1–6. https://doi.org/10.1109/VPPC.2017.8331026

Jonke P, Stöckl J, Miletic Z, Bründlinger R, Seitl C, Andrén F, Lauss G, Strasser T (2016) Integrated rapid prototyping of distributed energy resources in a real-time validation environment. In: 2016 IEEE 25th international symposium on industrial electronics (ISIE), pp 714–719. https://doi.org/10.1109/ISIE.2016.7744977

Khan A, Jarraya F, Gastli A, Ben-Brahim L, Hamila R, Rajashekara K (2017) Dual active full bridge implementation on Typhoon HIL for G2V and V2G applications. In: 2017 IEEE vehicle power and propulsion conference (VPPC), pp 1–6. https://doi.org/10.1109/VPPC.2017.8331023

Knezović K, Martinenas S, Andersen PB, Zecchino A, Marinelli M (2017) Enhancing the role of electric vehicles in the power grid: field validation of multiple ancillary services. IEEE Trans Transp Electrific 3(1):201. https://doi.org/10.1109/TTE.2016.2616864

Lian K, Lehn P (2005) Real-time simulation of voltage source converters based on time average method. IEEE Trans Power Syst 20(1):110. https://doi.org/10.1109/TPWRS.2004.831254

Liserre M, Sauter T, Hung JY (2010) Future energy systems: integrating renewable energy sources into the smart power grid through industrial electronics. IEEE Indust Electron Mag 4(1):18. https://doi.org/10.1109/MIE.2010.935861

Majstorovic D, Celanovic I, Teslic ND, Celanovic N, Katic VA (2011) Ultralow-latency hardware-in-the-loop platform for rapid validation of power electronics designs. IEEE Trans Indust Electron 58(10):4708. https://doi.org/10.1109/TIE.2011.2112318

Osório CRS, Miletic M, Zelic J, Majstorovic D, Gagrica O (2021) Advancements on real-time simulation for high switching frequency power electronics applications (invited paper). In: 2021 21st international symposium on power electronics (EE), pp 1–6. https://doi.org/10.1109/Ee53374.2021.9628306

Osório CRD, Koch GG, Pinheiro H, Oliveira RCLF, Montagner VF (2020) Robust current control of grid-tied inverters affected by LCL filter soft-saturation. IEEE Trans Indust Electron 67(8):6550. https://doi.org/10.1109/TIE.2019.2938474

Pallo N, Foulkes T, Modeer T, Fonkwe E, Gartner P, Pilawa-Podgurski RC (2017) Hardware-in-the-loop co-design testbed for flying capacitor multilevel converters. In: 2017 IEEE power and energy conference at Illinois (PECI), pp 1–8. https://doi.org/10.1109/PECI.2017.7935758

Rommel C (2019) HIL TESTED – powerful performance, functionality, and quality from model-based testing. Published by VDC Research. https://info.typhoon-hil.com/lp-white-paper-vdc-research-hil-tested

Salcedo R, Corbett E, Smith C, Limpaecher E, Rekha R, Nowocin J, Lauss G, Fonkwe E, Almeida M, Gartner P, Manson S, Nayak B, Celanovic I, Dufour C, Faruque M, Schoder K, Brandl R, Kotsampopoulos P, Ha TH, Davoudi A, Dehkordi A, Strunz K (2019) Banshee distribution network benchmark and prototyping platform for hardware-in-the-loop integration of microgrid and device controllers. J Eng 2019:5365

Typhoon HIL. (2023) Online documentation. www.typhoon-hil.com/documentation

Typhoon HIL. (2023a) Youtube channel. www.youtube.com/c/TyphoonHIL-com/featured

Typhoon HIL. (2023b) HIL fundamentals course. https://hil.academy/courses/hil-fundamentals/

Typhoon HIL. (2023c)Documentation – GDS oversampling. www.typhoon-hil.com/documentation/typhoon-hil-software-manual/concepts/gds_oversampling.html

Typhoon HIL. (2023e) Documentation – IO voltage levels. https://www.typhoon-hil.com/documentation/typhoon-hil-hardware-manual/hil4-6_series_user_guide/References/hil4-6_IO_voltage_levels.html

Typhoon HIL. (2023f) Documentation – IO timing. https://www.typhoon-hil.com/documentation/typhoon-hil-hardware-manual/hil4-6_series_user_guide/References/hil4-6_IO_timing.html

Vekić MS, Grabić SU, Majstorović DP, Čelanović IL, Čelanović NL, Katić VA (2012) Ultralow latency HIL platform for rapid development of complex power electronics systems. IEEE Trans Power Electron 27(11):4436. https://doi.org/10.1109/TPEL.2012.2190097

Xu L, Guerrero JM, Lashab A, Wei B, Bazmohammadi N, Vasquez J, Abusorrah AM (2021) A review of DC shipboard microgrids part i: power architectures, energy storage and power converters. IEEE Trans Power Electron 1. https://doi.org/10.1109/TPEL.2021.3128417

Zelic J, Novakovic L, Klindo I, Gruosso G (2020) Hardware in the loop framework for analysis of impact of electrical vehicle charging devices on distribution network. In: 2020 IEEE vehicle power and propulsion conference (VPPC), pp 1–5. https://doi.org/10.1109/VPPC49601.2020.9330863

Chapter 2
Control of Grid-Connected Inverter

Azra Malik and Ahteshamul Haque

Abstract The control of grid-connected inverters has attracted tremendous attention from researchers in recent times. The challenges in the grid connection of inverters are greater as there are so many control requirements to be met. The different types of control techniques used in a grid-connected inverter are discussed in detail in this chapter. In addition, a case study is also presented using the hardware setup of Typhoon HIL.

Keywords Control techniques · Grid-connected inverter · Synchronous reference frame · Real-time simulation · Typhoon HIL

2.1 Introduction

During the past few years, there has been an increased penetration of non-conventional distributed energy resources (DERs) into the conventional electricity distribution grids (Khan et al. 2020). This trend has witnessed an accelerated shift from low-voltage power networks to the smart micro-grid pattern with efficient and reliable interconnections of DERs at the point of common coupling (PCC). In this whole shift, power electronics converters play a major role in the transfer and control of generated power from one side to the other (Khan et al. 2019). They act as an interface between various electrical domains, for instance, DC–DC converters are utilized to supply to DC loads from the relevant DC sources at the required voltage and power. In a similar manner, DC–AC converters or inverters are utilized as an interface between DC generators like batteries, PV panels, etc., and AC receiving

---

A. Malik · A. Haque (✉)
Advance Power Electronics Research Lab, Department of Electrical Engineering, Jamia Millia Islamia (A Central University), New Delhi, India
e-mail: ahaque@jmi.ac.in

A. Malik
e-mail: azra1910177@st.jmi.ac.in

© The Author(s), under exclusive license to Springer Nature Singapore Pte Ltd. 2023
S. M. Tripathi and F. M. Gonzalez-Longatt (eds.), *Real-Time Simulation and Hardware-in-the-Loop Testing Using Typhoon HIL*, Transactions on Computer Systems and Networks, https://doi.org/10.1007/978-981-99-0224-8_2

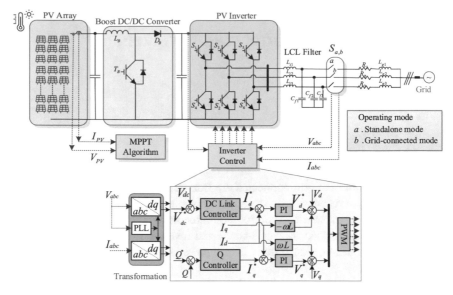

Fig. 2.1 Grid-connected PV system operation modes

ends like power grids, etc. Inverters are also divided into two different categories—voltage source and current source inverters (VSIs and CSIs) (Kouro et al. 2015). These names come from the fact that the respective converter output voltage and current are controlled independently. The major difference between the two topologies is that the VSIs are capable to perform voltage buck operation, whereas CSIs can only carry out voltage boost functions. For CSIs, three-phase configurations are considered more relevant than single-phase configurations. When the inverter functions as an integration between the DC source and the grid for efficient transfer and control of generated power, then it is termed a grid-connected inverter (Kurukuru et al. 2021). Overall, a grid-connected system works in different operation modes depending on the control switch states, which can be guided locally through the inverter or remotely through an operator (Yang et al. 2019). These operation modes are presented in Fig. 2.1 and are described below.

2.1.1 Standalone Mode

In this mode, the corresponding DER supplies to the load independently and it is disconnected from the grid (Khan et al. n.d.). The inverter in this mode is responsible for supplying to the AC loads joined at the PCC. The inverter also functions to maintain the voltage and frequency of the system at the suitable range as specified in the standards. During the fault events, the line breaker opens through the disconnection command issued by the respective protection devices (Bharath et al. 2018). There are various challenges accompanying a broad-scale utilization of standalone mode including voltage regulation, islanding-related issues, protection challenges, effect on the environment, and power quality (Bojoi et al. 2011). These challenges need to be addressed to gather more information on a large-scale adoption of standalone operation mode.

2.1.2 Grid-Connected Mode

In this mode, the inverter is connected to the grid at PCC and it transfers the generated power from the DC side to the AC side, i.e., grid and AC loads (Ahmed et al. 2011). The voltage reference is taken as per the grid side requirements for inverter controller. Furthermore, the inverter control is responsible for maintaining the frequency and power at the AC side. In this mode, synchronization is important and it is achieved through phase-locked loop (PLL) by the control algorithm (Bisht et al. 2020). Along with that, it keeps a track on harmonics and reduces the harmonics as per grid standards (Zmood and Holmes 2003). Inverter switches play a significant part in implementing the control technique.

2.1.3 Autonomous Mode

When grid-connected inverters intentionally separate themselves from the PCC, through opening the controlled switch, they operate autonomously. In this operation mode, they function as controlled voltage sources, which supply only to the local AC loads. Among all the discussed operating modes, grid-connected inverters have multiple roles to play like supplying to the local loads, DC and AC bus coupling, and delivering the generated energy to the grid, while following the prescribed regulated standards, for instance, IEEE 1547—2018 (Kazmierkowski and Malesani 1998). In addition, they support the grid through providing ancillary services like reactive power support, etc. For the safe operation of grid-connected systems, proper control design maintaining an optimum performance is crucial. It becomes more important when the grid is weak, since maintaining good power quality becomes a challenge. When DER penetration is increased, the power system may show low inertia which makes the voltage on the inverter side vulnerable to power variations. While DER's increased penetration into traditional power systems is a boon, there are some challenges and issues associated with its implementation (Han, et al. 2017). The impact of increased DERs over the power system transient stability and frequency stability are currently being explored. Though modern electric grid codes have adopted the compliance to reduce the impact caused by increased DERs diffusion (Xu et al. 2021). It includes the low-voltage ride-through (LVRT) requirements that mandate the solar and wind generators to continue connecting in the network for handling the voltage sag (Chandran et al. 2019). Similarly, high-voltage ride-through (HVRT) is also incorporated to cope up with the overvoltage profile seen in grid codes.

The ride-through functionality can be achieved by either utilizing external devices (e.g., flexible AC transmission system (FACT)) or inducing adjustment in the inverter control. Between the two, modification in control is a less complex and cost-effective method for achieving LVRT operation (Hasanien 2016). A droop control for LVRT operation is developed, which includes monitoring the deviation in the DC link voltage such that when the drop is observed, the maximum power point (MPP) tracker is changed to the LVRT mode of operation in the controller (Khan et al. n.d.). A similar dual current control is observed for supporting the grid during the fault events (Fatama et al. 2020). The control consists of handling the positive and negative sequence currents due to the fault condition in the inverter. In addition, it provides reactive power into the grid complying with the grid code regulations. Further, the FACT device can be used with inverter control to introduce reactive power depending on the requirements (Merabet 2017). Therefore, a coordinated reactive power control is implemented for ensuring LVRT functionality. Inverters can be either single-phase or three-phase depending upon the requirement. For grid-connected systems, single-phase inverters are advantageous since they have the capability to induce additional flexibility for controlling different line power flows. This capability can also be utilized for providing phase-wise voltage support. Active power plays an important role in phase balancing during difficult situations in the DER integration with the grid. Al-shetwi et al. (2017) Grid-connected inverters can be of various topologies and configurations including transformer-based and transformerless, for Photovoltaic (PV) systems, they can be string inverters, central inverters, multi-string inverters, etc. Further, there come numerous configurations under transformerless inverters including H-Bridge inverter, highly efficient and reliable inverter concept (HERIC) topology, H5, H6, etc.

2.2 Control Techniques Classification for Grid-Connected Inverters

For ensuring an efficient operation of the grid-connected system, with PV or wind generators, it is essential for inverters to have an optimum operation. An effective inverter operation can be achieved by applying proper inverter control (Ebrahimi et al. 2015). Inverter controllers help in maintaining the power factor through sinusoidal current injection towards the grid side (Zeb, et al. 2017). A fast and accurate design for inverter control is required in order to achieve a reliable operation of the whole system. The inverter control is of high significance and is further divided into two categories—(1) MPP control and (2) Inverter module control. MPP control is responsible for extracting maximum power from the generator's side, whether PV or wind. Similarly, inverter module consists of ensuring optimum grid synchronization and proper sine current injection towards grid. It also makes sure to provide effective real and reactive power flow and control of the DC-link voltage at the DC side. Both the controls are important for robust and efficient functionality of the whole system (Liu et al. 2020). The general control structure of inverter consists of two cascaded loops, one of them is an internal current control loop, controlling the grid current and the other is an outer voltage control loop, which controls the DC link voltage. The inner control loop should be fast enough for ensuring proper dynamic response and harmonics reduction within the required limits for distorted grid operation. The outer voltage control loop should be slower than the inner loop for an optimal power flow and better regulation (Ahmad, et al. 2018). This is required for establishing stability for a given controller in grid-connected inverter systems. These grid-connected inverter controllers can be of many types depending on the application and other requirements.

2.2.1 Linear Controllers

The linear controllers utilize conventional pulse width modulators (PMW), as compared to the non-linear controllers. They consider the overall system as linear, and further consider the dynamics and other features for proper control design (Rowan and Kerkman 1986). The analysis of these controllers is done through the general feedback control theory. The linear controllers have segregated voltage modulation and current error compensation parts. This helps in utilizing the benefits associated with open-loop modulation techniques (like optimal PWM, space vector PWM (SVPWM), and sinusoidal PWM (SPWM)) (Ramchand 2018). These benefits include a well-spaced harmonic continuum, constant switching frequency, open-loop testing of inverter, optimum DC-link utilization, and switching pattern. It consists of classical controllers, Proportional-resonant (PR) controllers, state-feedback controllers, and predictive controllers.

2.2.1.1 Classical Controllers

There are several classical controllers already presented in literature like proportional (P), proportional–integral (PI), proportional–integral–derivative (PID), etc. Among these, the most commonly employed controller is PI controller since it has the capability for efficient DC signal tracking. It is represented by a continuous-time expression as given below:

$$c(t) = K_P e(t) + K_I \int e(t). \tag{2.1}$$

The corresponding frequency domain representation can be given by the following expression:

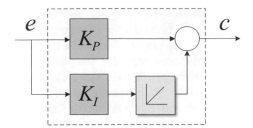

Fig. 2.2 PI controller equivalent block-diagram representation

$$G_{PI}(s) = K_P + \frac{K_I}{s} = K_P \left(\frac{1 + sT_I}{sT_I} \right) = K_P \left(1 + \frac{\tau_I}{s} \right). \tag{2.2}$$

Here, $T_I = K_P/K_I$ and $\tau_I = 1/T_I$, and based on the above representations, the standard block diagram implementation can be as shown in Fig. 2.2.

For attaining the desired closed-loop performance, proper K_P and T_I values are required. Depending on the order of the system and respective damping ratio and other system entities, the control parameter values can be designed according to the control objectives (Parvez et al. 2020). PI controller has been utilized with a successful closed-loop control for grid-connected inverter applications in the case of both PV and wind generators. For a three-phase grid-connected PV system, three PI compensators are utilized for generating the gate signals of switches for sinusoidal PWM (Dasgupta et al. 2011). Based on the PWM technique, the switching times for individual switches are affected. The demerit of this controller is that it involves inherent magnitude and phase tracking of the error signal.

The proportional resonant (PR) controller has a very similar working as that of PI controller. The difference lies in the operating reference frame, i.e., they are able to work in stationary reference frame, without any transformations. The proportional gain works for the dynamics in system considering phase and bandwidth, whereas the resonant term manifests an infinite gain at the given resonant frequency (Cai et al. 2014). PR controller takes the sinusoid frequency as the central frequency for tracking a signal sinusoidal in nature. Other than that, the integration is carried out in a varied manner for PR controller (Tarasantisuk et al. 2016). Here, the integrator integrates the frequencies nearer to the resonant frequency, as compared to PI controller (Li and Balog 2015). Hence, it does not involve any kind of stationary or phase-shift errors. For proper closed-loop control dynamics, it is desired that the damping ratio should be chosen more than or equal to one. The mathematical expression for a general PR controller is provided below:

$$G_{PR}(s) = K_P + \frac{2K_I(s)}{s^2 + \omega_0^2}. \tag{2.3}$$

Based on the above expression, the controller structure can be shown through the representation shown in Fig. 2.3.

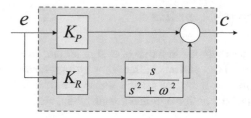

Fig. 2.3 PR controller block diagram

2.2.1.2 State-Feedback Controllers

For current error compensation in the grid-connected inverter control, one of the linear controllers can be a state-feedback controller, which is able to work in both stationary as well as synchronously rotating reference frame (Lee et al. 1992). The synthesis of the network is made based on the linear state-feedback theory. To ensure sufficient and proper damping, a feedback gain matrix is calculated through pole assignment utilization (Liu et al. 1998). The integral part of the controller is responsible for making sure that the static error is reduced to zero. However, it may cause the transient error to escalate to the point of unacceptable range. Therefore, it requires the feedback signals for both the disturbance and reference inputs to be summed in the feedback control law. It is observed that the state feedback controller provides superior performance as compared to the traditional PI controllers since the control technique provides dynamically correct compensation of the voltage signal.

2.2.1.3 Predictive Controllers

This kind of controller can be realized for a considered multivariable-based application. Due to the non-linearity feature, implementation of constraints becomes easier providing a faster dynamic response. It predicts the current error vector at the initiation of a single sampling time basis the actual error and the load-related parameters (Liao et al. 2021). Therefore, it is able to determine the generation of voltage vector for the next modulation sample, which will further help in minimizing the error in forecasting. Various predictive controllers for grid-connected PV systems have been proposed in literature like constant switching frequency-based predictive control, hybrid control with both predictive and hysteresis control, etc. Constant switching frequency-based control requires the switching frequency of inverter to be fixed and the current ripple is inconsistent. This control determines the voltage vector once for each sample period (Cortés et al. 2010). It enforces the current vector to follow the reference depending on the provided signal. Further, the determined voltage vector is executed in the modulation technique, including SVPWM, SPWM, etc. The disadvantage associated with this technique is that it does not ensure the peak current limit for inverter.

2.2.2 Non-Linear Controllers

These controllers are complex in terms of design and development as compared to the linear controllers. They perform extraordinarily as controllers for grid-connected systems as compared to the conventional linear controllers. They further include sliding mode, hysteresis, online optimized controllers, etc.

2.2.2.1 Sliding Mode Controller (SMC)

Among all the non-linear controllers, SMC algorithms are widely utilized owing to their improved and robust performance. The major benefit associated with SMC is that it is observed to be indifferent to parameter deviation and load disturbances (Cortes et al. 2009). Therefore, it is able to provide a non-varying steady-state response for grid-connected PV applications. However, the overall output of SMC is highly dependent on the appropriate sliding surface, which is sometimes difficult to define (Zheng et al. 2018). Further, the performance of SMC also relies on sampling time, thus, it may be affected due to distortion if the adequate sampling time is not identified. These disadvantages may highly affect the overall PV system efficiency in a negative way and therefore, it is recommended to consider all these facts before applying this technique.

2.2.2.2 Hysteresis Controllers

It is also one of the important controllers under non-linear controllers and has been employed for grid connected systems. It consists of creating an adaptive band to achieve an optimum and stable switching frequency (Davoodnezhad et al. 2014). Therefore, the output of hysteresis controller is found to be the individual switch states. Hence, it becomes important to consider the isolated neutral again for proper attaining proper control. The advantages of hysteresis controller include deficiency of tracking errors, insensitivity to load disturbances, lesser complexity, outstanding robustness, and fast dynamics. The limitations of this controller include load time constant and the switching speed, since the switching speed relies over the load parameters. When applied to three-phase systems, the comparator state change in one phase may impact the load voltage in other phases. Considering the three current errors as space vectors, this effect may be compensated (Chavali et al. 2022). This will establish numerous controller variants, termed space-vectors. Through suitable voltage vector selection, an optimum performance with better stability can be achieved.

2.2.2.3 Online Optimized Controllers

This controller performs optimization phenomena in real time and depends on complex real-time calculations using microprocessors. These controllers can be applied through either minimum switching frequency-based techniques or with field orientation-based techniques.

2.2.3 Adaptive and Robust Controllers

Adaptive controller provides the output response based on condition changes within the process (Castello et al. 2016). These controllers adjust the control action automatically depending on the operating conditions. They may manifest better performance, with insensitivity to parameter variations. However, they are considered to be computationally complex for implementation in grid-connected systems (Kumar and Tyagi 2021). On the other hand, robust control approach is adopted for designing the controller considering the uncertainties. These controllers are able to achieve stability along with robust performance, even for multi-variate systems. A proper criterion, clear description, and appropriate bounds must be defined to achieve desired robust control. H-infinity controllers under robust control solve multivariable problems by representing them as an optimization problem (Yang et al. 2020).

2.2.4 Intelligent Controllers

Through emulating the biological intelligence in human brain, intelligent controllers can be utilized to obtain automation. This is also applied for solving issues using the way human brain solves different issues (Bose 2008). A same approach is adopted to solve control problems. There are various controller classes under intelligent controllers as below.

2.2.4.1 Neural Network (NN) Controller

Neural Network (NN) is typically inspired by the biological neural network in the human brain. It consists of a connection of several artificial neurons making a network similar to the human nervous system (Meireles et al. 2003). These controllers are capable to attain better results using function mapping. Furthermore, the desired control objectives can be achieved through training in either online or offline mode. These intelligent controllers are fast and computationally efficient, therefore, they are considered highly preferable for grid-connected PV inverter control (Bose 2017).

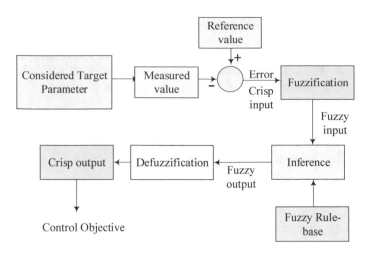

Fig. 2.4 A general block diagram for Fuzzy Logic (FL) controller

2.2.4.2 Fuzzy Logic (FL) Controller

In FL controllers, human knowledge is defined through some rules and implemented to modify and control system dynamics as per the desired control requirements (Hannan et al. 2019). The basic structure of FL control consists of four steps—fuzzification, rule base formation, inference, and de-fuzzification. For the first step, i.e., fuzzification, the given crisp data has to be fuzzified based on the rule base in the second step. These rules are formulated according to the application requirements (Hannan et al. 2015). These rules are assessed in the inference step and, based on that, decision is made as shown in Fig. 2.4. In the last step, the output, which is in fuzzy form has to be de-fuzzified to attain the equivalent crisp output. Therefore, properly defined control objectives are achieved.

2.3 Modeling of Inverters in Grid-Connected PV System

The system dynamics of an inverter and control structure can be represented through inverter modeling. It is an essential step towards attaining the inverter control objectives (Romero-cadaval et al. 2015). The overall process includes the reference frame transformation as an important process, where the control variables including voltages and currents in AC form, will be converted to DC form using Clark and park transformations. These transformations are applied to both one- and three-phase systems (Purba et al. 2019). Furthermore, the system modeling of inverter along with the L filter, PQ controller, voltage, and current controllers is established.

The grid-connected PV system control diagram for a three-phase inverter is depicted in Fig. 2.5. It involves the application of a cascaded control loop. The

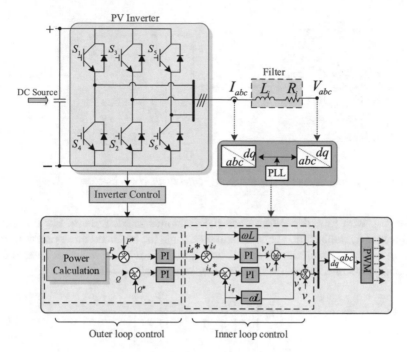

Fig. 2.5 Cascaded control loop for a three-phase inverter system

external loop consists of controlling the active and reactive power by PQ controller. It may also consist of indirect control through a DC-link voltage controller. This is generally applied to two-level grid-connected PV applications consisting of DC/DC converter for optimal power extraction (Strache et al. 2014). Further, the inner control loop is employed to control the injected current into the AC load. It aims to normalize the current output as per the reference fixed by the outer loop. The output voltage reference can be achieved with the help of inner current controller loop. This reference is utilized for PWM to produce the inverter output voltage through switching sequences in the inverter devices. In addition, a PLL is employed for the transformed variables to synchronize with the grid (Yu et al. 2019). A mathematical model can also be devised for modeling the three control loops-PQ, voltage, and current control loops.

2.3.1 Reference Frame Transformation

Three-phase to Stationary Reference Frame (abc to αβ)

From Fig. 2.6a, we can see that three-phase inverter consists of voltages V_a, V_b, and V_c as AC variables, and similarly, currents in three phases are other AC variables.

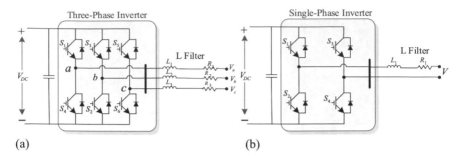

Fig. 2.6 a A general three-phase inverter, **b** A general single-phase inverter

They vary with time depending upon the dynamics in each phase. For translating the inverter dynamics into mathematical modeling, the balanced three-phase inverter system can be characterized through transformation into a two-phase system over stationary reference frame or $\alpha\beta$-reference frame as given below

$$\begin{bmatrix} t_\alpha \\ t_\beta \end{bmatrix} = \begin{bmatrix} 0 & \frac{1}{2} & \frac{-1}{2} \\ 0 & \frac{\sqrt{3}}{2} & \frac{-\sqrt{3}}{2} \end{bmatrix} \begin{bmatrix} t_a \\ t_b \\ t_c \end{bmatrix}, \tag{2.4}$$

where t_α and t_β are the variables in the stationary reference frame, whereas t_a, t_b, and t_c are the variables of the three-phase system. This transformation is popularly termed Clarke transformation.

Stationary reference frame to Synchronous reference frame ($\alpha\beta$ to dq)

The two-phase variables in the $\alpha\beta$ reference frame are actually AC variables rotating at the same speed as that of the three-phase variables. Therefore, it becomes difficult to apply classical controllers like PI controller, since it won't be able to accomplish zero-error tracking of the corresponding parameters. In that case, Park transformation can be applied. Park transformation refers to the conversion of stationary reference frame ($\alpha\beta$) parameters to synchronous reference frame parameters (dq) as given below

$$\begin{bmatrix} t_d \\ t_q \end{bmatrix} = \begin{bmatrix} \cos\theta & \sin\theta \\ -\sin\theta & \cos\theta \end{bmatrix} \begin{bmatrix} t_\alpha \\ t_\beta \end{bmatrix}, \tag{2.5}$$

where t_d and t_q are the parameters in the synchronous reference frame. Angle θ is the angular position for the given parameters. Quantities t_d and t_q are the DC quantities obtained through park transformation, which is considered as a preferable choice in the case of applying PI controllers.

Both above transformations were applied for three-phase systems. However, they may be applied for a single-phase inverter system as in Fig. 2.6b, by utilizing an orthogonal signal generator (OSG). This OSG creates an imaginary quantity, in quadrature with the original single-phase variable. Then, the two variables in αβ reference frame can be converted to dq reference frame with the help of Park transformation.

2.3.2 Inverter Cascaded Control

The two-stage three-phase inverter or DC/AC converter employs a cascaded control loop. The outer loop consists of controlling the active and reactive power through a PQ controller. It can also be controlled using DC-link voltage regulation. The objective of the inner control loop is to inject AC current control (Du et al. 2015). The reference for the output current is set by the outer control loop. Using the inner controller, the reference for output voltage is attained by applying inverse Clark and Park transformations. This voltage reference is provided to the PWM for producing the output voltage through proper switching sequences. A synchronization control unit, i.e., PLL is required for synchronizing the output with the grid. Modeling the entire system is important for proper controller (inner and outer) designs. A mathematical model of the whole system can be derived by modeling each control loop.

2.3.3 Modeling of the Outer PQ Loop

The outer PQ control loop is responsible for real and reactive power control with the help of instantaneous power theory (Kumar et al. 2020). The instantaneous real and reactive powers (P and Q) can be calculated in the dq reference frame as provided below

$$P = \frac{3}{2}\left(i_d v_d + i_q v_q\right), \tag{2.6}$$

$$Q = \frac{3}{2}\left(i_d v_q - i_q v_d\right). \tag{2.7}$$

For simplicity, supposing that the PLL is following the output voltage vector with the d-axis of the synchronous dq reference frame or $v_q = 0$, the transfer functions from the above calculations (ratio of real and reactive powers to dq output currents) can be found as given below:

$$\frac{P(s)}{i_d(s)} = \frac{3v_d(s)}{2}, \tag{2.8}$$

$$\frac{Q(s)}{i_q(s)} = \frac{-3v_d(s)}{2}. \tag{2.9}$$

For an ideal converter, the above calculations are used for open-loop control to regulate the real and reactive power. Nonetheless, a closed-loop control is preferred for better performance in the case of uncertainties (deviations in voltage and power losses) in the system.

2.3.4 Modeling of DC-Link Voltage Control Loop

An alternative to the PQ control loop is through the DC-link voltage regulation as the outer loop. If the DC-link voltage is maintained constant, it confirms the optimum power delivered from the DC side to the AC side. This approach is commonly applied in PV systems in the case of continuous supply through PV arrays. It is able to ensure optimum power delivery at the load end. For grid-tied applications, it is desired that the DC-link voltage should be more than the maximum grid voltage. For a loss-less power converter, input power and output power should be balanced as per the instantaneous power theory. This can be explained as

$$\text{Input DC power} = V_{DC}C_{DC}\frac{dV_{DC}}{dt}. \tag{2.10}$$

Input DC power = Output AC power

$$V_{DC}C_{DC}\frac{dV_{DC}}{dt} = \frac{3}{2}\left(i_d v_d + i_q v_q\right). \tag{2.11}$$

Assuming that the PLL is following the output voltage vector with the d-axis of the synchronous dq reference frame or $v_q = 0$, the transfer functions from the above calculations (ratio of DC-link voltage output and reactive powers to d-axis current) can be found as given below:

$$\frac{V_{DC}(s)}{i_d(s)} = \frac{3}{2}\frac{v_d(s)}{s V_{DC}C_{DC}}. \tag{2.12}$$

2.3.5 Modeling of Inner Current Control Loop

The objective of the current controller is to control the output AC current and in the process, it generates a reference for output voltage. As per the circuit in Fig. 2.5, applying Kirchohff's law, the three-phase inverter output current can be expressed below

$$L\frac{di_a}{dt} + Ri_a = v_{ia} - v_a, \tag{2.13}$$

$$L\frac{di_b}{dt} + Ri_b = v_{ib} - v_b, \tag{2.14}$$

$$L\frac{di_c}{dt} + Ri_c = v_{ic} - v_c. \tag{2.15}$$

where i_p, v_{ip}, and v_p are the current injected to the grid, inverter output voltage, and the AC load voltage, respectively, with p = a, b, c being the three phases. After applying the Clark and park transformations in the above equations, the following expressions are obtained:

$$L\frac{di_d}{dt} + Ri_d - \omega Li_q = v_{d1} - v_d, \tag{2.16}$$

$$L\frac{di_q}{dt} + Ri_q + \omega Li_d = v_{q1} - v_q, \tag{2.17}$$

where i_d and i_q represent the currents injected to the AC side on dq reference frame, and ω is the frequency of the system. Similarly, v_{d1}, v_{q1} represent the inverter output voltages in dq reference frame, and v_d, v_q are the output voltages or load voltages corresponding to d and q axes. From the above equations, the AC current to be injected can be regulated through the inverter output voltage control. However, there is a coupling between the d- and q-axis currents, which causees the control to be slightly complicated. The control dynamics are also affected by the load characteristics. Therefore, there is a need to modify the output voltage references through adding decoupling terms and, further, the modified current control modeling can be expressed as

$$L\frac{di_d}{dt} + Ri_d = v_{d1}^*, \tag{2.18}$$

$$L\frac{di_q}{dt} + Ri_q = v_{q1}^*. \tag{2.19}$$

From the above calculations, the d- and q-axis-related terms are found to be identical. Therefore, a single-axis analysis should be sufficient due to their similar dynamics. The plant transfer function can then be obtained as the ratio of the output current to the inverter output voltage.

$$\frac{i_d(s)}{v_{d1}^*(s)} = \frac{i_q(s)}{v_{q1}^*(s)} = \frac{1}{sL + R}. \tag{2.20}$$

2.4   Case Study: Single-Phase Grid-Connected PV Inverter Simulation Using Typhoon HIL-402

As discussed previously, a single-phase grid-connected PV inverter provides AC voltage and current, as required by the grid. To further verify this statement, this section provides a case study-related output results for an inverter. The overall schematic for single-phase PV inverter in the Typhoon HIL tool chain is shown in Fig. 2.7 and, further, the unmasked inverter components and various control loops are provided in Fig. 2.8.

When the grid-connected PV system is simulated in Typhoon HIL, the SCADA panel opens and various widgets can be used to see the desired signals as shown in Fig. 2.9. Overall sinusoidal voltage and current output to be injected to the grid are

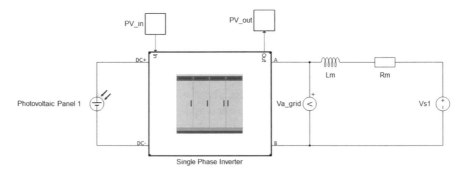

Fig. 2.7   Single-phase PV inverter schematic in Typhoon HIL tool chain

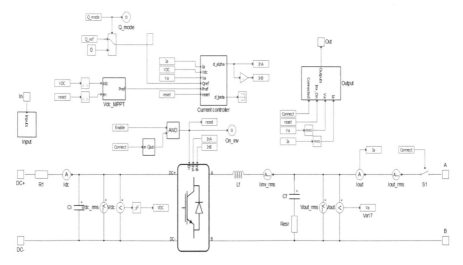

Fig. 2.8   Unmasked inverter control loops in Typhoon HIL schematic

shown in Fig. 2.10. It is clearly visible that the output voltage and current are AC waveforms at a frequency of 50 Hz, with voltage at amplitude 325 V (RMS value is 230 V) and current at amplitude 95A.

Further, the Typhoon HIL-402 device can also be utilized to obtain real-time simulation results over digital storage oscilloscope (DSO). The overall setup is provided in Fig. 2.11. Further, the inverter output voltage and current obtained over DSO are shown in Fig. 2.12. They are scaled outputs as can be visible from the figure. Hence, the results are verified through Typhoon HIL-402 setup.

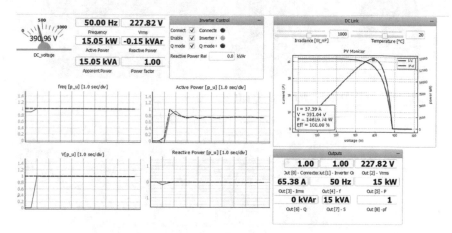

Fig. 2.9 The SCADA panel visuals after simulation

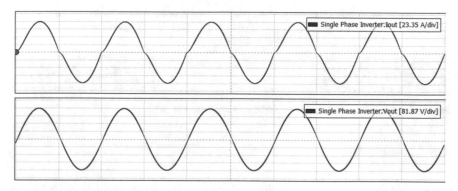

Fig. 2.10 The sinusoidal voltage and current obtained after simulation

Fig. 2.11 The overall system setup with the HIL-402 device

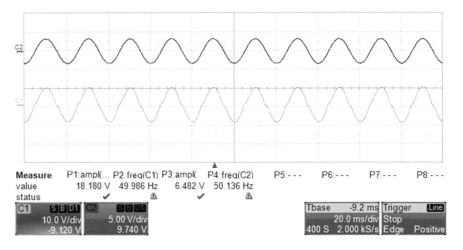

Fig. 2.12 The inverter output voltage and current obtained at DSO

2.5 Conclusion

Renewable energy sources including PV, winds, etc., are globally embraced at an increased rate. Their fast adoption has led to more utilization of power electronic converters worldwide for optimum power transfer and delivery. The advanced power electronics converter technology has provided significant favorable methods for interconnection with renewable energy generators, especially PV systems. This chapter talks about the importance of power converters, their modeling, and control functionalities in a grid-connected system. Majorly, it specifies the DC/AC converter or inverter mathematical analysis and various control approaches for performing proper power control and transfer. A basic control structure of a grid-connected three-phase inverter is detailed with PI control in the synchronous or dq reference frame. PI control provides minimum steady-state error with DC quantities, therefore stationary to synchronous frame transformation becomes important. The cascaded control consists of outer loop as the power control loop and the DC-link voltage control loop, and inner loop as the AC current control loop. This cascaded control along with each control loop modeling is able to regulate the DC voltage and AC current to be injected in an effective way. This kind of mathematical analysis can also be applied to other power electronics converters in order to achieve the desired control objectives. These investigations are further verified in a case study for single-phase grid-connected PV inverter simulation with the help of Typhoon HIL-402 device. The case study is able to show the relevance of the control and modeling.

References

Ahmad A et al (2018) Robust control of grid-tied parallel inverters using nonlinear backstepping approach. IEEE Access PP(8):1, https://doi.org/10.1109/ACCESS.2018.2875030

Ahmed KH, Massoud AM, Finney SJ, Williams BW (2011) A modified stationary reference frame-based predictive current control with zero steady-state error for LCL coupled inverter-based distributed generation systems 58(4):1359–1370

Al-shetwi AQ, Zahim M, Blaabjerg F (2018) Low voltage ride-through capability control for single-stage inverter-based grid-connected photovoltaic power plant. Sol Energy 159, 665–681, 2018, https://doi.org/10.1016/j.solener.2017.11.027

Bharath KVS, Haque A, Khan MA (2018) Condition monitoring of photovoltaic systems using machine learning techniques. 2018 2nd IEEE international conference power electron. Intell. Control Energy Syst. ICPEICES 2018, pp. 870–875, https://doi.org/10.1109/ICPEICES.2018.8897413

Bisht R, Bhattarai R, Subramaniam S, Kamalasadan S (2020) A novel synchronously rotating reference frame based adaptive control architecture for enhanced grid support functions of single phase inverters 9994(c):1–11, https://doi.org/10.1109/TIA.2020.2994879

Bojoi RI, Limongi LR, Roiu D, Tenconi A (2011) Enhanced power quality control strategy for single-phase inverters in distributed generation systems 26(3):798–806

Bose BK (2008) Neural network applications in power electronics and motor drives—an introduction and perspective. IECON Proc (Indus Electron Conf) 54(1):25–27, https://doi.org/10.1109/IECON.2008.4757921

Bose BK (2017) Artificial intelligence techniques in smart grid and renewable energy systems—some example applications. Proc IEEE 105(11):2262–2273. https://doi.org/10.1109/JPROC.2017.2756596

Cai H, Wei W, Peng Y, Hu H (2014) Fuzzy proportional-resonant control strategy for three-phase inverters in islanded micro-grid with nonlinear loads. IEEE international conference fuzzy system, pp 707–712, https://doi.org/10.1109/FUZZ-IEEE.2014.6891682

Castello J, Espi JM, Garcia-Gil R (2016) A new generalized robust predictive current control for grid-connected inverters compensates anti-aliasing filters delay. IEEE Trans Ind Electron 63(7):4485–4494. https://doi.org/10.1109/TIE.2015.2497303

Chandran P, Madhura RS, Roselyn JP, Devaraj D, Gopal V, Ravi A (2019) Development of intelligent fuzzy PQ-FRT control strategy for grid-connected solar PV system. IEEE International conference intelligent techology control Optimum. Signal Process. INCOS 2019, pp. 7–11, https://doi.org/10.1109/INCOS45849.2019.8951347

Cortés P, Wilson A, Kouro S, Rodriguez J, Member S (2010) Model predictive control of multilevel cascaded H-bridge inverters 57(8):2691–2699

Cortes D, Vázquez N, Alvarez-Gallegos J (2009) Dynamical sliding-mode control of the boost inverter. IEEE Trans Ind Electron 56(9):3467–3476. https://doi.org/10.1109/TIE.2008.2010205

Chavali RV, Dey A, Das B (2022) A hysteresis current controller PWM scheme applied to three-level NPC inverter for distributed generation interface. IEEE Trans Power Electron 37(2):1486–1495. https://doi.org/10.1109/TPEL.2021.3107618

Davoodnezhad R, Holmes DG, McGrath BP (2014) A novel three-level hysteresis current regulation strategy for three-phase three-level inverters. IEEE Trans Power Electron 29(11):6100–6109. https://doi.org/10.1109/TPEL.2013.2295597

Dasgupta S, Sahoo SK, Panda SK (2011) Single-phase inverter control techniques for interfacing renewable energy sources with microgrid-Part I: parallel-connected inverter topology with active and reactive power flow control along with grid current shaping. IEEE Trans Power Electron 26(3):717–731. https://doi.org/10.1109/TPEL.2010.2096479

Du Y, Lu DDC, Chu GML, Xiao W (2015) Closed-form solution of time-varying model and its applications for output current harmonics in two-stage PV inverter. IEEE Trans Sustain Energy 6(1):142–150. https://doi.org/10.1109/TSTE.2014.2360616

Ebrahimi M, Member S, Khajehoddin SA (2015) Fast and robust single-phase DQ current controller for smart inverter applications 8993(c):1–9, https://doi.org/10.1109/TPEL.2015.2474696

Fatama AZ, Khan MA, Kurukuru VSB, Haque A, Blaabjerg F (2020) Coordinated reactive power strategy using static synchronous compensator for photovoltaic inverters. Int Trans Electr Energy Syst 30(6):1–18. https://doi.org/10.1002/2050-7038.12393

Han Y et al (2017) Stability analysis of digital controlled single- phase inverter with synchronous reference frame voltage control 8993(c), https://doi.org/10.1109/TPEL.2017.2746743

Hannan MA, Ghani ZA, Hoque MM, Ker PJ, Hussain A, Mohamed A (2019) Fuzzy logic inverter controller in photovoltaic applications: Issues and recommendations. IEEE Access 7(c):24934–24955, https://doi.org/10.1109/ACCESS.2019.2899610

Hannan MA, Ghani ZA, Mohamed A, Uddin MN (2015) Real-Time testing of a fuzzy-logic-controller-based grid-connected photovoltaic inverter system. IEEE Trans Ind Appl 51(6):4775–4784. https://doi.org/10.1109/TIA.2015.2455025

Hasanien HM (2016) An adaptive control strategy for low voltage ride through capability enhancement of grid-connected photovoltaic power plants 31(4):3230–3237

Kazmierkowski MP, Malesani L (1998) Current control techniques for three-phase voltage-source PWM converters : a survey 45(5):691–703

Khan , Haque A, Kurukuru VSB (2019) Machine learning based islanding detection for grid connected photovoltaic system. 2019 international conferencepower electron. Control Autom. ICPECA 2019 - Proc., vol. 2019-Novem, no. 1, https://doi.org/10.1109/ICPECA47973.2019.8975614

Khan MA, Haque A, Bharath Kurukuru VS (2020) Reliability analysis of a solar inverter during reactive power injection. 9th IEEE international conference power electron. Drives Energy Syst. PEDES 2020, https://doi.org/10.1109/PEDES49360.2020.9379776

Khan MA, Haque A, Bharath KVS, Droop based low voltage ride through implementation for grid integrated photovoltaic system

Kouro S, Leon JI, Vinnikov D, Franquelo LG (2015) Grid-connected photovoltaic systems: an overview of recent research and emerging PV converter technology. IEEE Ind Electron Mag 9(1):47–61. https://doi.org/10.1109/MIE.2014.2376976

Kumar N, Saha TK, Dey J (2020) Multilevel Inverter (MLI)-based stand-alone photovoltaic system: modeling, analysis and control 14(1):909–915

Kurukuru VSB, Haque A, Khan MA, Sahoo S, Malik A, Blaabjerg F (2021) A review on artificial intelligence applications for grid-connected solar photovoltaic systems. Adv Renew Energy Syst (Part 1 2), pp 161–174, https://doi.org/10.1201/b18242-7

Kumar M, Tyagi B (2021) A robust adaptive decentralized inverter voltage control approach for solar PV and storage-based islanded microgrid. IEEE Trans Ind Appl 57(5):5356–5371. https://doi.org/10.1109/TIA.2021.3094453

Lee DC, Sul SK, Park MH (1992) High performance current regulator for a field-oriented controlled induction motor drive. Conference Rec. - IAS Annual Meeting (IEEE Ind. Appl. Soc., vol 1992-Janua, no 5, pp 538–544, https://doi.org/10.1109/IAS.1992.244349

Li X, Balog RS (2015) PLL-less robust active and reactive power controller for single phase grid-connected inverter with LCL filter, Conference Proceeding—IEEE appllication power electronics conference Expo. - APEC, vol 2015-May, May, pp 2154–2159, https://doi.org/10.1109/APEC.2015.7104647

Liao H, Zhang X, Ma Z (2021) Robust dichotomy solution-based model predictive control for the grid-connected inverters with disturbance observer. CES Trans Electr Mach Syst 5(2):81–89. https://doi.org/10.30941/cestems.2021.00011

Liu Q, Member S, Caldognetto T, Buso S (2020) Review and comparison of grid-tied inverter controllers in microgrids 35(7):7624–7639

Liu YH, Chen CL, Tu RJ (1998) A novel space-vector current regulation scheme for a field-oriented-controlled induction motor drive. IEEE Trans Ind Electron 45(5):730–737. https://doi.org/10.1109/41.720329

Merabet A (2017) Control system for dual-mode operation of grid-tied photovoltaic and wind energy conversion systems with active and reactive power injection. Saint Mary's Univ. Halifax, NS, Canada, no. August, 2017, https://doi.org/10.13140/RG.2.2.26509.44006

Meireles MRG, Almeida PEM, Simões MG (2003) A comprehensive review for industrial applicability of artificial neural networks. IEEE Trans Ind Electron 50(3):585–601. https://doi.org/10.1109/TIE.2003.812470

Parvez M, Fathi M, Elias M (2020) Comparative study of discrete pi and pr controls for single-phase UPS inverter 4:1–13, https://doi.org/10.1109/ACCESS.2020.2964603

Purba V, Johnson BB, Rodriguez M, Jafarpour S, Bullo F, Dhople SV (2019) Reduced-order aggregate model for parallel-connected single-phase inverters. IEEE Trans Energy Convers 34(2):824–837. https://doi.org/10.1109/TEC.2018.2881710

Ramchand R (2018) A new control strategy for single phase cascaded H bridge multilevel inverter in stationary reference frame with nonlinear loads

Rowan TM, Kerkman RJ (1986) A new synchronous current regulator and an analysis of current-regulated PWM inverters I(4)

Romero-cadaval E, François B, Malinowski M (2015) Grid-connected photovoltaic generation plants as alternative energy sources

Strache S, Wunderlich R, Heinen S (2014) A Comprehensive, quantitative comparison of inverter architectures for various PV Systems, PV cells, and irradiance profiles. IEEE Trans Sustain Energy 5(3):813–822. https://doi.org/10.1109/TSTE.2014.2304740

Tarasantisuk C, Kumsup S, Piyarat W, Witheepanich K (2016) Stationary frame current regulation using Proportional Resonant controller for single phase grid connected inverter. 2016 13th international conference electrical engineering computer telecommun infernational technology ECTI-CON 2016, no 2, pp 3–7, https://doi.org/10.1109/ECTICon.2016.7561350

Xu S, Xue Y, Chang L (2021) Review of power system support functions for inverter-based distributed energy resources- standards, control algorithms, and trends. IEEE Open J Power Electron 2:88–105. https://doi.org/10.1109/ojpel.2021.3056627

Yang L, Chen Y, Luo A, Huai K (2019) Stability enhancement for parallel grid-connected inverters by improved notch filter. IEEE Access 7:65667–65678. https://doi.org/10.1109/ACCESS.2019.2917533

Yang T, Cai Z, Xun Q (2020) Adaptive backstepping-based H∞ robust controller for photovoltaic grid- connected inverter. IEEE Access 8:17263–17272. https://doi.org/10.1109/ACCESS.2019.2962280

Yu C, Xu H, Liu C, Wang Q, Zhang X (2019) Modeling and analysis of common-mode resonance in multi-parallel PV string inverters. IEEE Trans Energy Convers 34(1):446–454. https://doi.org/10.1109/TEC.2018.2877911

Zeb K et al (2017) A comprehensive review on inverter topologies and control strategies for grid connected photovoltaic system 94:1120–1141, https://doi.org/10.1016/j.rser.2018.06.053

Zmood DN, Holmes DG (2003) Stationary frame current regulation of PWM inverters with zero steady-state error. IEEE Trans Power Electron 18(3):814–822. https://doi.org/10.1109/TPEL.2003.810852

Zheng L, Jiang F, Song J, Gao Y, Tian M (2018) A discrete-time repetitive sliding mode control for voltage source inverters. IEEE J Emerg Sel Top Power Electron 6(3):1553–1566. https://doi.org/10.1109/JESTPE.2017.2781701

Chapter 3
Grid-Forming Converter Control Techniques Implementation in Typhoon HIL

Jose Miguel Riquelme-Dominguez, Hugo Mendonça, and Francisco Gonzalez-Longatt

Abstract The concentration of greenhouse gases in the Earth's atmosphere is directly linked to the average global temperature on Earth. The climate emergency is a reality that requires enormous efforts from the energy sector to reduce greenhouse emissions. As the number of environmentally friendly sources of electricity is incremented, the electricity infrastructure is challenged by several aspects coming from the electrical behaviour of those sources. The environmentally friendly sources, like wind power and solar photovoltaic, use power electronic converters as interfaces to the grid, and them has a performance dramatically different from the synchronous machines used in the traditional power plants. As the number of old pollutant power plants connected to the power system increases to accommodate more and more generation coming renewable power generation, the operational security and resilience of the power system are negatively affected. Implementing novel control techniques on the power electronic converters has become a realistic option to cope with the adverse effects. One family of control technics known as "grid forming" looks like a promising mechanism to overcome some of the most important effects coming from the massive reduction in the number of synchronous generators connected to the power system. This chapter documents the implementation of different power electronic converter control strategies in the Typhoon HIL environment. The implemented controllers belong to the family of grid-forming techniques, and they are droop control, virtual synchronous machine, and synchronverter. To establish a fair comparison amonbehaviourree grid-forming converter control techniques and

J. M. Riquelme-Dominguez (✉) · H. Mendonça
Escuela Técnica Superior de Ingenieros Industriales, Universidad Politécnica de Madrid, Madrid, Spain
e-mail: jm.riquelme@upm.es

H. Mendonça
e-mail: hugo.rocha@upm.es

F. Gonzalez-Longatt
University of South-Eastern Norway, Porsgrunn, Norway
e-mail: fglongatt@fglongatt.org

Centre of Smart Grid, University of Exeter, Exeter, UK

evaluate their behaviour of them, a validation test has been carried out with each controller implementation connected to two isolated loads. Finally, a conclusion is given, revealing the strengths and weaknesses of each considered performance.

Keywords Grid-forming converters · Droop control · Virtual synchronous machine · Synchronverter · Typhoon HIL

3.1   Introduction

Significant changes are taking place in modern power systems. These changes touch on every aspect of the system, including how electricity is produced, the mechanism used to supply it, and how the end user uses the electricity.

In terms of electricity production, renewable energy generators such as wind and solar photovoltaics (PV), among others, are displacing conventional synchronous generators from the generation mix. Although these kinds of energy sources are considered environmentally friendly technologies, their output power highly depends on the primary renewable source, which is, to a certain extent, volatile and challenging due to dependence on the weather (Gonzalez-Longatt et al. 2021). In addition, the intermittent nature of renewable generators contrasts with the traditional operation of conventional synchronous generators. The operation of electricity distribution networks is also changing. In recent times, there has been a trend to optimising available resources in electricity transmission systems to take full advantage of distributed generation, such as small photovoltaic generators, microturbines, biomass, and fuel cells.

In fact, current trends suggest that the interaction between distribution system operators (DSOs) and transmission system operators (TSOs) is changing. All parties involved in the power system will profit from the complete interaction between TSOs and DSOs, which will release the potential of the present and foreseeable resources. Finally, consumers have become an active part of the power system due to the integration of embedded generation (EG), the widespread use of electric vehicles (EV), and electrical energy storage (EES). The emergence of the aforementioned technologies together with the change in consumer behaviour makes the latter a prosumer, an entity that both consumes and produces electricity. The unifying factor behind these advancements is the widespread adoption of power electronic converter interfaced technologies (Chamorro et al. 2021; Acosta et al. 2021). Nonconventional generation and storage technologies rely on power electronic converters (PECs) to supply more manageable electricity and to integrate energy storage and renewable resources in power systems.

The PEC is a critical component in integrating new low-carbon technologies (LCT), as it is an enabling technology that creates the necessary interface between several energy systems (Gonzalez-Longatt et al. 2021; Gorostiza and Gonzalez-Longatt 2020).

The growing use of PECs is reducing the number of synchronous generators (SGs) available in the power system, which has several implications from the point of view of system operation. Two significant difficulties have been cited in numerous research publications and projects (ENTSO-E 2021; Operator 2016): (i) low (or nonexistent) supply of total system rotational inertia and (ii) reduced and limited fault levels affecting short circuit ratio.

Different actors have identified the above-mentioned issues: transmission system operators, researchers, and manufacturers (IEEFA 2021). Several documents have reported some of the negative aspects associated with high penetration of PEC technology in power systems: lack of robustness (specifically during extreme overcurrent events and significant voltage drops), malfunction of the Phase-Locked Loop (PLL) to track extremely low voltage sags (Grid 2016), fault ride-through (FRT) failures, and unfavourable interactions. The IEEE's power system dynamic performance committee acknowledged, in April 2020 (Hatziargyriou et al. 2020), the necessity for new types of dynamic behaviour in electrical power systems with a high penetration of power electronic interfaced technologies. Due to the broad presence of PEC-interfaced technologies in bulk power systems, the classification and definition of power system stability phenomena were improved by adding new factors. Two new stability categories have been introduced (Hatziargyriou et al. 2020): (i) converter-driven stability and (ii) resonance stability.

The dynamic behaviour of renewable energy sources is radically different from conventional synchronous generators due to the presence of voltage-source converters (VSC), which serve as an interface with the grid (Hatziargyriou et al. 2020). The existence of high penetration of VSC in power systems has caused several local instabilities, known as a converter-driven instabilities. These instabilities can be caused, among others, by an incorrect design of the control settings or wrongly calibrated controllers. However, the substitution of conventional generation with PEC-based energy sources does not necessarily have a negative impact on the grid. Nevertheless, it is true that, as discussed above, it can be a source of instability.

On the other hand, if the controller settings are appropriately adjusted, PEC-based technologies can provide solutions to some of the problems of power systems, e.g., those derived from low inertia power systems. PEC-interfaced technologies that take the place of traditional synchronous generators can be equipped with controllers to react to unforeseen occurrences and system imbalances exceptionally quickly; in fact, PEC-interfaced technologies can do that much faster than mechanical synchronous machines. A PEC-dominated power system needs both short-term and long-term solutions to improve stability. One of these potential solutions is related to controlling the grid-side inverter based on VSCs using grid-forming control. The grid-forming control technique allows for supporting the grid operation since it emulates the behaviour of conventional synchronous generators. There are different alternatives for grid-forming control implementation.

A little summary of the main control techniques used to emulate the behaviour of a synchronous generator is shown in Fig. 3.1 (see more details at (Tamrakar et al. 2017)).

In this chapter, the authors present details of the modelling and implementation of three grid-forming control techniques in the digital real-time simulation framework of

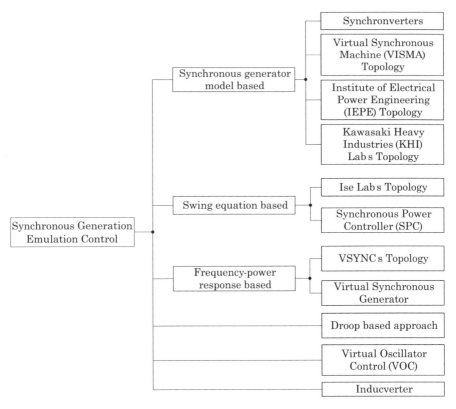

Fig. 3.1 Classification of different control strategies used for the implementation of synchronous generation emulation

Typhoon HIL. The grid-forming control techniques implemented in this chapter are Droop Control, Virtual Synchronous Machine (VSM), and Synchronverter (SynC).

This chapter is organised as follows: Sect. 3.2 presents the test system used to implement the grid-forming strategies and a description of the Droop, VSM, and SynC control techniques. Section 3.3 shows real-time simulation results considering the test system and the three grid-forming control techniques. Simulations have been carried out using the Virtual HIL functionality available in the Typhoon HIL Control Centre. Finally, Sect. 3.4 summarises the main points of this chapter.[1]

---

[1] No academic titles or descriptions of academic positions should be included in the addresses. The affiliations should consist of the author's institution, town/city, and country.

3.2 Modelling

The objective of this chapter is to present details of the modelling and implementation of three grid-forming control techniques in digital real-time simulation.

This section presents the modelling aspects of the three grid-forming control techniques considered in this chapter.

3.2.1 Test System

A straightforward but very extremely useful isolated test system is used for modelling purposes. The system is formed by a VSC-based power converter feeding two loads. Load 1 is a three-phase series RL branch, whereas Load 2 is a three-phase series RLC branch. An RLC filter is used as an interface between the VSC and Bus 1 to reduce harmonics due to switching. Every load is equipped with a circuit breaker (S_1, S_2) that allows to connect and disconnect them when required. The three-phase AC test system is depicted in Fig. 3.2, including the main parameters of the system and the main electrical variables involved.

The implementation of the test system in the Typhoon HIL Schematic Editor (.tse file) is shown in Fig. 3.3.

As depicted in Fig. 3.3, the test system is implemented with several Typhoon HIL blocks available in the library *core*: the "Three Phase Inverter" block used to model the VSC, "Resistor", "Inductor", and "Capacitor" blocks used to model the filter and the loads. Some measurement elements have also been included, such

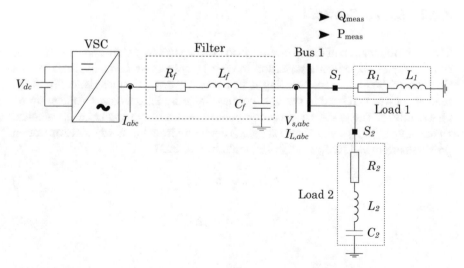

Fig. 3.2 Test System: A single VSC connected to two constant impedance loads

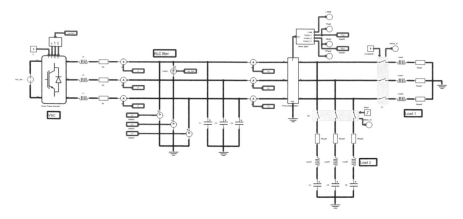

Fig. 3.3 Implementation of the isolated test systems inside the Typhoon HIL framework

as "Voltage Measurement", "Current Measurement", and a "Three-phase Metre" blocks. The latter is installed at Bus 1 to measure the loads' active and reactive power consumed. For the connection and disconnection of loads, blocks "Triple Pole Single Throw Contactor" are used as circuit breakers. Those circuit breakers can be controlled through "Step" signal blocks. In addition, "Probe" blocks need to be added to visualise signals in HIL SCADA during real-time simulations.

Once the test system has been presented, the following subsections introduce the three grid-forming control techniques.

3.2.2 Droop Control

The main foundation of droop control is to emulate the very well-known behaviour of a synchronous generator governor, in which any increase in the load results in a decrease of the frequency according to its frequency droop characteristic. Similarly, reactive power is related to the voltage magnitude by introducing a voltage droop characteristic. The grid-forming droop control uses a droop approach to calculate frequency (Δf_{droop}) and voltage (ΔV_{droop}) deviation from the steady-state operation point (Eberlein and Rudion 2020; Habibullah et al. 2021):

$$\Delta f_{droop} = m_p \Delta P \tag{3.1}$$

$$\Delta V_{droop} = m_q \Delta Q \tag{3.2}$$

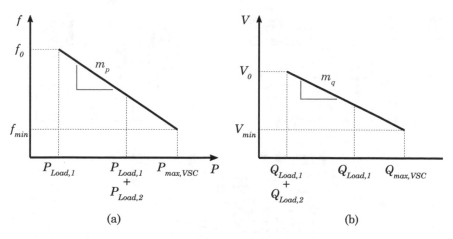

Fig. 3.4 Determination of m_p and m_q: **a** P-f droop and **b** Q–V droop

where m_p and m_q are the active and reactive power droop coefficients and ΔP and ΔQ are the low-pass filtered active and reactive power deviations from the steady-state operating point, respectively, the droop coefficients were determined according to Fig. 3.4.

As depicted in Fig. 3.4, the nominal frequency (f_0) and voltage (V_0) need to be specified for some load conditions (e.g. Load,1 and Load,2). Then, the minimum frequency (f_{min}) and voltage (V_{min}) are stated according to the maximum power of the VSC. In this chapter, the maximum active power of the VSC is 1.2 times the active power consumed by the loads, and the maximum reactive power is 1.2 times the reactive power consumed by Load 1. It is worth mentioning that the character of the Load 1 is inductive, while Load 2 is of a capacitive nature. For this reason, the total reactive power consumed by the two loads is lower than the one consumed by Load 1.

The following subsections show the model implementation of the droop control technique for a VSC-based power electronic converter.

3.2.2.1 Measurement and Reference System

In general, closed-loop control systems require sensors to perform their function. The droop control technique needs the appropriate variables in order to achieve its control action. Figure 3.5 shows the implementation of the necessary measurements for the grid-forming droop control. The input variables of the controller are the output current of the VSC (I_{a1}, I_{b1}, I_{c1}), the voltage of the filter capacitor (V_{a1}, V_{b1}, V_{c1}), and the output RLC filter current (I_{a2}, I_{b2}, I_{c2}). All of these variables are converted to $dq0$ rotating reference frame by the transformation block "abc to dq" from the library "core" of Typhoon HIL. As a result, the measurements of relevant voltages

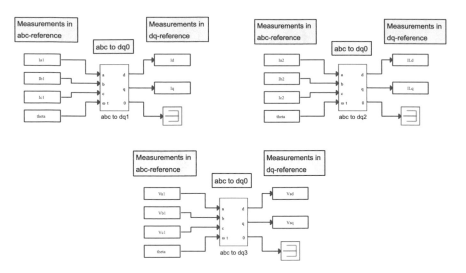

Fig. 3.5 Measurement acquisition in the synchronous rotating frame (*dq*-axis)

and currents (V_{sd}, V_{sq}, I_d, I_q, I_{Ld}, and I_{Lq}) are obtained in the rotating reference frame for further processing.

3.2.2.2   Active and Reactive Power Controller

The active and reactive power controller, which is one of the outer control loops, is used to regulate the voltage amplitude and frequency, respectively.

Once it has the required voltages and currents in the rotating frame, the instantaneous active (3.3) and reactive power (3.4) are calculated from the measured output voltage and current in the synchronous rotating frame (*dq*) by the following equations:

$$P_{meas} = \frac{3}{2}\left(V_{sd}I_{Ld} + V_{sq}I_{Lq}\right) \tag{3.3}$$

$$Q_{meas} = \frac{3}{2}\left(V_{sd}I_{Lq} + V_{sq}I_{Ld}\right) \tag{3.4}$$

The reference frequency ω_{set} and the reference voltage magnitude $V_{sd,set}$ come from the *P-f* and *Q–V* droop control (3.5) and (3.6). The primary control is expressed by the following equations:

$$\omega_{set} = \omega_0 + m_p(P_{set} - P_{meas}) \tag{3.5}$$

$$V_{sd,set} = V_0 + m_q(Q_{set} - Q_{meas}) \tag{3.6}$$

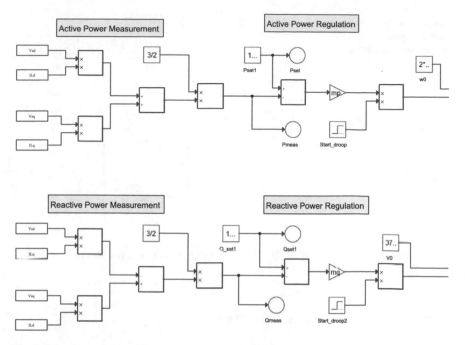

Fig. 3.6 Implementation of the power controllers in the droop control

Figure 3.6 depicts the implementation of the active and reactive power measurements in Typhoon HIL Schematic Editor with their input variables (V_{sd}, V_{sq}, I_{Ld}, I_{Lq}). These measurements are introduced in Eqs. (3.3) and (3.4) in order to obtain the measured active and reactive power (P_{meas}, Q_{meas}). The measured power is compared with the set power (P_{set}, Q_{set}) and the power deviation from the setpoint is obtained, which is introduced in (3.5) and (3.6). First, the power variation is multiplied by the respective droop coefficient and then added to the frequency and voltage setpoints to define the frequency reference (ω_{set}) and the voltage magnitude reference ($V_{sd,ref}$), respectively.

The reference frequency and the reference voltage magnitude are conducted separately through "Goto" blocks to the Active Power Synchronisation control and the Voltage and Current controllers, as explained in the following subsections.

3.2.2.3 Active Power Synchronisation

The active power synchronisation control serves to define the control angle (theta, $\theta = \omega t$) as depicted in Fig. 3.7. It takes the reference frequency as input through a "From" block in radians per second unit. Then, the reference frequency is integrated into an angle increment (Converters et al. 2010) by the "Integrator" block from the library "core". The result is the output signal θ. It is noteworthy that the Integrator

Fig. 3.7 Implementation of
the active power
synchronisation control

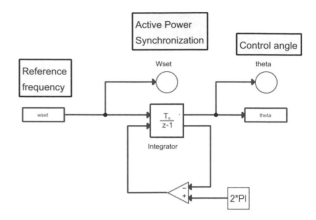

may be reset when the status of that block is greater than 2π, which corresponds to
one complete revolution.

3.2.2.4 Voltage and Current Controllers

The voltage and current controllers are known as the inner control loops. They are
formed by two pairs of cascaded PI-controllers, and their goal is the regulation of ac
voltages and currents at the output of the converter's filter. The control is made by
decoupling the reference voltage magnitude of each direct-quadrature axis, $V_{sd,ref}$ and
$V_{sq,ref}$. Figure 3.8 shows the implementation of the voltage and current controllers.

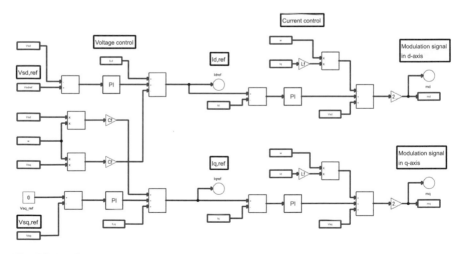

Fig. 3.8 Implementation of the voltage and current controllers

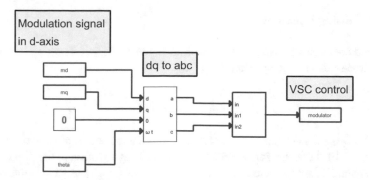

Fig. 3.9 Modulation signal generation for the control of the VSC

Figure 3.8 represents the voltage and current controllers implementation in Typhoon HIL Schematic Editor, including the structure of each individual controller, all feedback signals, and feed-forward terms. The voltage control receives the references of the voltage magnitude in the direct-quadrature axis ($V_{sd,ref}$, $V_{sq,ref} = 0$), and compares them against the actual value of those magnitudes (V_{sd}, V_{sq}). The error is sent to two PI-controllers that produce, after adding feedback and feed-forward terms, the reference current values for the current control, $I_{d,ref}$ and $I_{q,ref}$. The proportional-integral control is done using the "PID controller" block from the library "core", and it is characterised by the proportional and integral gains (K_{pv} and K_{iv}, respectively). The structure of the current control is similar to the one of the voltage control. It receives the references of the converter-side current magnitude in the direct-quadrature axis from the voltage control, $I_{d,ref}$ and $I_{q,ref}$. The error between the reference and the actual values (I_d and I_q) in each axis is sent to both PI-controllers (whose gains are K_{pc} and K_{ic}) that give, after adding feedback and feed-forward terms, the modulation signals in the d-q frame m_d and m_q.

Finally, the modulation signals (m_d, m_q) are first translated to "abc" reference frame with the block "dq to abc", and the output is concatenated with the block "Bus Join" from the library "core". The output of this block is the signal that governs the VSC, as depicted in Fig. 3.9.

3.2.3 Virtual Synchronous Machine

The Virtual Synchronous Machine (VSM) grid-forming technique is a more advanced control when compared to droop control. The novelty of the VSM is that it incorporates the well-known swing equation for the regulation of the active power generated by the VSC. It means that the rest of the control system, that is, the measurement system, the reactive power controller, the active power synchronisation, and the voltage and current controllers remain exactly the same as the ones for the droop control. The active power controller is the only component that is changed.

3.2.3.1 Swing Equation

As stated above, the active power regulation in the VSM grid-forming technique is done through the swing equation of an SG:

$$T_{acel} \frac{d\omega_{set}}{dt} = P_{set} - P_{meas} + K_d(\omega_0 - \omega_{set}) \tag{3.7}$$

where T_{acel} is the mechanical time constant, and K_d is the damping coefficient.

Figure 3.10 shows the implementation of the swing equation in the Typhoon HIL environment. It includes the measured active power (P_{meas}) at Bus 1; the nominal angular frequency (ω_0), and the active power command (P_{set}). Finally, the derivative of the reference frequency with respect to time is integrated to provide the reference frequency (ω_{set}).

As detailed in Sect. 3.2.3, the virtual rotor angular position of the VSM (θ) is given by the active power synchronisation control loop, and this angular position corresponds to the phase angle of the voltage induced by the VSM model.

The parameters of the VSM technique are defined for specific operating conditions of the VSC, as in the case of the droop control. The minimum angular frequency (ω_{min}) is set for the maximum active power of the VSC, $P_{max,VSC}$. The nominal frequency (ω_0) is set for specific load conditions, for example, $P_{set} = P_{Load,1}$. Therefore, K_d can be determined when the derivative of the reference frequency is zero:

$$K_d = \frac{P_{max,VSC} - P_{Load,1}}{\omega_0 - \omega_{min}} \tag{3.8}$$

The mechanical time constant is related to the maximum Rate of Change of Frequency (ROCOF) that the system can experiment with after an active power

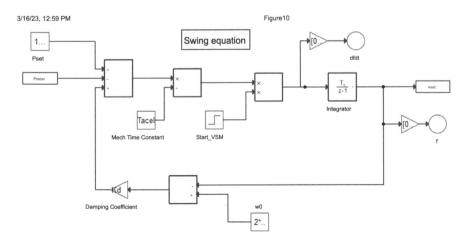

Fig. 3.10 Implementation of the swing equation in Typhoon HIL environment

unbalance. Just after the power imbalance, the angular frequency deviation is zero, and the derivative of the reference frequency only depends on T_{acel}, P_{set}, and P_{meas}. Thus, defining a maximum ROCOF, the mechanical constant keeps:

$$T_{acel} = \frac{P_{Load,1} - P_{max,VSC}}{\frac{d\omega_{set}}{dt}\Big|_{max}} = \frac{P_{Load,1} - P_{max,VSC}}{2\pi \frac{df_{set}}{dt}\Big|_{max}} \tag{3.9}$$

where $df_{set}/dt|_{max}$ is the maximum allowed ROCOF expressed in Hertz per second.

3.2.4 Synchronverter (SynC)

The Synchronverter is a grid-forming control that also mimics the behaviour of a synchronous generator, but it has the advantage that parameters such as inertia, damping, field inductance, and mutual inductance can be readily modified.

The different sections that make up the implementation of the Synchronverter are explained below.

3.2.4.1 Dynamic Swing Equations

The Synchronverter controller represents the dynamic swing equations of a synchronous generator by the equation of the frequency droop (3.10) and the equation of voltage droop (3.11) (Zhong and Weiss 2011).

$$T_{acel}\frac{d\omega_{set}}{dt} = T_{mech} - T_{elec} + D_p(\omega_0 - \omega_{set}) \tag{3.10}$$

$$K\frac{d(M_f i_f)}{dt} = Q_{set} - Q_{meas} + D_q(V_0 - V_{sd,set}) \tag{3.11}$$

The implementation of the dynamic swing equations in the Typhoon HIL environment is shown in Fig. 3.11, with the following input variables: the electromagnetic torque (T_{elec}), the reactive power (Q_{meas}) generated by the Synchronverter and the nominal voltage (V_0); the internal set values of the moment of inertia (T_{acel}), the integrator gain (K) to regulate the field excitation, the frequency droop coefficient (D_p), the voltage droop coefficient (D_q), the voltage setpoint ($V_{sd,set}$), the angular frequency setpoint (ω_0), the mechanical torque setpoint (T_{mech}) and the reactive power setpoint (Q_{set}); and the output variables of the reference angular frequency (ω_{set}) and reference field excitation ($M_f i_f$).

The mechanical torque (T_{mech}) can be obtained from the setpoint of active power (P_{set}) by dividing it by the nominal mechanical speed (ω_0). Again, the virtual rotor angle of the Synchronverter (θ) is given by the active power synchronisation control loop.

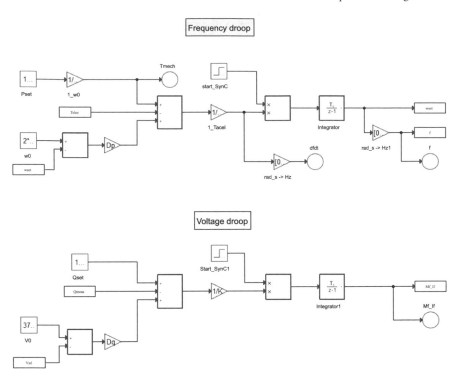

Fig. 3.11 Implementation of the dynamic swing equations in the Typhoon HIL environment

3.2.4.2 Trigonometric Functions

Trigonometric functions are necessary for the calculation of magnitudes as the active and reactive power measured at Bus 1, the electromagnetic torque provided by the VSC and the reference voltage at the output of the VSC filter. $\overline{\cos\theta}$ and $\overline{\sin\theta}$ are vectors defined as the three-phase angle difference with equal spacing of 120° or $2\pi/3$ in radians, as shown in Eqs. (3.12) and (3.13) (Fig. 3.12).

$$\overline{\cos\theta} = \begin{bmatrix} \cos\theta \\ \cos\left(\theta - \frac{2\pi}{3}\right) \\ \cos\left(\theta - \frac{4\pi}{3}\right) \end{bmatrix} \tag{3.12}$$

$$\overline{\sin\theta} = \begin{bmatrix} \sin\theta \\ \sin\left(\theta - \frac{2\pi}{3}\right) \\ \sin\left(\theta - \frac{4\pi}{3}\right) \end{bmatrix} \tag{3.13}$$

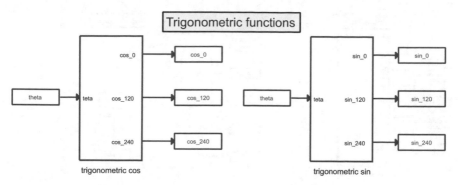

Fig. 3.12 Implementation of trigonometric functions in the Typhoon HIL environment

3.2.4.3 Active and Reactive Power Calculation

The measured active (P_{meas}) and reactive power (Q_{meas}) generated by the Synchronverter are given by Eqs. (3.14) and (3.15), where $\langle , \rangle$ denotes the inner product:

$$P_{meas} = \dot{\theta} M_f i_f \langle i, \overline{\sin \theta} \rangle \tag{3.14}$$

$$Q_{meas} = -\dot{\theta} M_f i_f \langle i, \overline{\cos \theta} \rangle \tag{3.15}$$

The implementation of the active and reactive power calculation in the Typhoon HIL environment are shown in Fig. 3.13. The inputs to the subsystems "Active Power" and "Reactive Power", which implements (3.14) and (3.15), are: the RLC filter output current ($I_{abc,2}$), the $\overline{\sin \theta}$ vector for the active power, the $\overline{\cos \theta}$ vector for the reactive power, the field excitation ($M_f\,i_f$), and the angular frequency reference (ω_{set}).

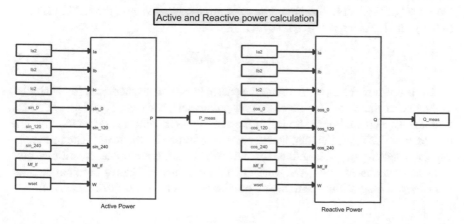

Fig. 3.13 Implementation of the active and reactive power equations in Typhoon HIL environment

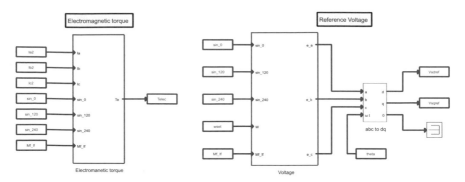

Fig. 3.14 Implementation of electromagnetic torque and the modulated voltage in the Typhoon HIL environment

3.2.4.4 Electromagnetic Torque and Control Signal of the VSC

The electromagnetic torque (T_{elec}) is the energy stored in the magnetic field of the machine, represented by Eq. (3.16):

$$T_{elec} = M_f i_f \langle i, \overline{\sin \theta} \rangle \qquad (3.16)$$

The implementation of the electromagnetic torque equation in the Typhoon HIL environment is shown in Fig. 3.14. The subsystem "Electromagnetic Torque" implements (3.16), being the input variables: the RLC filter output current ($i_2 = [i_{a2} \ i_{b2} \ i_{c2}]^T$), the $\overline{\sin \theta}$ vector, and the field excitation ($M_f i_f$).

Figure 3.14 also shows the calculation of the modulated voltage (e = [e_a e_b $e_c]^T$), which is first transformed from abc-frame to dq-frame, and then routed to the voltage and current controllers that generate the modulation signal of the VSC. The modulated voltage is calculated by Eq. (3.17) and implemented in the subsystem "Voltage" of Fig. 3.14. The input variables are the $\overline{\sin \theta}$ vector, the field excitation ($M_f i_f$), and the reference frequency (ω_{set}).

$$e = \dot{\theta} M_f i_f \overline{\sin \theta} \qquad (3.17)$$

The procedure followed in tuning the parameters of the Synchronverter is similar to the one of the VSM. In this case, four parameters, namely: T_{acel}, D_p, K and D_q need to be specified. For the tuning of the active power control parameters, $P_{max,VSC}$ is chosen as 1.2 times the total active power consumed by the loads. The minimum angular frequency (ω_{min}) is defined for that level of load. The nominal conditions refer to Load 1 connected to the VSC and Load 2 disconnected, with a nominal angular frequency of (ω_0). Thus, the damping coefficient can be determined by (3.18):

$$Dp = \frac{\dfrac{P_{max,VSC}}{\omega_{min}} - \dfrac{P_{Load,1}}{\omega_0}}{\omega_0 - \omega_{min}} \qquad (3.18)$$

The mechanical time constant is related to the maximum ROCOF that the system frequency can experiment immediately after an active power unbalance. This is because just after the power imbalance, the angular frequency deviation is zero, and the derivative of the reference frequency only depends on T_{acel}, T_{mech}, and T_{elec}. Thus, defining a maximum ROCOF, the mechanical constant keeps:

$$T_{acel} = \frac{\frac{P_{Load,1}}{\omega_0} - \frac{P_{max,VSC}}{\omega_{min}}}{\frac{d\omega_{set}}{dt}\Big|_{max}} = \frac{\frac{P_{Load,1}}{\omega_0} - \frac{P_{max,VSC}}{\omega_{min}}}{2\pi \frac{df_{set}}{dt}\Big|_{max}} \qquad (3.19)$$

where $df_{set}/dt|_{max}$ is the maximum allowed ROCOF expressed in Hertz per second. The tuning of Dq and K is identical to the one followed for T_{acel} and D_p.

3.3 Simulation and Results

This section shows the numerical results of the real-time simulations considering the proposed test system to illustrate the performance of the three implemented grid-forming control techniques. The models used for digital real-time simulation were implemented using the Typhoon HIL framework. The test system together with the controller models, was created using Typhoon HIL Schematic and then compiled and ran using the Typhoon HIL SCADA system.

The parameters of the test system and the three gird-forming control techniques are detailed in Tables 3.1, 3.2, 3.3, 3.4 and 3.5.

To check the correct implementation of the grid-forming control strategies, two scenarios have been proposed. In both scenarios, the VSC is feeding Load 1, and

Table 3.1 Main Parameters of the VSC

Description	Variable	Value and unit
Switching frequency	f_s	10 kHz
Rated DC link voltage	V_{dc}	1500 V
Filter resistance	R_f	2.1×10^{-3} Ω
Filter inductance	L_f	100×10^{-6} H
Filter capacitor	C_f	2500×10^{-6} F

Table 3.2 Main parameters of the loads

Description	Variable	Value and unit
Load 1 resistance	R_1	83×10^{-3} Ω
Load 1 inductance	L_1	137×10^{-6} H
Load 2 resistance	R_2	50×10^{-3} Ω
Load 2 inductance	L_2	68×10^{-6} H
Load 2 capacitance	C_2	13.55×10^{-3} F

Table 3.3 Main parameters of the droop control model

Description	Variable	Value
Active power droop coefficient	m_p	3.5910×10^{-6}
Reactive power droop coefficient	m_q	2.5675×10^{-5}
Voltage proportional gain	K_{pv}	1.6730
Voltage integrator gain	K_{iv}	374.59
Current proportional gain	K_{pc}	0.20
Current integrator gain	K_{ic}	4.14

Table 3.4 Main parameters of the VSM control model

Description	Variable	Value
Mechanical time constant	T_{acel}	3.9987×10^4
Damping coefficient	K_d	3.9987×10^5
Voltage proportional gain	K_{pv}	1.6730
Voltage integrator gain	K_{iv}	374.59
Current proportional gain	K_{pc}	0.20
Current integrator gain	K_{ic}	4.14

Table 3.5 Main parameters of the SynC control model

Description	Parameter	Value
Mechanical time constant	T_{acel}	125.20
Frequency droop coefficient	D_p	1.2521×10^3
Integrator gain	K	4.3285×10^5
Voltage droop coefficient	D_q	4.5214×10^4
Voltage proportional gain	K_{pv}	1.6730
Voltage integrator gain	K_{iv}	374.59
Current proportional gain	K_{pc}	0.20
Current integrator gain	K_{ic}	4.14

suddenly the Load 2 is connected to the system. The first case analyses the response of the grid-forming droop control considering a variation of the droop coefficient parameters. The second scenario compares the response of the three grid-forming techniques under the same event.

3.3.1 Analysis of Parameter Variation in Droop Control

The first scenario analyses the effect of varying the parameters of the droop control grid-forming technique. A pair of droop coefficients have been previously defined, and then they are modified to observe the response of the VSC control system. The first pair of droop parameters are selected to satisfy the following requirements: when only Load 1 is connected, the reference frequency is set to its nominal value ($\omega_0 = 120\pi$ rad/s), and the minimum frequency in Hertz is 59.5 Hz for the maximum active power of the converter (1.2 times the total active power consumed by the loads). In terms of reactive power, the peak nominal voltage at Bus 1 is defined as 400 V when the two loads are connected. The minimum voltage is set to 0.9 times the nominal voltage for the maximum reactive power of the VSC (1.2 times the reactive power consumed by Load 1). Figure 3.15 shows active power control performance in these conditions. It can be seen that the VSC delivers the requested active power ($P_{set} = 1.796$ MW) in less than 0.1 s, while the reference frequency is maintained at its nominal value, ω_0. At $t = 0.5$ s, switch S_2 is closed, and Load 2 is connected to the system. As a consequence, there is an active power imbalance of $|\Delta P| = 0.664$ MW, and the droop control reduces the reference frequency to 384 MHz. This situation is maintained until the end of the simulation.

The performance of the reactive power droop control is depicted in Fig. 3.16. It can be seen that the reactive power control is as fast as one of the active power, achieving the reactive power set ($Q_{set} = 1.118$ MVar) in less than 0.1 s. The voltage is set between the minimum voltage defined ($V_{min} = 360$ V) and its nominal value, according to the voltage droop characteristic: with just Load 1 connected, the voltage is set to $V_{set} = 371.28$ V. When Load 2 is connected at $t = 0.5$ s, its capacitive character almost compensates the reactive power consumed by Load 1, being the final reactive power consumed by loads of 1.5 kVar. The connection of Load 2 also implies an increment of the voltage at Bus 1, achieving the designed value of 400.0 V for that load conditions.

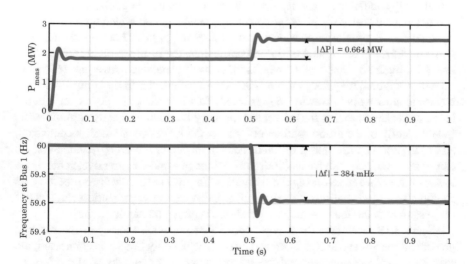

Fig. 3.15 Active power droop control performance after the connection of Load 2 at $t = 0.5$ s. Active power droop coefficient: $m_p = 3.591 \times 10^{-6}$

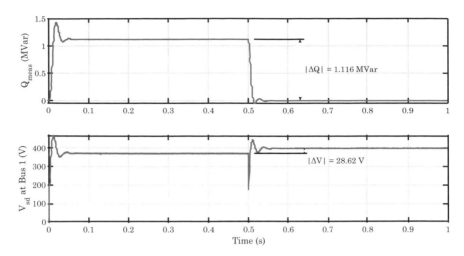

Fig. 3.16 Reactive power droop control performance after the connection of Load 2 at t = 0.5 s. Reactive power droop coefficient: $m_q = 2.5675 \times 10^{-5}$

These results confirm the correct implementation of several components of the model, including: the measurement system, the active power synchronisation control, the current and voltage controllers, and the active and reactive droop control. In the following, an analysis is carried out in which the droop parameters are scaled up and down to study the effects of these changes in the droop control performance. The initial droop coefficients were: $m_{p,1} = 3.591 \times 10^{-6}$ and $m_{q,1} = 2.5675 \times 10^{-5}$. These coefficients are scaled down by 0.8 to get $m_{p,2}$ and $m_{q,2}$, and scaled up by 1.2 to get $m_{p,3}$ and $m_{q,3}$. A simulation considering the connection of Load 2 at $t = 0.5$ s has been performed for the three pairs of droop coefficients, the results plotted in Figs. 3.17 and 3.18. Figure 3.17 shows the results in terms of generated active power and the reference angular frequency of the VSC. As depicted, the three control schemes behave identically before switch S_2 is closed. This is because the power set is equal to the power measured, so the reference frequency is maintained at its nominal value independently of the droop coefficients. At $t = 0.5$ s, the second load is connected, and each droop control has a different behaviour due to the droop parameters. As can be seen, the one with the larger droop coefficient is more aggressive, achieving a lower value of the reference angular frequency. This means that the droop coefficients should be selected carefully because a great value can cause protection elements of the system to react due to some magnitude exceeding a threshold.

Figure 3.18 represents the droop control performance in terms of the reactive power and the voltage at Bus 1. As can be seen, when only Load 1 is connected, the reactive power measurement and the reactive power set coincide, so the reference voltage is the same for the three droop schemes. At $t = 0.5$ s, the second load is connected, and the control system with the greater droop parameters is more aggressive, achieving a reference voltage of 405.5 V.

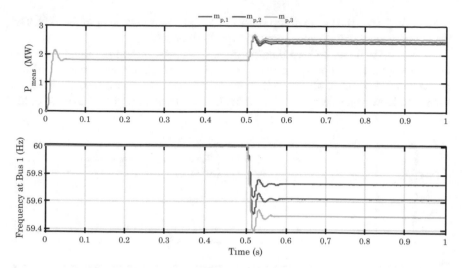

Fig. 3.17 Droop control performance after the connection of Load 2 at t = 0.5 s. Three different active power droop coefficients were considered

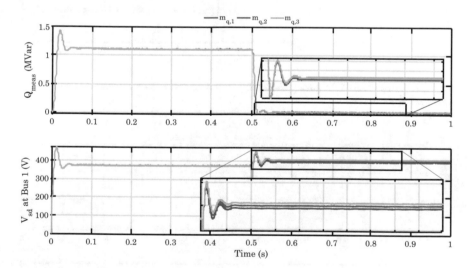

Fig. 3.18 Droop control performance after the connection of Load 2 at t = 0.5 s. Three different reactive power droop coefficients were considered

3.3.2 Comparison of the Three Grid-Forming Strategies

The three strategies have been compared in the same scenario. Again, the VSC is feeding Load 1, and Load 2 is suddenly connected to Bus 1 at $t = 0.5$ s. The VSM and the Synchronverter strategies have been designed for a minimum angular frequency of $f_0 = 59.7$ Hz, whereas in droop control, this value has been kept to 59.5 Hz. The results of the simulations are plotted in Figs. 3.19 and 3.20.

In Fig. 3.19, it can be seen that the three strategies present the same evolution in the first part of the simulation due to the initial load conditions. When Load 2 is connected at $t = 0.5$ s, all the techniques increment the active power delivered to the loads. However, as can be seen in the reference frequency plot, one of the droop

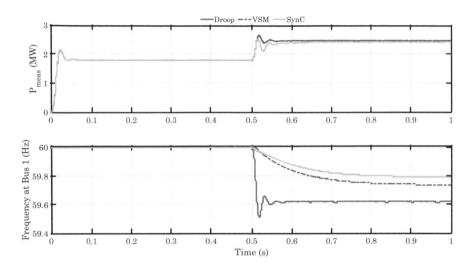

Fig. 3.19 Comparison of the three grid-forming techniques after the connection of Load 2 at t = 0.5 s. Active power control

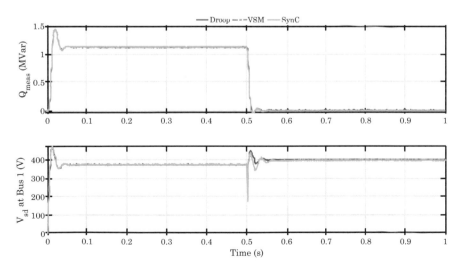

Fig. 3.20 Comparison of the three grid-forming techniques after the connection of Load 2 at t = 0.5 s. Reactive power control

control is much more aggressive than the other two due to the linear characteristic of the droop strategy. As a result, the reference frequency falls slower in the VSM and SynC strategies by incorporating the dynamic swing equation, which considers the derivative of that magnitude. In this simulation, it is shown that the incorporation of the electromagnetic torque in the swing equation is preferable to have slower decay of the reference frequency, as depicted in Fig. 3.19.

Figure 3.20 depicts the evolution of the reactive power measured and the reference voltage at Bus 1 for the three simulations. As shown, the results in this case are quite similar. In fact, the results of the droop control and the VSM strategy are identical as the VSM technique has been implemented with a droop control for the reactive power controller. The Synchronverter technique presents similar results with a greater oscillation when the power imbalance occurs due to the control of the electromagnetic torque instead of the active power, as shown in Fig. 3.19.

3.4   Conclusions

In this chapter, three grid-forming control techniques for VSC have been implemented in the Typhoon HIL environment. The test system consisted of an isolated three-phase system formed by the converter and two loads. The implementation of the control techniques implied several subsystems and control loops, including the measurement system, the current and voltage controllers, the active power synchronisation control, and the active and reactive power controllers. Throughout this chapter, all of these components have been presented in detail, both the fundamental equations and their modelling in the Typhoon HIL Schematic Editor. Finally, simulation results show the correct functioning of the grid-forming techniques. An analysis of the droop control coefficients was presented. The comparison between the three control techniques showed that VSM and SynC provide inertia to the system in contrast to droop control. The main difference between VSM and SynC is related to the inclusion of the dynamic swing equations of the synchronous generator in terms of torque.

References

Acosta MN, Gonzalez-Longatt F, Topić D, Andrade MA (2021) Optimal microgrid-interactive reactive power management for day-ahead operation. Energies 14(5):1275. https://doi.org/10.3390/en14051275
Australian Energey Market Operator, "Black System South Australia 28 September 2016," 2017. Accessed: Mar. 30, 2021. [Online]. Available: www.aemo.com.au
Chamorro HR et al (2021) Data-driven trajectory prediction of grid power frequency based on neural models. Electronics 10(2):151. https://doi.org/10.3390/electronics10020151
Converters V, Zhang L, Harnefors L, Member S, Nee H, Member S (2010) Power-synchronization control of grid-connected 25(2):809–820

Eberlein S, Rudion K (2021) Small-signal stability modelling, sensitivity analysis and optimisation of droop controlled inverters in LV microgrids. Int J Electr Power Energy Syst 125:106404, https://doi.org/10.1016/j.ijepes.2020.106404

Gonzalez-Longatt F, Acosta MN, Chamorro HR, Rueda JL (2021) Power converters dominated power systems. In: Gonzalez-Longatt F, Rueda JL (eds) Modelling and simulation of power electronic converter controlled power systems in PowerFactory, Switzerland: Springer Nature Switzerland AG

Gorostiza FS, Gonzalez-Longatt F, (2020) Deep reinforcement learning-based controller for SOC management of multi-electrical energy storage system. IEEE Trans Smart Grid, 1–1, https://doi.org/10.1109/TSG.2020.2996274

Hatziargyriou N et al (2020) Stability definitions and characterisation of dynamic behavior in systems with high penetration of power electronic interfaced technologies, https://doi.org/10.1109/JSYST.2015.2444893

Habibullah M, Gonzalez-Longatt F, Acosta MN, Chamorro HR, Rueda JL, Palensky P (2021) On short circuit of grid-forming converters controllers: a glance of the dynamic behaviour

High penetration of power electronic interfaced power sources and the potential contribution of grid forming converters european network of transmission system operators for electricity ENTSO-E technical group on high penetration of power electronic interfaced power sources. Accessed: Mar. 30, 2021. [Online]. Available: www.entsoe.eu

IEEFA, "Australia's Opportunity To Plan Ahead for a Secure Zero-Emissions Electricity Grid Towards Ending the Reliance on Inertia for Grid Stability," 2021

National Grid, "System Operability Framework 2016," UK Electr. Transm., no. November, pp. 68–72, 2016, Accessed: Mar. 30, 2021. [Online]. Available: www.nationalgrid.com/sof

Tamrakar U, Shrestha D, Maharjan M, Bhattarai B, Hansen T, Tonkoski R (2017) Virtual inertia: current trends and future directions. Appl Sci 7(7):654. https://doi.org/10.3390/app7070654

Zhong QC, Weiss G (2011) Synchronverters: inverters that mimic synchronous generators. IEEE Trans Ind Electron 58(4):1259–1267. https://doi.org/10.1109/TIE.2010.2048839

Chapter 4
Model Predictive Control for Grid-Connected Converters with Typhoon HIL

Daniel M. Lima, Dimas A. Schuetz, Felipe B. Grigoletto, Fernanda Carnielutti, Humberto Pinheiro, Luiz A. Maccari Jr., Vinícius F. Montagner, and Caio R. D. Osório

Abstract The growing use of distributed energy resources is driving several improvements in power electronics and their control strategies, especially regarding voltage-source converters, which are crucial to integrate these resources into the electric grid. Control strategies such as classical linear proportional-integral, proportional-resonant, state feedback and deadbeat are generally employed for these applications. However, these strategies usually do not take into account nonlinearities such as control action saturation and current limitations. To solve these issues, Model Predictive Control (MPC) has become a very powerful alternative for controlling grid-connected converters (GCCs), allowing to encompass in the control design different linear and nonlinear constraints. Among the MPC controllers, the Finite Control Set MPC (FCS-MPC) is an attractive solution for controlling GCCs. In FCS-MPC, an optimization problem is formulated with a cost function that expresses the control objectives, such as current reference tracking, capacitor voltage regulation, minimization of losses and common-mode voltages. Besides, FCS-MPC can be implemented with one voltage vector per sampling period, or with a switching sequence, characterizing the Modulated MPC. In this context, this chapter will present different MPC strategies for GCCs and how they can be implemented, tested and validated using the Typhoon

---

D. M. Lima · L. A. Maccari Jr.
Federal University of Santa Catarina - UFSC, Blumenau, Brazil
e-mail: daniel.lima@ufsc.br

D. A. Schuetz · F. Carnielutti (✉) · H. Pinheiro · V. F. Montagner
Federal University of Santa Maria - UFSM, Santa Maria, Brazil
e-mail: fernanda.carnielutti@ufsm.br

F. B. Grigoletto
Federal University of Pampa - UNIPAMPA, Alegrete, RS, Brazil
e-mail: grigoletto@gmail.com

C. R. D. Osório
Typhoon HIL, Novi Sad, Serbia
e-mail: caio.osorio@typhoon-hil.com

© The Author(s), under exclusive license to Springer Nature Singapore Pte Ltd. 2023   75
S. M. Tripathi and F. M. Gonzalez-Longatt (eds.), *Real-Time Simulation and Hardware-in-the-Loop Testing Using Typhoon HIL*, Transactions on Computer Systems and Networks, https://doi.org/10.1007/978-981-99-0224-8_4

HIL platform and the Test-Driven Design (TDD) approach. In power electronics, the TDD can be used to address the performance of GCCs, and also provides a tool to benchmark different implementations of current controllers in a fair way. In this chapter, TDD will be used in order to test, validate and compare the performance of MPC controllers for three-phase GCCs with LCL filters. TDD was carried out using Python scripts and the Typhoon HIL platform, testing the current controllers under different steady-state and transient conditions.

Keywords Grid-connected converters · Hardware-in-the-loop · Model predictive control · Single resonant controller · Test-driven design

4.1 Introduction

Due to concerns about climate change and environmental issues, it is important to increase the share of clean energies in the global energy matrix. In order to achieve this goal, renewable energy systems, such as wind, solar and distributed generation, have experienced a fast development in the last decades (Willis and Scott 2000; Blaabjerg et al. 2006; Guerrero et al. 2010). In this scenario, grid-connected converters (GCCs) play a fundamental role. In the grid-following mode, the current control of the GCCs allows regulating the active and reactive power flow between the primary source and the grid, ensuring synchronization with the voltage at the point of common coupling (PCC) and providing grid currents in compliance with stringent grid code standards, as, for instance, IEEE1547 Std. (IEEE 2011). However, this control loop has to overcome some important challenges, such as operation under grid voltage harmonics, unbalances and faults, uncertainty on the grid impedance at the PCC and limitations in the control signal. In order to cope with these tasks, different control strategies are utilized in the literature. Among them, one can cite the classical linear strategies given in the synchronous reference frame, such as the proportional-integral (SRFPI) controller and the proportional resonant (PR) controller (Dannehl et al. 2010; Teodorescu et al. 2006). Although these classical strategies allow acceptable performances, they present difficulties to incorporate, in the design stage, the optimization of the performance indices, constraints in the control signal and also parameter uncertainties.

In the direction of providing optimal performance under constrained control action, the Model Predictive Control (MPC) has become an important alternative in power electronics, as, for instance, in machine drives, electrical vehicles, multilevel converters and renewable generation systems (Panten et al. 2016; Nauman and Hasan 2016; Young et al. 2016; Sultana et al. 2017; Rodriguez and Kazmierkowski 2013). With this technique, the constrained optimal control action can be computed and implemented in real time through available microcontrollers. MPCs for power converters can be divided into two main types: Continuous Control Set (CCS-MPC) and Finite Control Set (FCS-MPC) (Vazquez et al. 2017; Ferreira and Gonzatti 2018; Falkowski and Sikorski 2018a, b; Lekouaghet et al. 2018). Many MPC algorithms use a continuous set control input, that is, the control input can assume any real value between some arbitrary range. The FCS-MPC takes a different approach by limiting the possible number of control input values to a finite set. In this type of MPC algorithm, an optimal control input that belongs to the set of all possible switching states of the power converter is computed at every sampling period, but this results in a variable switching frequency pattern (Kouro et al. 2009; Rodriguez et al. 2013; Vazquez et al. 2017; Karamanakos et al. 2020; Scoltock et al. 2013; Kakosimos and Abu-Rub 2018; Alam et al. 2019; Aguilera et al. 2013; Karamanakos and Geyer 2020; Geyer 2016). The main advantage is the possibility of easily insert constraints in the control law to achieve robust performance and stability against parametric uncertainties and variations. However, the main drawback of the FSC-MPC is the variable switching frequency, which can cause problems such as vibrations, electrical resonances, acoustic noise and widespread harmonic content with low-order components leading to difficulties in the design of the power filter. On the other hand, CCS-MPC controllers have as their main characteristic the use of a PWM modulator. The control law is obtained as a continuous signal and is modulated as a PWM signal. As a result, this approach presents a fixed switching frequency and, consequently, fixed harmonic content on the voltages generated by the GCC. However, it brings difficulties when it is necessary to introduce restrictions on the control law, as, for example, to cope with robustness specifications or saturation (Camacho and Bordons 2004; Rossiter 2005).

In the context of grid-connected applications, the GCC control system is generally implemented in a Digital Signal Processor (DSP) or a similar system. This control system should be properly tested and validated for various operational conditions (OCs), including faults in the grid, strong and weak grids, disturbances, etc. before connecting the inverter to the grid. These conditions are usually described in Technical Standards and Grid Codes of Power System Operators, with which commercial GCCs must comply; otherwise, problems may arise in the operation of the power system. For example, in 2016, a fire caused a major system disturbance in the Southern California Area (Corporation 2017). During this occurrence, known as the Blue Cut Fire, faults in the power system were detected, and the worst event was a line-to-line fault with a phase jump and voltage sag. The phase jump caused an abrupt phase shift in the voltages, which was erroneously interpreted by the synchronization algorithms of the PV inverters as a frequency deviation from the nominal 60 Hz. The actual frequency dropped to 59.86 Hz, but the synchronization algorithms detected 57 Hz. As a result, the PV inverters started to trip and disconnect from the grid, causing a cascaded event where more than 2GW of solar power was lost in a single day.

Looking back at the events of the Blue Cut Fire, would it be possible to have prevented the loss of such a huge amount of PV power? Would the extensive test and validation of the control systems under various operation conditions have been able to detect such a misbehaviour of the synchronization algorithms before deploying the inverters in the field?

Extensive testing of inverter control software and firmware in order to comply with the Grid Codes and avoid events such as the Blue Cut Fire is generally a demanding process. Also, implementing all these different OCs in a real laboratory environment can be difficult and, in some cases, dangerous. Laboratory setups for emulating grid conditions are usually expensive and limited, in the sense that they are not able to emulate the behaviour of various phenomena that can occur in real power systems. A virtual automated test environment can accelerate this process, since it can safely simulate any desired OCs, which lowers the design/redesign costs and time-to-market of new GCC systems.

The Test-Driven Design (TDD) is a methodology from the software area that can be adapted to many problems, including GCC test and design. The core idea of the TDD is to verify specific functionalities of the software by performing a series of automated tests in a software unit (a small part of the code) (Janzen and Saiedian 2005, 2008; Williams et al. 2003; Jeffries 2007). In the case of a GCC, each of its control blocks can be subjected to various tests, following the TDD methodology, to ascertain, for example, if grid standards are met. The TDD is also capable of providing a fair methodology to benchmark different implementations of controllers for GCCs. In this chapter, the TDD will be used in order to test, validate and compare the performance of Model Predictive Controllers for three-phase GCCs with LCL filters. The TDD was carried out using Python scripts and the Typhoon HIL platform to test the current controllers under different steady-state and transient conditions.

The remainder of this chapter is divided as follows. The system modelling and control algorithms are presented in Sects. 4.2 and 4.3, respectively. TDD with Typhoon HIL is presented in Sect. 4.4, and the results are shown in Sect. 4.5. Finally, conclusions are given in Sect. 4.6.

4.2 System Model

In this section, a discrete-time state-space model of a grid-connected three-phase two-level inverter with an LCL filter will be derived, which will be used in the next sections. This type of system is common in the literature and its modelling is well-known, hence, the equations will be presented here in a straightforward manner. The reader can use (Teodorescu et al. 2011) for an in-depth analysis.

The three-phase two-level converter and LCL filter are shown in Fig. 4.1, and they are connected to the grid via the point of common coupling (PCC). In this representation of the system, the output voltages of the inverter are u_{ab} and u_{bc}, the converter-side currents are i_{1a}, i_{1b} and i_{1c}, the voltages across the filter capacitors are v_{ab} and v_{bc}, and the grid-side currents are i_{2a}, i_{2b} and i_{2c}. The DC-link voltage is assumed to be constant.

In this chapter, we assume that the system is balanced and that there is no path for the zero sequence currents. In the stationary reference frame (Clarke Transform), the system in Fig. 4.1 is represented by two decoupled single-phase systems ($\alpha\beta$ coordinates):

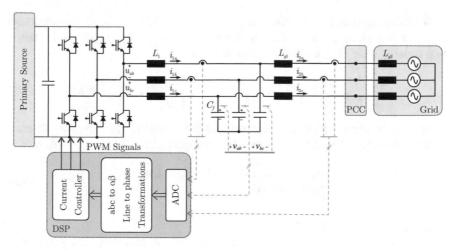

Fig. 4.1 Three-phase two-level gird-connected inverter with LCL filter

$$\begin{bmatrix} \dot{\mathbf{x}}_\alpha \\ \dot{\mathbf{x}}_\beta \end{bmatrix} = \begin{bmatrix} \mathbf{A}_c & 0 \\ 0 & \mathbf{A}_c \end{bmatrix} \begin{bmatrix} \mathbf{x}_\alpha \\ \mathbf{x}_\beta \end{bmatrix} + \begin{bmatrix} \mathbf{B}_c & 0 \\ 0 & \mathbf{B}_c \end{bmatrix} \begin{bmatrix} u_\alpha \\ u_\beta \end{bmatrix} + \begin{bmatrix} \mathbf{F}_c & 0 \\ 0 & \mathbf{F}_c \end{bmatrix} \begin{bmatrix} w_\alpha \\ w_\beta \end{bmatrix} \tag{4.1}$$

where $\mathbf{x}_j = \begin{bmatrix} i_{1j}, v_{cj}, i_{2j} \end{bmatrix}^T$, u_j and w_j (with $j = \alpha, \beta$) are, respectively, the system states, the control inputs and the disturbances. The matrices are defined as

$$\mathbf{A}_c = \begin{bmatrix} 0 & \frac{-1}{L_1} & 0 \\ \frac{1}{C_f} & 0 & \frac{-1}{C_f} \\ 0 & \frac{1}{L_2} & 0 \end{bmatrix}, \mathbf{B}_c = \begin{bmatrix} \frac{1}{L_1} \\ 0 \\ 0 \end{bmatrix}, \mathbf{F}_c = \begin{bmatrix} 0 \\ 0 \\ \frac{-1}{L_2} \end{bmatrix}, \tag{4.2}$$

where L_1 is the converter-side filter inductance, C_f is the filter capacitance and L_{g1} is the grid-side filter inductance. The grid impedance is considered mainly inductive and has an estimated value of L_{g2}, and, finally, $L_2 = L_{g1} + L_{g2}$.

A discretized model is needed for the proposed controllers, which are given in the next section. Thus, using an adequate discretization method, such as Euler and Zero-Order Hold (ZOH), and a constant sampling period T_s, the following discretized state-space model is obtained from (4.1):

$$\begin{cases} \mathbf{x}(k + 1) = \mathbf{A}\mathbf{x}(k) + \mathbf{B}\mathbf{u}(k) + \mathbf{F}\mathbf{w}(k), \\ \mathbf{y}(k) = \mathbf{C}\mathbf{x}(k) \end{cases} \tag{4.3}$$

where k denotes the sampling instant, $\mathbf{x} = \begin{bmatrix} \mathbf{x}_\alpha^T, \mathbf{x}_\beta^T \end{bmatrix}^T$, $\mathbf{u} = \begin{bmatrix} u_\alpha, u_\beta \end{bmatrix}^T$ and $\mathbf{w} = \begin{bmatrix} w_\alpha, w_\beta \end{bmatrix}^T$.

Matrices $\mathbf{A} \in \mathfrak{R}^{6 \times 6}$, $\mathbf{B} \in \mathfrak{R}^{6 \times 2}$ and $\mathbf{F} \in \mathfrak{R}^{6 \times 2}$ are obtained from the discretization process. Matrix $\mathbf{C}$ is defined according to the control system specifications.

It is also important to consider a one-sample time delay in the model, since the control action computed at k will only be applied at $k + 1$. This avoids issues related to a variable delay. The model (4.3) then becomes

$$\begin{bmatrix} \mathbf{x}(k+1) \\ \mathbf{u}(k) \end{bmatrix} = \begin{bmatrix} \mathbf{A} & \mathbf{B} \\ \mathbf{0} & \mathbf{0} \end{bmatrix} \begin{bmatrix} \mathbf{x}(k) \\ \mathbf{u}(k-1) \end{bmatrix} + \begin{bmatrix} \mathbf{0} \\ \mathbf{I} \end{bmatrix} \mathbf{u}(k) + \begin{bmatrix} \mathbf{F} \\ \mathbf{0} \end{bmatrix} \mathbf{w}(k) \qquad (4.4)$$

where $\mathbf{I}$ is an identity matrix of dimension 2 for a three-phase three-wire inverter controlled in the $\alpha\beta$ coordinate system. In a compact form, (4.3) can be written as

$$\begin{cases} \bar{\mathbf{x}}(k+1) = \bar{\mathbf{A}}\bar{\mathbf{x}}(k) + \bar{\mathbf{B}}\mathbf{u}(k) + \bar{\mathbf{F}}\mathbf{w}(k), \\ \mathbf{y}(k) = \bar{\mathbf{C}}\bar{\mathbf{x}}(k) \end{cases} \qquad (4.5)$$

with $\bar{\mathbf{x}}(k) = \left[\mathbf{x}(k)^T, \mathbf{u}(k-1)^T\right]^T$.

In the next section, the model derived here will be used in order to design different current controllers, with an emphasis on Model Predictive Control.

4.3   Control Algorithms

The current controllers used in this chapter will be presented here, which comprise two different Model Predictive Control (MPC) algorithms and a Proportional Resonant (PR) controller. It is important to observe that, in this chapter, only the current control loop of the GCC will be described, in order to simplify the analysis. However, the HIL simulations also take into account other important control loops, such as the Phase-Locked Loop (PLL) for grid synchronization.

4.3.1   Model Predictive Control (MPC)

Model Predictive Control is one of the most used advanced control techniques in the industry (Camacho and Bordons 2004). However, MPC is not a single algorithm but a family of algorithms that has some common basic concepts. A typical MPC algorithm uses a model of the system to predict its future behaviour, and then computes the control action through an optimization that is carried out at each sampling time T_s using the information of the references, control inputs and predictions. Depending on the type of model, optimization problem and how this problem is solved, different MPC algorithms are defined. One essential part of MPC is its associated cost function J, where the user can specify the tuning parameters of the algorithm. A basic cost function for a Single-Input Single-Output (SISO) system is given below

$$J = \delta \sum_{i=N_1}^{N_2} (y_{\text{ref}}(k+i) - y(k+i)) + \lambda \sum_{i=1}^{N_u} \Delta u(k+i-1), \qquad (4.6)$$

where y is the output, y_{ref} is the output reference, $\Delta u(k) = u(k) - u(k-1)$ is the control effort and N_1, N_2, N_u, δ and λ are, respectively, the initial and final prediction horizon, the control horizon, the future error weight and control effort weight. The prediction horizon defines the future time window to be considered in the control problem, the control horizon determines when the controller can act in the future time window and the weights are adjusted to make the controller prioritize one of the terms of the cost function in relation to the others, where a higher value for a given weight means higher importance. The user can change the tuning parameters to achieve the desired closed-loop response. The first term of (4.6) is the predicted future reference tracking error and the second one the control effort. By choosing the weights adequately, the user can prioritize the reference tracking or the dynamic response of the controller. A relative low value of λ will produce a faster closed-loop response. It is worth noting, however, that (4.6) is just an example; different MPC algorithms can use different cost functions that have more (or less) terms with different purposes.

The MPC algorithms obtain the control action to be applied in the process through the minimization of J while taking into account process constraints, such as the maximum voltage that can be synthesized. For more detailed information, see Camacho and Bordons (2004). The MPC algorithms used in this chapter are presented below.

4.3.1.1 Finite Control Set Model Predictive Control (FCS-MPC)

As explained before, the FCS-MPC limits the possible number of control input values to a finite set. This characteristic is useful to directly control the switching states of an inverter, that is, without any modulation (Karamanakos et al. 2020; Karamanakos and Geyer 2020; Geyer 2016). Hence, the finite number of possible control actions are the allowed switching vectors of the inverter. In the case of the three-phase two-level inverter in Fig. 4.1, there are 7 distinct control input vectors available, which are given by the pairs $(u_{\alpha j}, u_{\beta j})$, with $j = 1, \ldots, 7$. Assuming, this constraint on the control inputs, that the prediction horizon is $N_1 = N_2 = N > 1$ (due to the implementation delay), and that the control action $\mathbf{u}_j(k)$ is kept constant within this horizon ($N_u = 1$), the prediction at $k + N$ is given by

$$\bar{\mathbf{x}}_j(k+N) = \bar{\mathbf{A}}^N \bar{\mathbf{x}}_j(k) + \mathbf{B}_p \mathbf{u}_j(k) + \mathbf{F}_p \mathbf{W}_p, \qquad (4.7)$$

for each possible input vector j, where

$$\mathbf{u}_j(k) = \begin{bmatrix} u_{\alpha j} & u_{\beta j} \end{bmatrix}^T$$
$$\mathbf{B}_p = \left(\bar{\mathbf{A}}^{N-1} + \bar{\mathbf{A}}^{N-2} + \cdots + \bar{\mathbf{A}}^0 \right) \bar{\mathbf{B}}$$
$$\mathbf{F}_p = \begin{bmatrix} \bar{\mathbf{A}}^{N-1} \mathbf{F} \cdots \bar{\mathbf{A}} \mathbf{F} & \mathbf{F} \end{bmatrix} \tag{4.8}$$
$$\mathbf{W}_p{}^T = \begin{bmatrix} \mathbf{w}^T(k) \cdots \mathbf{w}^T(k+N-1) \end{bmatrix}.$$

Now, consider the output vector given by

$$\mathbf{y}_j(k+N) = \bar{\mathbf{C}} \bar{\mathbf{x}}_j(k+N). \tag{4.9}$$

With the predicted output given by the previous equation, it is possible to compute the future tracking error

$$\mathbf{e}_j(k+N) = \mathbf{y}_{\text{ref}}(k+N) - \mathbf{y}_j(k+N), \tag{4.10}$$

where $\mathbf{y}_{\text{ref}}$ is the reference vector.

To obtain the optimal control input, the FCS-MPC first computes the weighted Euclidian norm of the future error, which is also called the cost function, for each vector j:

$$\left\| \mathbf{e}_j(k+N) \right\|_{\mathbf{M}}^2 \triangleq \left\| \mathbf{y}_{\text{ref}}(k+N) - \mathbf{y}_j(k+N) \right\|_{\mathbf{M}}^2$$
$$\triangleq \left\| \mathbf{y}_{\text{ref}}(k+N) - \mathbf{f} - \bar{\mathbf{C}} \mathbf{B}_p \mathbf{u}_j(k) \right\|_{\mathbf{M}}^2 \tag{4.11}$$

where $\mathbf{M}$ is diagonal and positive definite. $\mathbf{M}$ is a tuning parameter that must be chosen by the user and works as a weighting factor for the different outputs of the optimization problem. Also,

$$\mathbf{f} = \bar{\mathbf{C}} \left[\bar{\mathbf{A}}^N \bar{\mathbf{x}}_j(k) + \mathbf{F}_p \mathbf{W}_p{}^T \right].$$

Then, the optimal control input to be applied is the one which generates the lowest norm of the error according to (4.11).

The choice of the tuning parameters N and $\mathbf{M}$ will depend on the system and control specifications. For the case of power regulation using GCCs with LCL filters, the controlled variables are usually the grid-side currents, but since this requires a model with a higher dimension, a longer prediction horizon is needed, which, in turn, increases the computational burden and stability problems associated with model uncertainties and disturbance estimation (Panten et al. 2016).

To avoid this issue, a simplified model will be used where the LCL filter is approximated by an L-filter. In this case, the converter-side currents are the states and the capacitor voltages are considered as disturbances.

Applying this to model (4.1) results in the following simplified system matrices:

$$\mathbf{A}_c = \frac{-1}{L_1}\begin{bmatrix} 1 & 0 \\ 0 & 1 \end{bmatrix}, \mathbf{B}_c = \frac{1}{L_1}\begin{bmatrix} 1 & 0 \\ 0 & 1 \end{bmatrix}, \mathbf{F}_c = \frac{-1}{L_1}\begin{bmatrix} 1 & 0 \\ 0 & 1 \end{bmatrix} \qquad (4.12)$$

where the states are given by $\mathbf{x}_\alpha = i_{1\alpha}$ and $\mathbf{x}_\beta = i_{1\beta}$, u_α and u_β are the control inputs, and $w_\alpha = v_{c\alpha}$ and $w_\beta = v_{c\beta}$ are the disturbances. The tuning parameters are chosen as $N = 2$ and $\mathbf{M} = \mathbf{I}$.

The FCS-MPC algorithm can be summarized in the following steps, given the tuning parameters N and $\mathbf{M}$, at each sampling k:

(1) obtain the disturbance prediction vector $\mathbf{W}_p$;
(2) obtain the reference vector $\mathbf{y}_{\mathrm{ref}}(k + N)$;
(3) for each possible control input vector j, compute the cost function (4.11) using (4.7);
(4) At $k + 1$, implement the control input vector j that results in the lowest value for the cost function (4.11).

4.3.1.2 Space-Vector Modulated Model Predictive Control (SVM²PC)

The Space Vector Modulated Model Predictive Control (SVM²PC), first proposed in Osorio et al. (2021), differently from the FCS-MPC, combines control and space-vector modulation (SVM) in a single optimization problem. As discussed previously, the FCS-MPC algorithm computes the value of the cost function for all possible switching states and then applies the control action that results in the lowest cost function value, which is constant inside a sampling period.

The SVM²PC uses an approach where the duty cycles of each possible sector are included in the optimization problem, thus, an optimal switching sequence is obtained.

This approach solves some issues of the FCS-MPC, such as variable switching frequency.

With SVM²PC, an SVM is used to define the switching states of each inverter semiconductor switch. Thus, the control input vector belongs to a finite $\mathbf{u} : \mathfrak{R}^2 \rightarrow \Omega = \{\mathbf{v}^0, \mathbf{v}^1, \ldots, \mathbf{v}^7\}$, which is implemented within a sample period following a symmetrical switching sequence, such as

$$V_j = \{\underbrace{\mathbf{v}_{j0}}_{0.5d_{j0}}, \underbrace{\mathbf{v}_{j1}}_{0.5d_{j1}}, \underbrace{\mathbf{v}_{j2}}_{d_{j2}}, \underbrace{\mathbf{v}_{j1}}_{0.5d_{j1}}, \underbrace{\mathbf{v}_{j0}}_{0.5d_{j0}}\}, \qquad (4.13)$$

where $(\mathbf{v}_{j0}, \mathbf{v}_{j1}, \mathbf{v}_{j2})$ is the set of the three nearest voltage vectors that comprise a given sector S_j in the space vector, with $j = 1, \ldots, 6$. The switching sequence with five vectors (4.13) is shown in Fig. 4.2, along with the evolution of the states within a sampling time T_s. With each switching vector $\mathbf{v}_{ji}$, there is an associated duty cycle d_{ji}, in a way that

Fig. 4.2 Implementation of
a switching sequence V_j and
the state dynamics within a
sampling time T_s

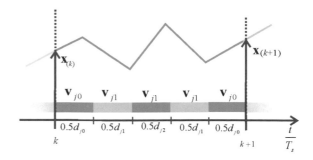

$$\sum_{i=0}^{2} d_{ji} = 1, \quad \text{with } d_{ji} \geqslant 0. \tag{4.14}$$

Assuming that the GCC depicted in Fig. 4.1 is a two-level converter, and that the control input $\mathbf{u}(k)$ is implemented using the switching sequence (4.13), the system model (4.5) can be approximated using the average voltage vector being synthesized, resulting in

$$\bar{\mathbf{x}}_j(k+1) = \bar{\mathbf{A}}\bar{\mathbf{x}}(k) + \bar{\mathbf{B}}\,\mathbf{v}_j(k)\,\mathbf{d}_j(k) + \bar{\mathbf{F}}\mathbf{w}(k), \tag{4.15}$$

with $\mathbf{v}_j = \begin{bmatrix} \mathbf{v}_{j1}, & \mathbf{v}_{j2} \end{bmatrix}$ and $\mathbf{d}_j = \begin{bmatrix} d_{j1}, & d_{j2} \end{bmatrix}^T$.

Therefore, assuming a prediction horizon $N_1 = N_2 = N > 1$ due to the implementation delay, and that the control action $\mathbf{u}_j(k)$ is kept constant within this horizon ($N_u = 1$), the prediction at $k + N$ is given by

$$\bar{\mathbf{x}}_j(k+N) = \bar{\mathbf{A}}^N\bar{\mathbf{x}}_j(k) + \mathbf{B}_p\,\mathbf{u}_j(k) + \mathbf{F}_p\mathbf{W}_p, \tag{4.16}$$

with the same definitions of (4.7), except that

$$\mathbf{u}_j(k) = \mathbf{v}_j(k)\,\mathbf{d}_j(k). \tag{4.17}$$

Now, the output and error vectors are defined as in (4.9) and (4.10), respectively, with the same considerations. Replacing (4.16) and (4.9) into (4.10), the error vector, after some manipulations, can be written as

$$\mathbf{e}_j(k+N) \stackrel{\Delta}{=} d_{j1}\,\mathbf{h}_{j1} + d_{j2}\,\mathbf{h}_{j2} + \mathbf{h}_3, \tag{4.18}$$

with

$$\mathbf{h}_{j1} = -\bar{\mathbf{C}}\mathbf{B}_p\,\mathbf{v}_{j1}(k)$$
$$\mathbf{h}_{j2} = -\bar{\mathbf{C}}\mathbf{B}_p\,\mathbf{v}_{j2}(k) \tag{4.19}$$
$$\mathbf{h}_3 = \mathbf{y}_{\text{ref}}(k+N) - \bar{\mathbf{C}}\bar{\mathbf{A}}^{N-1}\,\mathbf{x}_j(k) - \bar{\mathbf{C}}\mathbf{F}_p\,\mathbf{W}_p.$$

From (4.18), the cost function can be written as

$$\left\|\mathbf{e}_j(k+N)\right\|^2 \triangleq \left\|\mathbf{H}_j\,\mathbf{d}_j(k) + \mathbf{h}_3\right\|_{\mathbf{M}}^2 \tag{4.20}$$

where

$$\mathbf{H}_j = \begin{bmatrix} \mathbf{h}_{j1} & \mathbf{h}_{j2} \end{bmatrix} \tag{4.21}$$

and the diagonal positive definite matrix $\mathbf{M}$ represents the output weighting parameters of the optimization problem.

The objective of the SVM^{2}PC algorithm is to find the sector S_j and duty cycle vector $\mathbf{d}_j(k)$ which minimizes the cost function (4.20) under the constraints given in (4.14), i.e. the optimal values for the proposed optimization problem. As shown in Osorio et al. (2021), it is possible to write this problem as a constrained convex optimization which has a global optimal solution.

Also note that it is possible to choose different prediction horizons and output vectors.

Following the same reasoning presented in Sect. 4.3.1.1, the SVM^{2}PC algorithm will be used to control the converter-side current with $N = 2$ and $\mathbf{M} = \mathbf{I}$. Given that $\mathbf{h}_{j1}$ and $\mathbf{h}_{j2}$ can be computed off-line for each sector S_j by using (4.19), as explained in Osorio et al. (2021), the SVM^{2}PC algorithm has the following steps (Osorio et al. 2021) at each sampling time k:

(1) obtain the disturbance prediction vector $\mathbf{W}_p$;
(2) obtain the reference vector $\mathbf{y}_{\text{ref}}(k+N)$;
(3) compute the value of $\mathbf{h}_3$ in (4.19);
(4) for each sector S_j:

- compute the duty cycles d_{j1} and d_{j2} using (4.22):
- if $d_{j1} + d_{j2} \geq 1$, compute (4.24) with $\mathbf{A}_0$ and b_0 from (4.23);
- if $d_{j1} \leq 0$, compute (4.24) with $\mathbf{A}_1$ and b_1 from (4.23);
- if $d_{j2} \leq 0$, compute (4.24) with $\mathbf{A}_2$ and b_2 from (4.23);

(5) compute $d_{j0} = 1 - d_{j1} - d_{j2}$;

- if $d_{ji} > 1$, set $d_{ji} = 1$ and the other two duty cycles equal to zero;

(6) compute the cost function (4.20) for each sector S_j;
(7) select the optimal duty cycles $\mathbf{d}^*$ and sector S^* which results in the lowest value for cost function (4.20);

(8) at $k + 1$, implement the control input vector through the symmetrical sequence in (4.13), the three switching vectors of sector S^* and duty cycles $\mathbf{d}^*$.

$$\mathbf{d}_j = -\left(\mathbf{H}_j{}^T \mathbf{M} \mathbf{H}_j\right)^{-1} \mathbf{H}_j{}^T \mathbf{M} \mathbf{h}_3. \tag{4.22}$$

$$\begin{cases} \mathbf{A}_0 = \begin{bmatrix} 1 & 1 \end{bmatrix}, & b_0 = -1 \\ \mathbf{A}_1 = \begin{bmatrix} -1 & 0 \end{bmatrix}, & b_1 = 0 \\ \mathbf{A}_2 = \begin{bmatrix} 0 & -1 \end{bmatrix}, & b_2 = 0 \end{cases} \tag{4.23}$$

$$\begin{bmatrix} \mathbf{d}_j \\ \mu_i \end{bmatrix} = -\begin{bmatrix} \mathbf{H}_j{}^T \mathbf{M} \mathbf{H}_j & 0.5\mathbf{A}_i^T \\ \mathbf{A}_i & 0 \end{bmatrix}^{-1} \begin{bmatrix} \mathbf{H}_j{}^T \mathbf{M} \mathbf{h}_3 \\ b_i \end{bmatrix}. \tag{4.24}$$

4.3.2　Proportional Single Resonant Controller

The Proportional Resonant (PR) current controller has been widely applied to GCCs (Teodorescu et al. 2006, 2011). In its classical form, the control system assumes the structure presented in Fig. 4.3, which is a state-feedback approach with additional resonant states.

The PR has a very large gain at a specific frequency, allowing reference tracking and disturbances rejection at this frequency. The damped (non-infinite gain) discrete resonant controller used here has the following difference equation:

$$\xi_i(k+2) - 2e^{-aT_s}\cos(\omega T_s)\xi_i(k+1) + e^{-2aT_s}\xi_i(k) = y_{\text{ref},i}(k) - y_i(k), \tag{4.25}$$

where the subscript i indicates the resonant controller for some output i, $y_{\text{ref},i}$ is the desired reference for the output y_i, ξ_i is the resonant controller state and a, for each angular frequency ω, depends on the damping ratio ζ, that is

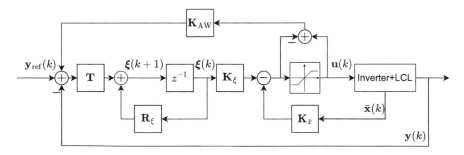

Fig. 4.3 Control system structure of the Resonant controller

$$a = \frac{\omega \zeta}{\sqrt{1 - \zeta^2}}.$$

A state-space realization of (4.25) is given by

$$\begin{bmatrix} \xi_i(k+1) \\ \xi_i(k+2) \end{bmatrix} = \begin{bmatrix} 0 & 1 \\ -e^{-2aT_s} & 2e^{-aT_s}\cos(\omega T_s) \end{bmatrix} \begin{bmatrix} \xi_i(k) \\ \xi_i(k+1) \end{bmatrix} + \begin{bmatrix} 0 \\ 1 \end{bmatrix} (y_{\mathrm{ref},i}(k) - y_i(k))$$

$$(4.26)$$

or, in a compact form, with $i = \alpha, \beta$,

$$\boldsymbol{\xi}_i(k+1) = \mathbf{R}_i \boldsymbol{\xi}(k) + \mathbf{T}_i(y_{\mathrm{ref},i}(k) - y_i(k)). \tag{4.27}$$

In this chapter, the PR will be used to control the grid-side currents, thus, there will be one resonant controller at the fundamental frequency for each grid-side current ($i_{g\alpha}$ and $i_{g\beta}$). As a result, it will be called here a single resonant controller. It is worth mentioning that it is possible to add more resonant frequencies to improve, for example, harmonic rejection, but this is not performed in this chapter. For more information, please refer to Teodorescu et al. (2006, 2011).

Combining the resonant controllers for both $\alpha\beta$ coordinates, we have

$$\boldsymbol{\xi}(k+1) = \mathbf{R}_\xi \boldsymbol{\xi}(k) + \mathbf{T}(\mathbf{y}_{\mathrm{ref}}(k) - \mathbf{y}(k)), \tag{4.28}$$

where $\mathbf{y} = \begin{bmatrix} i_{g\alpha}, & i_{g\beta} \end{bmatrix}^T$, and

$$\mathbf{R}_\xi = \begin{bmatrix} \mathbf{R}_\alpha & 0 \\ 0 & \mathbf{R}_\beta \end{bmatrix}, \quad \mathbf{T} = \begin{bmatrix} \mathbf{T}_\alpha \\ \mathbf{T}_\beta \end{bmatrix}.$$

Aiming to systematize the controller design, let us join the controller and the plant states into a single vector. Therefore, the state-space representation of the combined system becomes

$$\begin{cases} \begin{bmatrix} \bar{\mathbf{x}}(k+1) \\ \boldsymbol{\xi}(k+1) \end{bmatrix} = \begin{bmatrix} \bar{\mathbf{A}} & 0 \\ -\mathbf{T}\bar{\mathbf{C}} & \mathbf{R}_\xi \end{bmatrix} \begin{bmatrix} \bar{\mathbf{x}}(k) \\ \boldsymbol{\xi}(k) \end{bmatrix} + \begin{bmatrix} \bar{\mathbf{B}} \\ 0 \end{bmatrix} \mathbf{u}(k) + \begin{bmatrix} 0 \\ \mathbf{T} \end{bmatrix} \mathbf{y}_{\mathrm{ref}} + \begin{bmatrix} \bar{\mathbf{F}} \\ 0 \end{bmatrix} \mathbf{w}(k), \\ \mathbf{y}(k) = \begin{bmatrix} \bar{\mathbf{C}} & 0 \end{bmatrix} \begin{bmatrix} \bar{\mathbf{x}}(k) \\ \boldsymbol{\xi}(k) \end{bmatrix} \end{cases}$$

$$(4.29)$$

where $\bar{\mathbf{C}}$ is adequately chosen so that $\mathbf{y}$ is equal to the grid-currents, as defined in (4.28).

The next step is to find the state feedback gains $\mathbf{K}_\xi$ and $\mathbf{K}_{\bar{x}}$. In this chapter, the Discrete Linear Quadratic Regulator (DLQR), which is an optimal control design method, will be used with (4.29) to obtain the feedback gains. The use of the DLQR will result in the feedback control law shown below

$$\mathbf{u}(k) = -\mathbf{K}_{\text{LQR}} \begin{bmatrix} \bar{\mathbf{x}}(k) \\ \boldsymbol{\xi}(k) \end{bmatrix} = -\begin{bmatrix} \mathbf{K}_{\bar{x}}, & \mathbf{K}_{\xi} \end{bmatrix} \begin{bmatrix} \bar{\mathbf{x}}(k) \\ \boldsymbol{\xi}(k). \end{bmatrix}. \qquad (4.30)$$

Also, to avoid issues with the windup of the resonant states, an anti-windup technique will be used with gain $\mathbf{K}_{\text{AW}}$, as shown in Fig. 4.3.

For this controller, the tuning parameters are

$$\mathbf{Q}_x = \begin{bmatrix} 100\mathbf{I}_8 & 0 \\ 0 & 0.001\mathbf{I}_4 \end{bmatrix}, \quad \mathbf{Q}_u = \mathbf{I}_2, \quad \mathbf{K}_{\text{AW}} = 0.8\mathbf{I}_2, \qquad (4.31)$$

where $\mathbf{I}_j$ is an identity matrix with dimension j, and $\mathbf{Q}_x$ and $\mathbf{Q}_u$ are diagonal matrices used to weight the states and the inputs in the DLQR algorithm.

These controllers, as well as the three-phase two-level GCC with LCL filter, will be implemented in the Typhoon HIL platform in order to verify and compare their performances for different grid conditions.

4.4 Typhoon HIL and Test-Driven Design

Test-Driven Design (TDD) is a methodology for software development in which the designer creates tests for a given application, code, firmware, etc. before it is actually fully implemented in practice (Janzen and Saiedian 2005, 2008; Kumar and Bansal 2013). Usually, the application is referred to as Device Under Test (DUT). By means of the TDD, software development is performed recursively, as shown in Fig. 4.4.

The main steps in the implementation of a TDD are (i) creation of a test or a set of tests for the specific application; (ii) running the tests; (iii) if the application passes all tests and does not need additional features, the process ends; (iv) if one or more

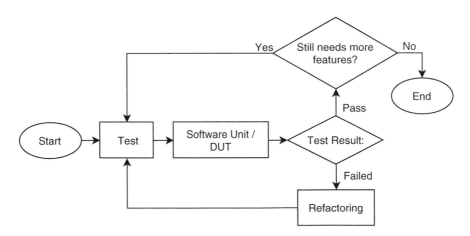

Fig. 4.4 Basic TDD flowchart. Figure adapted from Magnago et al. (2019)

test fails, the application goes through a redesign stage to fix the issues, being tested until it passes in all tests (Jeffries 2007).

As a result, the TDD is a powerful tool that can also be applied for testing, developing and validating controllers for GCCs; for example, in this chapter, the MPC current controller is the DUT considered for the TDD. The inverter topology and the grid-side filter are chosen by the designer, according to the application. As an example, let us consider here the three-phase two-level GCC with LCL filter, as described in Sect. 4.2, where either the grid-side or inverter-side currents can be controlled. In this case, there are also measurements of the voltages of the capacitors of the LCL filter and the line-to-line voltages. The performance evaluation of the designed MPC current controller can consider different operational conditions (OCs), where the grid characteristics, current references and other parameters are varied. Many other characteristics can be included in the OCs, such as different types of faults, voltage harmonics, changes in the grid impedance, grid frequency variations, voltage disturbances and many others (Teodorescu et al. 2007).

Finally, performance indexes are defined prior to the execution of the tests, so as to quantitatively evaluate the obtained results.

Using Typhoon HIL, the TDD can be implemented as automated test scripts written in Python, using the Typhoon Test IDE platform. The inverter, LCL filter, utility grid, sensors, and the other elements of the power stage are modelled with high-fidelity in the Schematic Editor and emulated in real-time using a Typhoon HIL simulator. The controller is implemented either in a C block in the Schematic Editor, or externally by means of a DSP. In this chapter, the MPC current controllers will be implemented in a C block. The schematic is then compiled, the automated test scripts in Python are run in the Typhoon Test IDE environment. Once the set of tests are complete, the results are presented in the Allure platform and as an automatically generated PDF report that contains the test parameters, OCs, results and performance indexes. In the beginning of the report, a summary of the test results is presented that provides a straightforward evaluation of the overall controller performance. Also, more detailed results for each test are presented in specific tables and plots. In Fig. 4.5, a table is presented where the settling times (performance index) for the SVM^{2}PC described in Sect. 4.3.1.2 are shown; a more detailed description of the TDD for the controllers presented in Sect. 4.3 will be shown in the next section. With the TDD, it is possible to compare the performance of different controllers for various OCs, where the green and red cells in the tables indicate whether the performance requirements were met or not, respectively, by the current controller.

Steady-state and transient performance indexes are used to evaluate the current controllers. The steady-state ones are computed from the grid-side currents over one cycle of the fundamental frequency of the grid voltages, and are

1. THD(%): Average of the total harmonic distortion of the grid-side currents of each phase;
2. $E_{N2}(A)$: Average current tracking error calculated as the 2-norm of the grid current vector in $\alpha\beta$ coordinates;

3.4 Settling time of the current tracking error. (Max = 10.0 ms)

Table 15 - t_{set} (ms) under Type A fault

	0.1	0.55	1.0
-90.0°	4.13	4.18	4.53
0.0°	4.13	3.62	0.0
90.0°	4.14	4.33	5.57

Table 16 - t_{set} (ms) under Type B fault

	0.1	0.55	1.0
-90.0°	298.24	298.26	299.36
0.0°	297.85	297.2	0.0
90.0°	298.65	298.4	298.36

Table 17 - t_{set} (ms) under Type C fault

	0.1	0.55	1.0
-90.0°	297.85	298.44	297.75
0.0°	297.94	297.71	0.0
90.0°	298.23	297.8	298.59

Table 18 - t_{set} (ms) under Type E fault

	0.1	0.55	1.0
-90.0°	297.71	298.62	298.08
0.0°	297.68	298.86	0.0
90.0°	297.58	297.82	298.96

Fig. 4.5 SVM^{2}PC: colour-graded table showing the settling time (performance index) for different grid faults

3. $E_{N2max}(A)$: Maximum current tracking error, calculated as the maximum value of the 2-norm of the current vector in the $\alpha\beta$ coordinates;
4. Current harmonics: Peak value of the grid-side current harmonics as a percentage of the amplitude of the fundamental component.

The transient indexes are calculated for both the grid-side and inverter-side currents considering their responses against faults in the grid:

1. $t_{set}(s)$: Time interval from the beginning of the fault until the instant when the current tracking error is lower than a predefined value;
2. $i_{gM}(A)$: Maximum Root-Mean Square (RMS) value of the grid-side currents, calculated by means of a half-cycle fundamental moving average for each grid-side current;
3. $i_{cM}(A)$: Maximum peak of the inverter-side currents.

For example, Fig. 4.6 shows the time-domain waveforms for a test that failed, that is, a Type C Fault with $0.55\angle-90°$ for the SVM^{2}PC. From top to bottom, the waveforms displayed are the grid phase voltages, inverter-side currents, RMS value

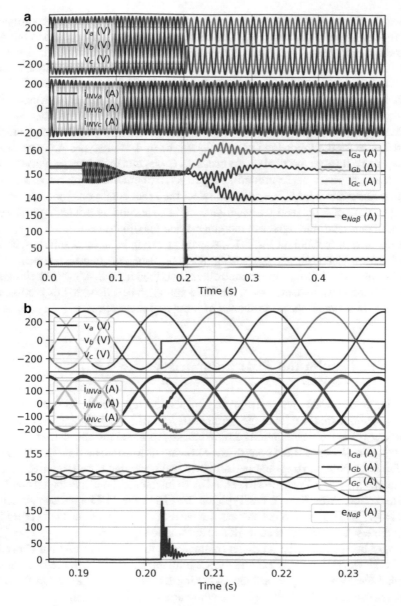

Fig. 4.6 SVM^{2}PC: time-domain plots for Type C Fault with $0.55\angle{-90°}$. **a** Complete plot; **b** Transient response plot

of the grid-side currents and the norm of current tracking error in the $\alpha\beta$ coordinates. Figure 4.6a presents the complete capture time, while Fig. 4.6b gives a zoom over the transient response. The system is also tested under weak grid conditions, defined from the short-circuit ratio, which then defines the Thevenin equivalent impedance at the PCC. The inverter-side current reference is gradually increased and the THDs of the grid-side currents and voltages are calculated. Values of THDs above a predefined threshold limit are detected and reported, indicating a possible instability.

The current controller behaviour is characterized using the performance indexes listed before for the following OCs: Faults on the grid of the types A (symmetrical three-phase), B (asymmetrical phase-ground), C (asymmetrical phase-phase) and E (asymmetrical phase-phase-ground), as well as harmonics on the grid voltage and stability under weak grid. The user can select the exact conditions for each test, for example, type of fault (and its magnitude and phase), the amplitude and phase of grid harmonics, current reference increments and maximum value.

A detailed description of the TDD, as well as example projects and files, can be found at https://hil.academy/courses/digital-control-of-grid-tied-converters/.

In the next section, the current controllers described in Sect. 4.3 will be simulated using Typhoon HIL and, after that, the TDD approach described in this section will be used to benchmark the performance of these current controllers.

4.5   HIL Results and TDD Comparisons

In order to illustrate the performance of the current controllers described in Sect. 4.3, simulation results in the Virtual HIL (VHIL) were obtained and are presented here. The main simulation parameters are presented in Table 4.1, and they are the same for all three current controllers, allowing us to compare their performances in a fair way. First, the transient performance and harmonic rejection will be evaluated manually, using Typhoon HIL SCADA to interact and observe the results from the simulation. After that, in the Typhoon Test IDE, the TDD approach described in Sect. 4.4 will be used to benchmark the performance of these current controllers using the test automation framework and automatic reporting.

It must be noted that the sampling frequency of the three controllers is chosen as 20 kHz. However, the SVM^2PC and PR are implemented with asymmetrical regular sampled PWM, resulting in a fixed switching frequency of 10 kHz. On the other hand, the standard FCS-MPC operates with variable switching frequency.

4.5.1   Transient Performance

First, the performance response of the FCS-MPC controller under strong and weak grid conditions, respectively, are shown in Figs. 4.7 and 4.8. In these figures, a step in the current reference from 0A to the nominal current of 214A is applied. For the

Table 4.1 Simulation parameters

Parameter	Value
AC voltage	Three phase 220V @60 Hz
Strong grid	SCR/X/R : 1000/2
Weak grid	SCR/ X/R: 2/2
Inverter rated power	100 kVA
DC bus voltage	800 V
Inverter-side inductors	500 μH
LCL filter capacitor	100 μF
Grid-side inductors	50 μH
Sampling frequency	20 kHz
Switching frequency (SVM^{2}PC and PR)	10 kHz

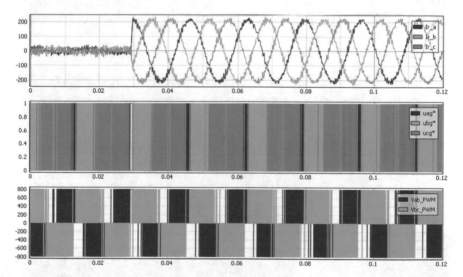

Fig. 4.7 Transient performance of the FCS-MPC algorithm under strong grid conditions

FCS-MPC, a damping resistor equal to 0.25Ω is added in series to the capacitor filter. It can be seen that, as expected, the FCS-MPC not only has a fast dynamic response for both weak and strong grid conditions, but also presents a high ripple on the currents injected into the grid.

In the sequence, the results of the SVM^{2}PC algorithm are presented in Fig. 4.9 for strong grid, and in Fig. 4.10 for weak grid conditions. As the FCS-MPC, the SVM^{2}PC has a fast dynamic response for both weak and strong grid conditions. Moreover, it is possible to see that the control actions saturate when the converter crosses the overmodulation region. Also, it is possible to note that the grid-side currents of the FCS-MPC shown in Figs. 4.7 and 4.8 present a higher harmonic content than the ones obtained with the SVM^{2}PC technique, due to the fact that, in the FCS-MPC, only

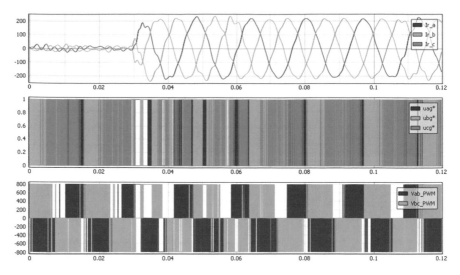

Fig. 4.8 Transient performance of the FCS-MPC algorithm under weak grid conditions

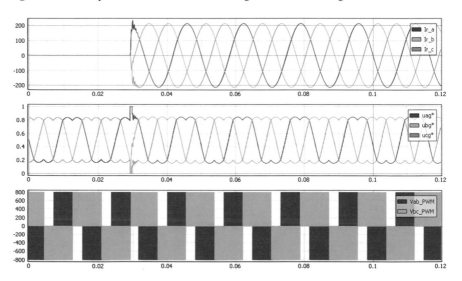

Fig. 4.9 Transient performance of the SVM^{2}PC algorithm under strong grid conditions

one vector is applied in each sampling period T_s, while the SVM^{2}PC implements a switching sequence.

Finally, the performance of the single resonant controller is presented in Figs. 4.11 and 4.12 for strong and weak grid conditions, respectively. The results reveal the convergence to the new current reference with a slower transient response when compared to the SVM^{2}PC and FCS-MPC controllers.

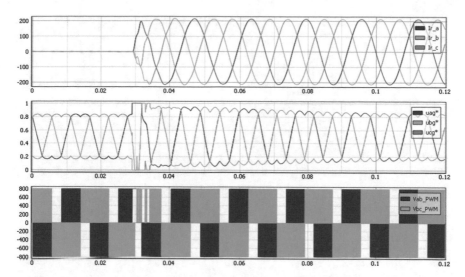

Fig. 4.10 Transient performance of the SVM^{2}PC algorithm under weak grid conditions

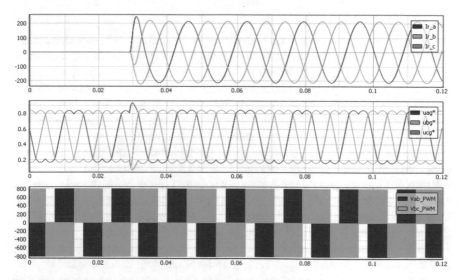

Fig. 4.11 Transient performance of the single resonant controller under strong grid conditions

4.5.2 Harmonic Rejection Performance

Now, let us analyse the harmonic rejection performance of each controller in the presence of 5% of 5th order harmonic, 3% of 7th order harmonic and 1% of 11th order harmonic in the grid voltages. First, the performance of the FCS-MPC, for both weak and strong grids, respectively, is shown in Figs. 4.13 and 4.14. It is possible to see

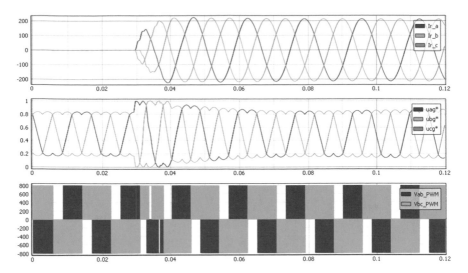

Fig. 4.12 Transient performance of the single resonant controller under weak grid conditions

that the grid voltage harmonics are reproduced in the grid currents, especially under weak grid conditions. This can be attributed to the fact that the FCS-MPC needs an adequate estimation of the disturbances, i.e. the voltages across the filter capacitors, in order to operate properly. In this implementation, this estimation is obtained simply by the rotation of the positive sequence component of the filter voltages. In this sense, the performance of the FCS-MPC with distorted grid conditions could be improved by using a better method for the estimation of the disturbances.

In the same way, the results for the SVM^{2}PC under weak and strong grid conditions, respectively, are presented in Figs. 4.15 and 4.16. Similar to the FCS-MPC, it can be noted that the grid voltage harmonics are also reproduced in the grid currents, especially under weak grid conditions. The same remarks concerning the improvement of the disturbance estimation for the FCS-MPC can be made for the SVM^{2}PC; the only difference is that the disturbances now are the grid voltages.

Finally, the harmonic rejection performance of the resonant controller under weak grid conditions is shown in Fig. 4.17, and under strong grid conditions in Fig. 4.18. These results show a similar performance to the other compared techniques, but with less distortion for the currents under a weak grid. The harmonic rejection performance of the resonant controller can be improved by the addition of multiple resonant controllers tuned to the harmonic components that should be rejected.

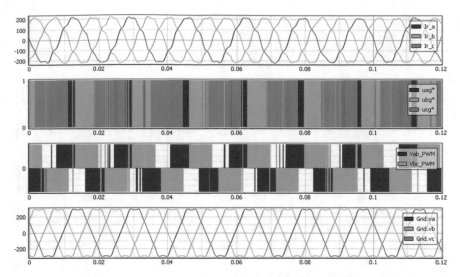

Fig. 4.13 Harmonic rejection performance of the FCS-MPC algorithm under weak grid conditions

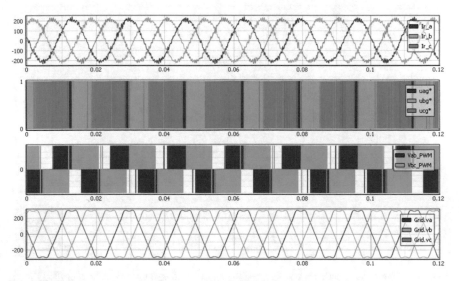

Fig. 4.14 Harmonic rejection performance of the FCS-MPC algorithm under strong grid conditions

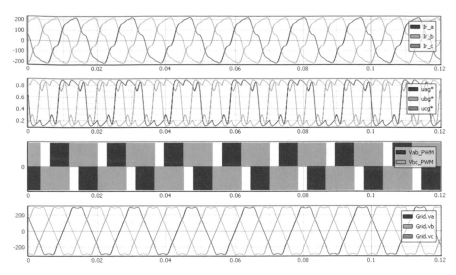

Fig. 4.15 Harmonic rejection performance of the SVM^{2}PC algorithm under weak grid conditions

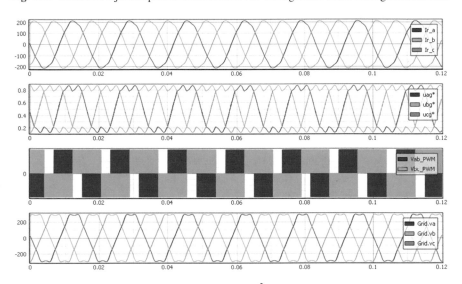

Fig. 4.16 Harmonic rejection performance of the SVM^{2}PC algorithm under strong grid conditions

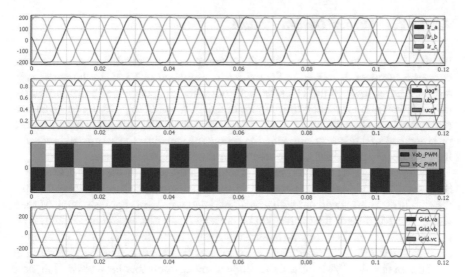

Fig. 4.17 Harmonic rejection performance of the single resonant controller under weak grid conditions

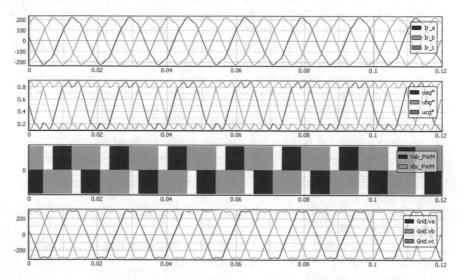

Fig. 4.18 Harmonic rejection performance of the single resonant controller under strong grid conditions

4.5.3   Test-Driven Design Results

Finally, in order to benchmark and fairly compare the performances of the three current controllers described in Sect. 4.3 and simulated in VHIL in the above subsections, let us now run the TDD for each controller independently. It is important to mention that, in this chapter, the purpose of the TDD is not to prove that a given current controller is better than the other, but solely to show how different controllers can be compared in a fair way, allowing the designer to analyse the results and redesign the controller appropriately, so as to obtain the desired performance.

The tests for the TDD, OCs and parameters are the same as the ones described in Sect. 4.4. The grid faults and harmonic tests are performed with a strong grid, while the stability test is carried out with a weak grid and volt-var support. In total, 51 tests were run, and some of the results will be presented and discussed here. In Fig. 4.19, from top to bottom, the overview of the TDD results for FCS-MPC, SVM^{2}PC and single resonant controller, respectively, are shown as presented in the Allure platform. It can be seen that the SVM^{2}PC and the single resonant controller have passed the same number of tests, although, as will be discussed in the following, the tests in which the controllers passed and failed are different, as well as their individual performances. The FCS-MPC, due to its simple implementation, has a degraded performance when compared to the SVM^{2}PC and the single resonant current controllers.

To complement the Allure results, a PDF report is also automatically generated. In the report, initially, a table summarizing the test results is presented for the fault conditions on the grid. As described in Sect. 4.4, the tests are divided into Steady State and Transient, and are evaluated for four different types of faults in the grid. The test summaries for the FCS-MPC, SVM^{2}PC and single resonant current controllers, respectively, are shown in Figs. 4.21, 4.20 and 4.22. The metrics shown in the table are the ones enumerated in Sect. 4.4, and the thresholds used for the TDD were chosen as (these values can be modified by the users to suit their design needs)

1. THD(%): Average value of the THD of the grid-side currents of the three phases. Max = 2.0%.
2. $E_{N2}(A)$: Average current tracking error, represented as the 2-norm of $\alpha\beta$ current vector. Max = 3.5 A.
3. $E_{N2max}(A)$: Maximum current tracking error, represented as the maximum value of the 2-norm of the $\alpha\beta$ current vector. Max = 8.0 A.
4. $t_{set}(s)$: Settling time of the current tracking error. Max = 10.0 ms.
5. $i_{gM}(A)$: Maximum RMS value of the grid-side current of the three phases. Max = 250 A.
6. $i_{cM}(A)$: Maximum RMS value of the converter-side current of the three phases. Max = 355 A.

The THD and stability test summaries are shown in Figs. 4.24, 4.23 and 4.25 for the performance of the FCS-MPC, SVM^{2}PC and single resonant current controllers, respectively. As a result, it can be noted that Figs. 4.21, 4.20, 4.22, 4.24, 4.23 and 4.25 present an overview of the performances of the controllers. If the designer wishes

Fig. 4.19 Overview of the TDD results for FCS-MPC, SVM^{2}PC and single resonant controller

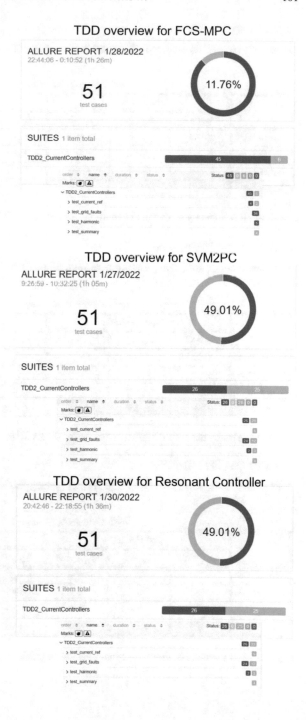

1.1 Fault tests summary

Maximum settling time compared to post fault simulation time interval: 100%.

Table 1 - Report summary

	Steady State			Transient		
	THD(%)	$E_{N2}(A)$	$E_{N2max}(A)$	$t_{set}(ms)$	$Ig_M(A)$	$Ic_M(A)$
Type A Fault	Failed	Failed	Failed	Failed	OK	OK
Type B Fault	Failed	Failed	Failed	Failed	OK	OK
Type C Fault	Failed	Failed	Failed	Failed	OK	OK
Type E Fault	Failed	Failed	Failed	Failed	OK	OK

Fig. 4.20 Test summary of grid fault conditions for the FCS-MPC current controller

1.1 Fault tests summary

Maximum settling time compared to post fault simulation time interval: 100%.

Table 1 - Report summary

	Steady State			Transient		
	THD(%)	$E_{N2}(A)$	$E_{N2max}(A)$	$t_{set}(ms)$	$Ig_M(A)$	$Ic_M(A)$
Type A Fault	OK	OK	OK	OK	OK	OK
Type B Fault	OK	Failed	Failed	Failed	OK	OK
Type C Fault	OK	Failed	Failed	Failed	OK	OK
Type E Fault	OK	Failed	Failed	Failed	OK	OK

Fig. 4.21 Test summary of grid fault conditions for the SVM^{2}PC current controller

1.1 Fault tests summary

Maximum settling time compared to post fault simulation time interval: 99%.

Table 1 - Report summary

	Steady State			Transient		
	THD(%)	$E_{N2}(A)$	$E_{N2max}(A)$	$t_{set}(ms)$	$Ig_M(A)$	$Ic_M(A)$
Type A Fault	OK	OK	OK	Failed	OK	Failed
Type B Fault	OK	OK	OK	Failed	OK	OK
Type C Fault	OK	OK	OK	Failed	OK	OK
Type E Fault	OK	OK	OK	Failed	OK	OK

Fig. 4.22 Test summary of grid fault conditions for the single resonant current controller

1.2 Harmonic test summary

Harmonics that did not comply: 11, 13, 23, 25, 29, 31

1.3 Current reference steps summary

Table 2 - Current reference stability test (A).

	21.4	68.48	115.56	162.64	209.72	256.8
THDi (%)	60.64	17.0	8.13	4.87	3.5	2.71
THDv (%)	15.51	13.94	11.66	11.21	10.38	9.71

Fig. 4.23 Test summary of THD and stability tests for the FCS-MPC current controller

1.2 Harmonic test summary

Harmonics that did not comply: 11, 13

1.3 Current reference steps summary

Table 2 - Current reference stability test (A).

	21.4	68.48	115.56	162.64	209.72	256.8
THDi (%)	2.25	1.66	1.11	0.81	0.63	0.51
THDv (%)	0.53	1.17	1.33	1.37	1.37	1.37

Fig. 4.24 Test summary of THD and stability tests for the SVM^{2}PC current controller

1.2 Harmonic test summary

Harmonics that did not comply: 5, 7, 11, 13

1.3 Current reference steps summary

Table 2 - Current reference stability test (A).

	21.4	68.48	115.56	162.64	209.72	256.8
THDi (%)	2.21	1.34	0.88	0.65	0.51	0.42
THDv (%)	0.52	0.92	1.01	1.04	1.05	1.08

Fig. 4.25 Test summary of THD and stability tests for the single resonant current controller

or needs to have a more detailed analysis of the behaviour of the controller for a specific test, this information can be accessed either in Allure or in the PDF report. However, by analysing Figs. 4.21, 4.20, 4.22, 4.24, 4.23 and 4.25, some conclusions can already be drawn.

First, regarding the fault conditions on the grid, we can see that the FCS-MPC fails in all tests, while the SVM^{2}PC passes in all THD tests, but fails the current error tests and, finally, the single resonant controller passes all the steady-state tests.

The poor performance of the FCS-MPC derives from the fact that only one voltage vector is implemented in a sampling period T_s, and the controlled variable is the converter-side current. It must be stressed that a compensation of the reactive power of the LCL filter capacitors was included in the FCS-MPC algorithm, in order to provide adequate references for the converter-side currents. To further improve its performance, a modified FCS-MPC with reduction of the tracking error could be implemented in order to provide better current tracking capability, the converter-side inductors of the LCL filter could be increased, the switching frequency could also be increased or the grid-side currents could be controlled. The SVM^{2}PC, on the other hand, needs an adequate estimation of the disturbances, i.e. the grid voltages, in order to operate properly. Since, in this example, this estimation is obtained by the rotation of the positive sequence component of the grid voltages, not considering the negative sequence, the unbalances are not properly addressed. This could be solved by improving the estimation of the disturbance. Lastly, the improved behaviour of the resonant controller is due to the fact that, as described in Sect. 4.3, it is implemented as two decoupled controllers, one for the α and one for the β axis. As a result, for asymmetrical faults, the controller has a better performance, as it is able to deal with the unbalanced phase voltages.

Now, considering the THD, the FCS-MPC does not comply with a high number of harmonic components due to the higher current ripple and tracking error when compared to the SVM^{2}PC and the single resonant controller. In these tests, the SVM^{2}PC has a better performance than the other controllers. The single resonant controller was tuned to track only the fundamental component of the current; a multiple resonant controller, designed to reject specific harmonics, would yield a better performance. Finally, the stability tests for weak grid conditions demonstrate that the FCS-MPC presents an indication of instability for some values of the current reference. This may be attributed to the fact that the converter-side current reference depends on the voltages of the filter capacitors, in order to compensate for its reactive currents at the fundamental frequency. On the other hand, the SVM^{2}PC and the single resonant controller are stable under weak grid conditions.

Let us now take a look at some specific test results, in order to further analyse the performance of the three current controllers. As an example, consider a type A fault with $0.55\angle-90°$, where the FCS-MPC failed in all tests but the $i_{gM}(A)$ and $i_{cM}(A)$ ones, the SVM^{2}PC passed in all the tests, and the single resonant controller failed in the $i_{cM}(A)$ and the settling time. The grid phase voltages for the considered fault condition are shown in Fig. 4.26. The complete responses of the FCS-MPC can be seen in the TDD results presented in Figs. 4.27, 4.28 and 4.29. The waveforms displayed in Figs. 4.27 and 4.29, are, from top to bottom, the grid phase voltages, inverter-side currents, RMS value of the grid-side currents and the norm of current tracking error in the $\alpha\beta$ coordinates. As presented in Fig. 4.20, the FCS-MPC has a poor performance for fault conditions, and this can be seen in the TDD waveforms. The currents injected in the grid have large ripples, as depicted in Fig. 4.28, having a negative impact both on the THD and the current tracking error. From Fig. 4.28, we can see that the dynamic response of the FCS-MPC controller is quite fast; however, the calculated settling time was equal to 296.96 ms, making the controller fail the

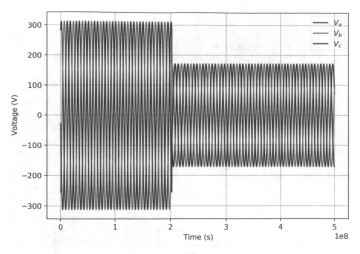

Fig. 4.26 Grid phase voltages for a type A fault with $0.55\angle-90°$

test. This discrepancy is due to the fact that the currents of the FCS-MPC have a high ripple, and the Python script thus interprets that the controller has not reached a steady state in the predefined time window used to calculate the settling time. The norm of the tracking error (the blue curve), shown in Fig. 4.29, lies outside of the defined threshold, represented by the red curve; this behaviour again can be explained as a result of the current ripple. On the other hand, the maximum values of $i_{gM}(A)$ and $i_{cM}(A)$ are within the predefined limits. As previously mentioned, these results could be improved by modifying the FCS-MPC in order to include a compensation for the tracking error to provide better current tracking, controlling the grid-side currents or increasing either the converter-side inductor or the switching frequency. For example, Fig. 4.30 shows the norm of the current tracking error for a case where the converter-side inductor was increased from 500 μH to 1500 μH. We can see that the error has significantly decreased, but at the expense of a large inductance. The metrics for the FCS-MPC were computed as (these values were extracted from the detailed tables presented in the PDF report)

1. THD: 5.23%
2. E_N: 11.67 A
3. E_{N2max}: 28.16 A
4. t_{set}: 296.96 ms
5. i_{gM}: 156.66 A
6. i_{cM}: 243.3 A.

The second controller that was tested was the SVM^{2}PC controller. A zoom of the transient response and the steady-state current tracking error, respectively, are presented in Figs. 4.31, 4.32 and 4.33. The same waveforms described for the FCS-MPC are shown. Especially in Fig. 4.33, it can be seen that the norm of the current

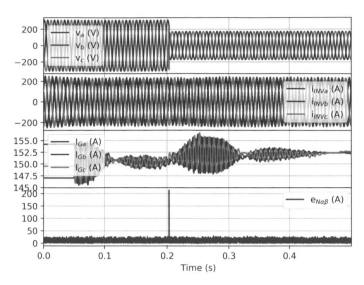

Fig. 4.27 Complete transient response of the FCS-MPC current controller for a type A fault with $0.55\angle-90°$

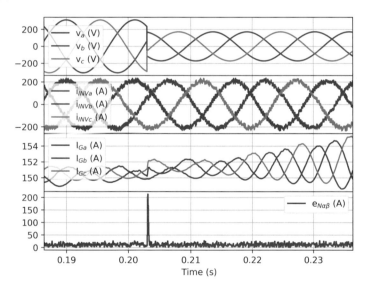

Fig. 4.28 Zoom of the transient response of the FCS-MPC current controller for a type A fault with $0.55\angle-90°$

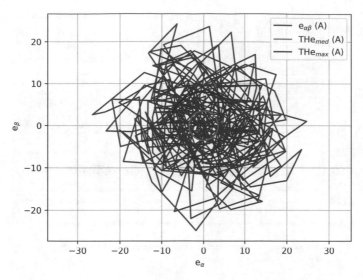

Fig. 4.29 Norm of the current tracking error of the FCS-MPC current controller for a type A fault with $0.55\angle{-}90°$

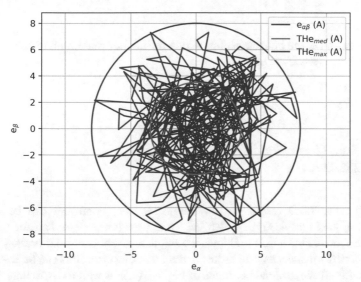

Fig. 4.30 Norm of the current tracking error of the FCS-MPC current controller for a type A fault with $0.55\angle{-}90°$ and with an increased converter-side inductor

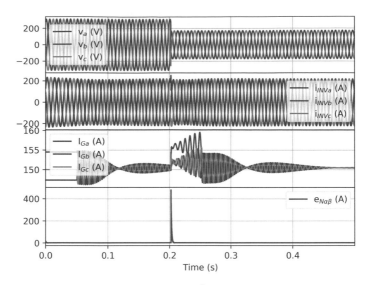

Fig. 4.31 Complete transient response of the SVM^{2}PC current controller for a type A fault with $0.55\angle-90°$

tracking error, i.e. the blue curve, lies within the defined maximum threshold in red. In quantitative terms, we have

1. THD: 0.21%
2. E_N: 1.23 A
3. E_{N2max}: 1.59 A
4. t_{set}: 4.33 ms
5. i_{gM}: 159.54
6. i_{cM}: 252.92 A.

Finally, the TDD results for the single resonant controller can be seen in Figs. 4.34, 4.35 and 4.36, in which the same waveforms described for the FCS-MPC are shown. For these conditions, the single resonant controller presents a good harmonic performance, as well as low current tracking error; this can be seen by the good quality of the grid-side currents of Fig. $i_{cM}(A)$, as well as by the norm of the current tracking error (blue curve) lying inside the defined threshold (red curve) in Fig. 4.36. However, the controller presents a slower dynamic response when compared to the SVM^{2}PC, which is expected. This can be quantified in the larger value of the settling time for the single resonant controller. The performance indexes for the single resonant current controller are

1. THD: 0.5%
2. E_N: 0.98 A
3. E_{N2max}: 2.36 A

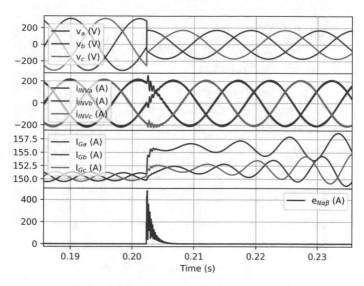

Fig. 4.32 Zoom of the transient response of the SVM^{2}PC current controller for a type A fault with $0.55\angle-90°$

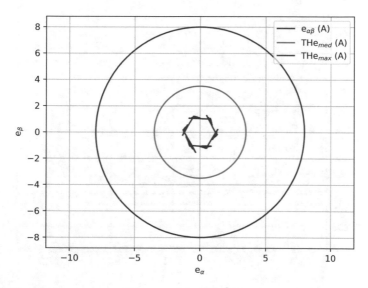

Fig. 4.33 Norm of the current tracking error of the SVM^{2}PC current controller for a type A fault with $0.55\angle-90°$

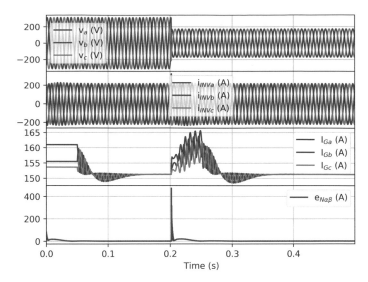

Fig. 4.34 Complete transient response of the single resonant current controller for a type A fault with $0.55\angle{-90}°$

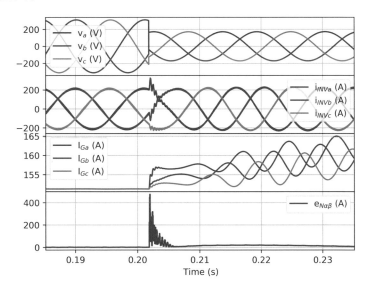

Fig. 4.35 Zoom of the transient response of the single resonant current controller for a type A fault with $0.55\angle{-90}°$

4. t_{set}: 40.23 ms
5. i_{gM}: 165.66 A
6. i_{cM}: 326.16 A.

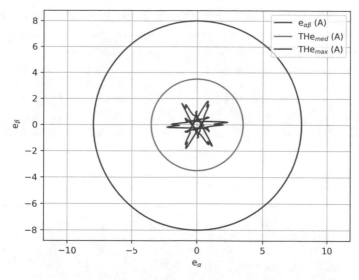

Fig. 4.36 Norm of the current tracking error of the single resonant current controller for a type A fault with $0.55\angle-90°$

Finally, let us compare the performance of the three current controllers with respect to the harmonic content in steady state. The goal of this test is to demonstrate the capability of the current controller under test to reject disturbances resulting from harmonics in the grid voltage. The tests are carried out with a strong grid and with the reference equal to the nominal value of the current. The tables with the odd harmonic content for, respectively, the FCS-MPC, SVM^{2}PC and the single resonant current controllers are presented in Fig. 4.37a, b and c. In the tables, the lines indicate the amplitude and phase of the grid background voltage harmonics, while the columns represent the order of the grid current harmonics. The FCS-MPC has some high-order harmonic components due to the high current ripple and tracking error, but has a better performance than the resonant controller when the low-order harmonics are considered. It can be seen that the SVM^{2}PC has the best harmonic performance. The single resonant controller, as the name indicates, was designed to track only the fundamental component of the current; this performance could be improved by using a multiple resonant controller designed to reject specific harmonic components.

4.6   Conclusions

This chapter described a Test-Driven Design methodology applied to current controllers for GCCs. The TDD allows the benchmarking of controllers through a series of automated tests that put them through a wide range of OCs, which mimic possible real-life scenarios. Hence, by choosing suitable performance indices and OCs,

Table 27 - Amplitude of the odd grid side current harmonic normalized with respect to the fundamental (%).

	3	5	7	9	11	13	15	17	19	21	23	25
0, 0	0.17	0.17	0.96	0.16	0.04	0.3	0.22	0.22	0.7	0.3	0.68	0.81
5, 0	0.34	0.81	0.4	0.48	1.18	0.59	0.21	0.32	0.5	0.17	0.47	0.75
10, 0	0.76	1.91	0.68	0.89	1.28	1.84	0.89	0.71	0.37	0.86	0.13	0.25
20, 0	0.18	2.47	1.65	0.24	2.99	2.09	0.68	0.41	0.66	0.45	0.3	0.2
30, 0	0.03	3.25	2.63	0.25	4.45	2.86	0.21	0.97	0.74	0.21	1.05	0.24

(a) FCS-MPC

Table 27 - Amplitude of the odd grid side current harmonic normalized with respect to the fundamental (%).

	3	5	7	9	11	13	15	17	19	21	23	25
0, 0	0.0	0.1	0.08	0.0	0.05	0.04	0.0	0.03	0.03	0.0	0.02	0.02
5, 0	0.0	0.69	0.62	0.0	1.0	0.74	0.0	0.03	0.03	0.0	0.02	0.02
10, 0	0.0	1.33	1.2	0.0	1.96	1.45	0.0	0.03	0.03	0.0	0.02	0.02
20, 0	0.0	2.61	2.36	0.0	3.89	2.87	0.0	0.03	0.02	0.0	0.02	0.02
30, 0	0.0	3.89	3.52	0.0	5.82	4.29	0.0	0.03	0.02	0.0	0.02	0.02

(b) SVM2PC

Table 27 - Amplitude of the odd grid side current harmonic normalized with respect to the fundamental (%).

	3	5	7	9	11	13	15	17	19	21	23	25
0, 0	0.0	0.37	0.24	0.0	0.12	0.08	0.0	0.05	0.04	0.0	0.02	0.02
5, 0	0.0	1.37	1.2	0.0	1.01	0.8	0.0	0.04	0.03	0.0	0.02	0.02
10, 0	0.0	2.41	2.21	0.0	1.94	1.54	0.0	0.04	0.02	0.0	0.01	0.01
20, 0	0.0	4.49	4.21	0.0	3.8	3.02	0.0	0.02	0.01	0.0	0.01	0.01
30, 0	0.0	6.53	6.18	0.0	5.63	4.49	0.0	0.03	0.03	0.0	0.04	0.04

(c) Single resonant

Fig. 4.37 Amplitude of the odd grid-side current harmonic normalized with respect to the fundamental (%): **a** SVM^{2}PC, **b** FCS-MPC and **c** Single resonant current controller

the behaviour of the inverter with the current controller can be rapidly and safely assessed, showing the user if a redesign is necessary or not. The TDD also allows an easy and fair comparison among different current controllers.

In this chapter, the TDD was applied to three different controllers: the SVM^{2}PC, a FCS-MPC and a single resonant controller. The SVM^{2}PC has shown superior performance under balanced faults at the grid side. However, under unbalanced faults at the grid side, the performance of the SVM^{2}PC can be improved with a better prediction of the disturbances. The performance of the FCS-MPC could be improved by (i) using a modified FCS-MPC with better steady-state estimation, (ii) increasing the converter-side inductor of the LCL filter or the switching frequency or (iii) controlling the grid-side currents. Finally, the TDD also indicates that the performance of the single resonant controller could be improved by using a multiple resonant controller, designed to reject specific harmonics.

The framework presented in this chapter, using Typhoon HIL for the TDD implemented as automated tests written in Python, can be readily extended to other controllers applied to grid-connected converters and also to other applications in power electronics. This indicates a great potential for improvement in the product development stage, increasing safety and reliability, and reducing cost and time, while keeping high standards in the control systems validation.

Acknowledgements The authors would like to thank Typhoon HIL, especially their engineers Henrique Magnago and Tiarles Guterres, who were responsible for the implementation of the automated Python scripts for the TDD, and UFSM-PPGEE for the support required to obtain the results presented in this chapter. This study was financed in part by the Coordenação de Aperfeiçoamento de Pessoal de Nível Superior - Brasil (CAPES), and by the Conselho Nacional de Desenvolvimento Científico e Tecnológico - Brasil (CNPq).

References

Aguilera RP, Lezana P, Quevedo DE (2013) Finite-control-set model predictive control with improved steady-state performance. IEEE Trans Indust Inf 9(2):658. https://doi.org/10.1109/TII.2012.2211027

Alam KS, Akter MP, Xiao D, Zhang D, Rahman MF (2019) Asymptotically stable predictive control of grid-connected converter based on discrete space vector modulation. IEEE Trans Indust Inf 15(5):2775. https://doi.org/10.1109/TII.2018.2876274

Blaabjerg F, Teodorescu R, Liserre M, Timbus A (2006) Overview of control and grid synchronization for distributed power generation systems. IEEE Trans Indust Electron 53(5):1398

Camacho E, Bordons C (2004) Model predictive control. Springer, Berlin

Dannehl J, Fuchs F, Hansen S, Thøgersen P (2010) Investigation of active damping approaches for PI-based current control of grid-connected pulse width modulation converters with LCL filters. IEEE Trans Indust Appl 46(4):1509. https://doi.org/10.1109/TIA.2010.2049974

Falkowski P, Sikorski A (2018a) Finite control set model predictive control for grid-connected AC-DC converters with LCL filter. IEEE Trans Indust Electron 65(4):2844. https://doi.org/10.1109/TIE.2017.2750627

Falkowski P, Sikorski A (2018b) Comparative analysis of finite control set model predictive control methods for grid-connected AC-DC converters with LCL filter. In: Proceedings of the IEEE 27th international symposium industrial electronics (ISIE), pp 193–200. https://doi.org/10.1109/ISIE.2018.8433673

Ferreira SC, Gonzatti RB, Pereira RR, da Silva CH, da Silva LB, Lambert-Torres G (2018) Finite control set model predictive control for dynamic reactive power compensation with hybrid active power filters. IEEE Trans Indust Electron 65(3):2608

Geyer T (2016) Model predictive control of high power converters and industrial drives. Wiley

Guerrero J, Blaabjerg F, Zhelev T, Hemmes K, Monmasson E, Jemei S, Comech M, Granadino R, Frau J (2010) Distributed generation: toward a new energy paradigm. Indust Electron Mag IEEE 4(1):52. https://doi.org/10.1109/MIE.2010.935862

IEEE (2011) IEEE:1547 standard for interconnecting distributed resources with electric power systems

Janzen D, Saiedian H (2005) Test-driven development concepts, taxonomy, and future direction. Computer 38(9):43. https://doi.org/10.1109/MC.2005.314

Janzen D, Saiedian H (2008) Does test-driven development really improve software design quality? Softw IEEE 25:77. https://doi.org/10.1109/MS.2008.34

Jeffries R, Melnik G (2007) Guest editors' introduction: Tdd-the art of fearless programming. Softw, IEEE 24:24. https://doi.org/10.1109/MS.2007.75

Kakosimos P, Abu-Rub H (2018) Predictive control of a grid-tied cascaded full-bridge NPC inverter for reducing high-frequency common-mode voltage components. IEEE Trans Indust Inf 14(6):2385. https://doi.org/10.1109/TII.2017.2768585

Karamanakos P, Geyer T (2020) Guidelines for the design of finite control set model predictive controllers. IEEE Trans Power Electron 35(7):7434. https://doi.org/10.1109/TPEL.2019.2954357

Karamanakos P, Liegmann E, Geyer T, Kennel R (2020) Model predictive control of power electronic systems: methods, results, and challenges. IEEE Open J Ind Appl 1:95. https://doi.org/10.1109/OJIA.2020.3020184

Kouro S, Cortes P, Vargas R, Ammann U, Rodriguez J (2009) Model predictive control-a simple and powerful method to control power converters. IEEE Trans Indust Electron 56(6):1826. https://doi.org/10.1109/TIE.2008.2008349

Kumar S, Bansal SK (2013) Comparative study of test driven development with traditional techniques

Lekouaghet B, Boukabou A, Lourci N, Bedrine K (2018) Control of PV grid connected systems using mpc technique and different inverter configuration models. Electric Power Syst Res 154:287

Magnago H, Guterres T, Carnielutti F, Massing J, Vieira R, Pinheiro H (2019) A test driven design approach to benchmark current controllers for grid-tied inverters. In: 2019 20th workshop on control and modeling for power electronics (COMPEL), pp 1–8. https://doi.org/10.1109/COMPEL.2019.8769672

NAER Corporation (2017) 1,200 MW fault induced solar photovoltaic resource interruption disturbance report

Nauman M, Hasan A (2016) Efficient implicit model-predictive control of a three-phase inverter with an output LC filter. IEEE Trans Power Electron 31(9):6075. https://doi.org/10.1109/TPEL.2016.2535263

Osório CRD, Schuetz DA, Koch GG, Carnielutti F, Lima DM Jr, LAM, Montagner VF, Pinheiro H, (2021) Modulated model predictive control applied to lcl-filtered grid-tied inverters: a convex optimization approach. IEEE Open J Ind Appl 2:366. https://doi.org/10.1109/OJIA.2021.3134585

Panten N, Hoffmann N, Fuchs FW (2016) Finite control set model predictive current control for grid-connected voltage-source converters with LCL filters: a study based on different state feedbacks. IEEE Trans Power Electron 31(7):5189. https://doi.org/10.1109/TPEL.2015.2478862

Rodriguez J, Kazmierkowski MP, Espinoza JR, Zanchetta P, Abu-Rub H, Young HA, Rojas CA (2013) State of the art of finite control set model predictive control in power electronics. IEEE Trans Indust Inf 9(2):1003. https://doi.org/10.1109/TII.2012.2221469

Rodriguez J, Kazmierkowski MP, Espinoza JR, Zanchetta P, Abu-Rub H, Young HA, Rojas CA (2013) State of the art of finite control set model predictive control in power electronics. IEEE Trans Indust Inf 9(2):1003. https://doi.org/10.1109/TII.2012.2221469

Rossiter JA (2005) Model-based predictive control: a practical approach. Control series. CRC Press, Boca Raton

Scoltock J, Geyer T, Madawala UK (2013) A comparison of model predictive control schemes for mv induction motor drives. IEEE Trans Indust Inf 9(2):909. https://doi.org/10.1109/TII.2012.2223706

Sultana WR, Sahoo SK, Sukchai S, Yamuna S, Venkatesh D (2017) A review on state of art development of model predictive control for renewable energy applications. Renew Sustain Energy Rev 76:391. https://doi.org/10.1016/j.rser.2017.03.058

Teodorescu R, Blaabjerg F, Liserre M, Loh P (2006) Proportional-resonant controllers and filters for grid-connected voltage-source converters. Electr Power Appl, IEE Proc 153(5):750

Teodorescu R, Liserre M, Rodriguez P (2007) Grid converters for photovoltaic and wind power systems Wiley-IEEE Press

Teodorescu R, Liserre M, Rodriguez P (2011) Grid converters for photovoltaic and wind power systems, vol 29. Wiley

Vazquez S, Rodriguez J, Rivera M, Franquelo LG, Norambuena M (2017) Model predictive control for power converters and drives: advances and trends. IEEE Trans Indust Electron 64(2):935. https://doi.org/10.1109/TIE.2016.2625238

Vazquez S, Rodriguez J, Rivera M, Franquelo LG, Norambuena M (2017) Model predictive control for power converters and drives: advances and trends. IEEE Trans Indust Electron 64(2):935. https://doi.org/10.1109/TIE.2016.2625238

Williams L, Maximilien E, Vouk M (2003) Test-driven development as a defect-reduction practice. In: 14th international symposium on software reliability engineering, 2003. ISSRE 2003., pp 34–45. https://doi.org/10.1109/ISSRE.2003.1251029

Willis H, Scott WG (2000) Distributed power generation: planning and evaluation. In: Power engineering (Willis). Taylor & Francis

Young HA, Perez MA, Rodriguez J (2016) Analysis of finite-control-set model predictive current control with model parameter mismatch in a three-phase inverter. IEEE Trans Indust Electron 63(5):3100. https://doi.org/10.1109/TIE.2016.2515072

Chapter 5
Grid-Connected Multilevel Converter with Optimal Programmed PWM and Virtual Synchronous Machine

Felipe B. Grigoletto, Dimas A. Schuetz, Jonas R. Tibola, Fernanda Carnielutti, and Humberto Pinheiro

Abstract Optimal Programmed Pulse-Width Modulation (OP-PWM) and its predecessor Selective Harmonic Elimination (SHE) are strong candidates to modulate high-power medium-voltage inverters, since they result in reduced switching losses for a given power quality requirement. SHE can lead to high amplitude harmonic components located at frequencies close to the eliminated ones requiring a bulky filter or increased commutation losses. To overcome these limitations, this chapter proposes the definition of the commutation angles of an OP-PWM from an optimization problem that considers both the limit of the current Total Harmonic Distortion (THD) an its individual low-order harmonic amplitude. On the other hand, primary controllers for grid-connected converters, such as the Virtual Synchronous Machines (VSM) have at their output the amplitude and frequency of the voltage to be synthesized by the converter which, matches with the input of an OP-PWM. This chapter aims to develop an optimized modulation technique for grid-connected converters in accordance with the limits of current harmonic content of the IEEE 1547 standard. Finally, real-time operation of a grid-connected three-phase neutral point clamped converter is carried out in the Hardware-in-the-Loop Typhoon (HIL) 402 to demonstrate the performance of the proposed approach.

F. B. Grigoletto (✉)
Federal University of Pampa, Alegrete, RS, Brazil
e-mail: grigoletto@gmail.com

D. A. Schuetz · J. R. Tibola · F. Carnielutti · H. Pinheiro
Federal University of Santa Maria - UFSM, Santa Maria, Brazil
e-mail: dimasschuetz@gmail.com

J. R. Tibola
e-mail: jrtibola@gmail.com

F. Carnielutti
e-mail: fernanda.carnielutti@gmail.com

H. Pinheiro
e-mail: humberto.ctlab.ufsm.br@gmail.com

© The Author(s), under exclusive license to Springer Nature Singapore Pte Ltd. 2023 117
S. M. Tripathi and F. M. Gonzalez-Longatt (eds.), *Real-Time Simulation and Hardware-in-the-Loop Testing Using Typhoon HIL*, Transactions on Computer Systems and Networks, https://doi.org/10.1007/978-981-99-0224-8_5

Keywords Grid-connected inverter · Hardware-in-the-loop · Multilevel
converter · Selective harmonic elimination · Optimal programmed pulse-width
modulation · Virtual synchronous machine

5.1 Introduction

High-power converters operating at medium or high voltages generally employ low
switching frequencies to reduce the switching losses (Holmes and Lipo 2003). In this
scenario, multilevel inverters play an important role due to significant advantages
when compared to two-level inverters. Among these advantages are the good quality
of the output voltages, reduced voltage over the semiconductor devices, and reduced
common mode voltage (Vijeh et al. 2019). Furthermore, grid connected generation
systems must comply with current harmonic limits from grid codes and standards. To
achieve these goals, a suitable design of control strategies, modulation techniques,
and filtering have to be taken into account.

Several Pulse-Width Modulation (PWM) modulation strategies have been proposed in the literature (Leon et al. 2016, 2017). Basically, they can be classified
in: Carrier-Based (CB-PWM) (Grigoletto and Pinheiro 2011; McGrath and Holmes
2022), Space Vector (SV-PWM) (Grigoletto and Pinheiro 2009; Jayakumar et al.
2021; Schuetz et al. 2021) and Pre-Programmed (PP-PWM) (Turnbull 1964; Patel
and Hoft 1973; Enjeti et al. 1990). CB-PWM is a simple and effective way to produce
a switched voltage waveform by comparing a desired reference voltage with a high-
frequency carrier. The simple design and implementation are outstanding advantages
of this strategy; however, the method may be not flexible when it comes to improving the harmonic spectrum or the switching losses. On the other hand, SV-PWM
uses vector representations of the switching states to synthesize a desired voltage.
This strategy presents flexibility to design the switching sequences in order to deal
with losses reduction and quality of output voltages. However, the complexity of the
algorithms may require microcontrollers with high processing capacity.

When it comes to PP-PWM techniques, the switching angles are defined in order
to reach an objective such as losses reduction, elimination of specific harmonic components, etc. Among PP-PWM strategies, the Selective Harmonic Elimination (SHE)
is an attractive candidate which aims to eliminate specific output voltage harmonic
components while synthesizing a desired fundamental component (Turnbull 1964;
Patel and Hoft 1973; Enjeti et al. 1990). In this technique, the switching angles are
determined offline and stored in look-up-tables.

It is important to mention that grid codes limit the individual harmonic components
and the Total Harmonic Distortion (THD) of the currents injected into the grid. By
means of the conventional SHE modulation it is possible to eliminate some predefined
low-order harmonics. However, the presence of significant harmonic components in
a frequency range near the eliminated ones can be noted, which may compromise
the compliance with the standards. In order to overcome this issue, the number
of switching angles can be increased, but the complexity of the algorithm is also

increased, as well as the number of commutations and, consequently, the switching losses are incremented (Dahidah et al. 2015).

Recently, SHE techniques have been proposed with different optimization objectives. In Agelidis et al. (2006), a SHE algorithm is described whose objective function is defined in order to find a solution that results in the minimal THD value. However, this algorithm does not consider the non-eliminated harmonic components. Conversely, Napoles et al. (2008) proposes a technique that aims to mitigate each harmonic component and minimize the THD. In this approach, passive filters tuned at desired harmonic frequencies are used to comply with the standard limits for grid-connected converters. As a disadvantage, this technique does not consider the output filter gain in the search algorithm of the switching angles.

Compared to SHE strategies, the mitigation of harmonic components instead of its elimination allows to meet the grid code requirements with lower filter size (Napoles et al. 2010). Although the low-order harmonics are not eliminated by the mitigation techniques, they are limited according to maximum values. On the other hand, high-order harmonics, that are not considered in the conventional SHE modulation, are also minimized to comply with the grid code limits. Usually, modulation for the mitigation of harmonics is formulated as an optimization problem, where the THD is also minimized (Franquelo et al. 2007).

Regarding SHE modulation, an important challenge in this technique is the search of the switching angles, where nonlinear equations must be solved. The complexity of this problem increases according to the number of angles and nonlinear equations involved. As an alternative to the well-known numeric Newton-Raphson (NR) method, intelligent algorithms have been employed due to their strong abilities to solve complex optimization problems and find the global optimal solution (Li et al. 2022; Memon et al. 2018).

One example of intelligent algorithms applied to SHE modulation is the Particle Swarm Optimization (PSO). This method computes the positions of each particle and global optimal position of all particles aiming to guide them to the optimal solution (Sadoughi et al. 2022; Jiang et al. 2022). Other examples are Genetic Algorithms (GA), Differential Evolution (DE) and Bee Algorithms (BA), which can be implemented with the common objective of finding the optimal solution from a defined objective function (Memon et al. 2018). In addition, Artificial Neural Network (ANN) algorithms can be also applied to SHE modulation; however, they need a high amount of data for the learning process (Chabni et al. 2017).

Although the SHE strategies have been widely applied to electrical machine drives, some authors have expanded their concepts for grid-tied converters (Tibola et al. 2011; Zhao et al. 2016). These applications require proper closed-loop control and synchronization methods. On the other hand, grid-connected converters can be controlled by emulating the behavior of synchronous generators (Beck and Hesse 2007; Zhong and Weiss 2011; Qing-Chang et al. 2014). As advantages of this methods, it can be cited the presence of inertia and the sharing of active and reactive power with the electrical power system. Moreover, the virtual synchronous generator intrinsically presents a frequency loop which can simplify the implementation of the OP-PWM.

In this context, this chapter presents an optimal programmed PWM strategy coordinated with a virtual synchronous machine. The innovation of this chapter is outlined as (i) An OP-PWM where the harmonic current content is minimized and is kept under the limits of the IEEE 1519 standard; (ii) The proposed modulation strategy presents a superior performance in terms of harmonic distortion compared to SHE-PWM for an equivalent number of switching angles and output filter; (iii) The closed-loop control by means of he Virtual Synchronous machine concept provides the synchronization angle required for the OP-PWM technique that supports the overall implementation of the grid-connected converter.

This chapter is organized as follows: Sect. 5.2 introduces the SHE modulation, Sect. 5.3 presents a detailed description of the proposed OP-PWM algorithm for the mitigation of individual harmonic components and THD minimization. Section 5.4 presents the closed-loop control structure based on the concept of Virtual Synchronous Machine for grid-tied converters. Furthermore, the design and implementation of the modulator is presented. Section 5.5 shows the main Hardware-in-the-Loop (HIL) results and, finally, the conclusions are presented in Sect. 5.6.

5.2 Selective Harmonic Elimination Modulation

Let us consider a voltage waveform, periodic in the time domain t with quarter-wave symmetry. This function can be described by the Fourier series as follows:

$$v(\omega t) = \sum_{h=1}^{\infty} a_h \sin(h\omega t), \tag{5.1}$$

where a_h is given by

$$a_h = \frac{4}{h\pi} \sum_{n=1}^{\infty} (-1)^{n+1} \cos(h\alpha_n), \tag{5.2}$$

and α_n corresponds to the switching angles.

Figure 5.1 shows a unipolar symmetrical PWM pulse pattern generated from the SHE modulation, which uses seven switching angles for a quarter period.

The selective harmonic elimination (SHE) modulation uses N switching angles to eliminate $(N-1)$ individual harmonic components in the output voltage (Patel and Hoft 1973). For the selective harmonic elimination problem, the set of nonlinear equations (5.6) must be solved (Patel and Hoft 1973):

$$\frac{4}{\pi} \left[\cos(\alpha_1) - \cos(\alpha_2) + \cdots + (-1)^{n+1} \cos(\alpha_N) \right] = m_i \tag{5.3}$$

$$\frac{4}{5\pi} \left[\cos(5\alpha_1) - \cos(5\alpha_2) + \cdots + (-1)^{n+1} \cos(5\alpha_N) \right] = 0 \tag{5.4}$$

$$\frac{4}{7\pi} \left[\cos(7\alpha_1) - \cos(7\alpha_2) + \cdots + (-1)^{n+1} \cos(7\alpha_N) \right] = 0 \tag{5.5}$$

$$\vdots \qquad\qquad \vdots \qquad\qquad \vdots$$

$$\frac{4}{H\pi} \left[\cos(H\alpha_1) - \cos(H\alpha_2) + \cdots + (-1)^{n+1} \cos(H\alpha_N) \right] = 0, \tag{5.6}$$

where n is the n-th switching angle defined by $n = 1, 2, \ldots N$ and h is the h-th harmonic component to be eliminated, where $h = 1, 5, 7, \ldots H$. Moreover, m_i is the modulation index defined by the ratio between the fundamental voltage component and the dc-link voltage. In order to eliminate H harmonic components, N switching angles are needed, where $H = N - 1$. In order to ensure a valid solution, the angles α_n must satisfy the condition (5.7):

$$0 < \alpha_1 < \alpha_2 < \alpha_3 < \cdots < \alpha_N < \frac{\pi}{2}. \tag{5.7}$$

The nonlinear equations of (5.6) can be solved by numeric methods as, for example, Newton-Raphson. Other solutions can be obtained by computational tools, such as the $fmincon$ solver, available in MATLAB® software. Usually, different solutions exist for (5.6); however it is important to note that harmonic components at frequencies higher than the eliminated components are not taken into account in the conventional formulation of SHE. For grid-tied converters, these harmonic voltage components can result in harmonic current components that exceed the limits defined by the grid-connection standards, such as the IEEE 1547 shown in Table 5.1.

It is important to mention that the limits presented in Table 5.1 are given as a percentage value of the nominal current in a period of 15–30 min. Moreover, even harmonics must be limited to 25% of the odd harmonics (IEEE Standard 2018).

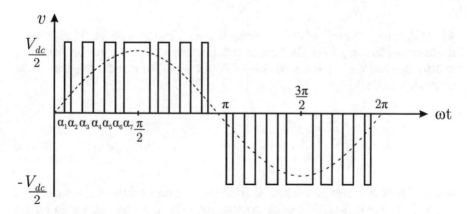

Fig. 5.1 PWM switching pattern with seven switching angles of quarter period

Table 5.1 Individual harmonic current distortion and THD limits of the IEEE 1547 standard (IEEE Standard 2018)

Harmonic order (h)	$\delta_h \cdot 100\ (\%)$
$h < 11$	4
$11 \leq h < 17$	2
$17 \leq h < 23$	1,5
$23 \leq h < 35$	0,6
$35 \leq h$	0,3
THD	**5**

5.3 Optimal Programmed PWM for Harmonic Mitigation and THD Minimization

This chapter proposes that the harmonic current components are kept below the limit established by IEEE 1547. In addition, the optimization algorithm searches a solution that minimizes the THD and complies with the limits of the IEEE 1547.

The definition of the switching angles can be treated as an optimization problem, that is:

$$\min_{\alpha_n} \text{THD}_i(\alpha) \text{ subjected to} \begin{cases} f_{\text{THD}_i}(\alpha_n) \leq 0 \\ f_1(\alpha_n) = 0 \text{ and } f_h(\alpha_n) \leq 0 \\ 0 < \alpha_1 < \cdots < \alpha_N < \frac{\pi}{2} \end{cases} \quad (5.8)$$

In this way, the THD of a voltage is written by

$$\text{THD}_v = \sqrt{\sum_{h=2}^{\infty} \left(\frac{V_h}{V_1}\right)^2}, \quad (5.9)$$

where V_h is the amplitude of the individual harmonic components and V_1 represents the fundamental component. On the other hand, the amplitude of the current harmonic components and the THD$_i$ can be defined by means of the transfer function of the output filter, that is.

$$f_{\text{THD}_i}(\alpha_n) = \sqrt{\sum_{h=2}^{\infty} \left(\frac{I_h}{I_1}\right)^2} = \frac{1}{V_1|G(j\omega 1)|}\sqrt{\sum_{h=2}^{\infty} [V_h|G(j\omega h)|]^2}, \quad (5.10)$$

where I_h is the amplitude of the individual harmonic component of the output current and I_1 is the fundamental harmonic component. LCL filters are often used for the grid-connection of power converters due to their third-order characteristic, which results in high attenuation of the harmonic components. Considering an LCL filter, the transfer function that relates the grid current i_g and the converter PWM voltage v_{in} is given by

$$G(s) = \frac{i_g(s)}{v_{in}(s)} = \frac{sCR_d + 1}{s^3CL_1L_2 + s^2CR_d(L_1 + L_2) + s(L_1 + L_2)}, \tag{5.11}$$

where C is the capacitance of the output filter, R_d is the damping resistance and L_1 e L_2 are the inductances. Considering $s = j\omega$ for sinusoidal systems, the filter gain as a function of the harmonic order is $|G(j\omega)|$. Therefore, this gain can be used to define the upper bound of the harmonic current components according to the standard. As the limit for THD_i is 5%, according to Table 5.1, the following constrains must be met:

$$f_{\text{THD}_i}(\alpha_n) = \text{THD}_i - 0.05 \cdot I_{nom} \leq 0, \tag{5.12}$$

where I_{nom} is the nominal output current. The following constrains are derived in order to search for the switching angles that satisfy the individual harmonic limits:

$$\left[\cos(\alpha_1) + \cdots + (-1)^{n+1}\cos(\alpha_N)\right] = m_i \frac{\pi}{4} \tag{5.13}$$

$$\left[\cos(5\alpha_1) + \cdots + (-1)^{n+1}\cos(5\alpha_N)\right] \leq \frac{5\pi\delta_5 I_{nom}}{4|G(5 \cdot j\omega)|V_{dc}} \tag{5.14}$$

$$\left[\cos(7\alpha_1) + \cdots + (-1)^{n+1}\cos(7\alpha_N)\right] \leq \frac{7\pi\delta_7 I_{nom}}{4|G(7 \cdot j\omega)|V_{dc}} \tag{5.15}$$

$$\vdots \qquad\qquad \vdots \qquad\qquad \vdots \tag{5.16}$$

$$\left[\cos(H\alpha_1) + \cdots + (-1)^{n+1}\cos(H\alpha_N)\right] \leq \frac{H\pi\delta_h I_{nom}}{4|G(H \cdot j\omega)|V_{dc}}. \tag{5.17}$$

Note that in (5.17) the values of H are not related to N as stated for the conventional SHE. In this work, it was established that $H = 50$ and $N = 7$. Therefore, in a general way, the inequalities of (5.17) can be expressed by

$$f_h(\alpha_n) = \sum_{n=1}^{N}(-1)^{n+1}\cos(h\alpha_n) - \frac{h\pi\delta_h I_{nom}}{4|G(h \cdot j\omega)|V_{dc}} \leq 0. \tag{5.18}$$

Due to the quarter-wave symmetry, the even harmonic components are zero. On the other hand, the components multiple of three are canceled for three-phase three-wire systems.

5.4 Closed-Loop Grid-Connected Converter

The grid-connection is carried out by a three-level neutral point clamped converter according to the block diagram presented in Fig. 5.2. Furthermore, this figure illustrates the modulation strategy associated with the virtual synchronous machine con-

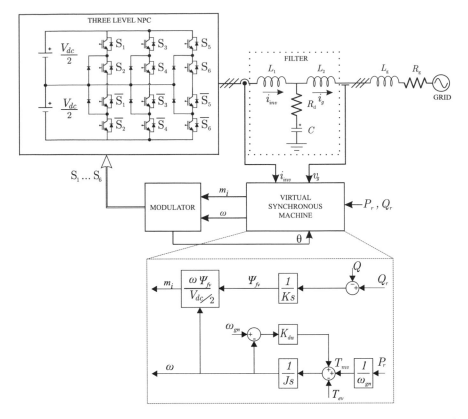

Fig. 5.2 Block diagram of the grid-connected converter with the proposed OP-PWM strategy and the virtual synchronous machine

trol concept. The control scheme uses the measurements of the grid voltage and output currents to provide the modulation index m_i and synchronization angle θ. Once m_i and θ are defined, the switching angles α_n are obtained from look-up tables stored offline. Once these angles are found, the modulator executes the logic operations to generate the drive signals for the power switches.

5.4.1 Virtual Synchronous Machine (VSM)

The virtual synchronous machine (VSM) is a concept where the grid-connected inverter emulates the behavior of an electromechanical synchronous machine. This technique allows to explore the good characteristics of synchronous machines, such as easy share of active and reactive power, as well as grid voltage and frequency support. In this way, the mechanic and electric equations of the system are derived in order to mimic the behavior of a synchronous machine (Zhong and Weiss 2011).

Let us consider a thee-phase synchronous machine, where the flux can be written as

$$\Phi_{abc} = L i_{abc} + M_f i_f , \qquad (5.19)$$

where M_f is the mutual inductance among the rotor and stator windings, i_{abc} is a vector of the stator currents and i_f is the field current. Furthermore:

$$\Phi_{abc} = \begin{bmatrix} \Phi_a \\ \Phi_b \\ \Phi_c \end{bmatrix} ; \quad i_{abc} = \begin{bmatrix} i_a \\ i_b \\ i_c \end{bmatrix} ; \quad M_f = \begin{bmatrix} M_f \cos(\theta) \\ M_f \cos(\theta - \frac{2\pi}{3}) \\ M_f \cos(\theta - \frac{4\pi}{3}) \end{bmatrix} , \qquad (5.20)$$

and R_s is the equivalent resistance of the stator windings. From (5.19), the terminal voltage is given by

$$\begin{aligned} \mathbf{v}_{abc} &= -R_s i_{abc} - \frac{d\Phi}{dt} \\ &= -R_s i_{abc} - L_s \frac{d i_{abc}}{dt} + e_{abc} \end{aligned} \qquad (5.21)$$

The electromotive force $\mathbf{e}_{abc}$ due to the rotor movement is:

$$\mathbf{e}_{abc} = -\frac{d M_f}{dt} i_f - \frac{d i_f}{dt} M_f \qquad (5.22)$$

since $d\theta_g/dt = \omega_g$, $d i_f/dt = 0$, $M_f i_f = \Psi_{fv}$ it can be written:

$$\mathbf{e}_{abc} = \begin{bmatrix} \omega_g \Psi_{fv} \sin(\theta) \\ \omega_g \Psi_{fv} \sin(\theta - \frac{2\pi}{3}) \\ \omega_g \Psi_{fv} \sin(\theta - \frac{4\pi}{3}) \end{bmatrix} \qquad (5.23)$$

The swing equation describes the mechanical dynamics of the machine as

$$J \frac{d\omega}{dt} = T_{mv} - T_{ev} - K_{dw}(\omega_{gn} - \omega), \qquad (5.24)$$

where J is the moment of inertia, K_{dw} is the constant damping factor of the machine, which is related to the frequency oscillation.

On the other hand, the electromagnetic torque T_{ev} is dependent of the electric variables and acts as the feedback in the loop:

$$T_{ev} = \frac{e_a i_a + e_b i_b + e_c i_c}{\omega}. \qquad (5.25)$$

The mechanic torque T_{mv} is directly related to the active power reference, such as

$$T_{mv} = \frac{P_{ref}}{\omega_{gn}}. \tag{5.26}$$

The amplitude of the voltage produced by the VSM is associated with the reactive power equation:

$$\Psi_{fv} = \frac{Q - Q_{ref}}{sK} \tag{5.27}$$

where, K is the constant related with amplitude oscillation, Q_{ref} is the reference reactive power and Q is the reactive expressed as

$$Q = \omega \mathbf{i}_{abc} \mathbf{M}_f i_f \tag{5.28}$$

According to the description, the behavior of a VSM is emulated by the grid-connected converter. More details about the virtual synchronous machine concept can be found in Zhong and Weiss (2011). The information of amplitude and frequency of the synchronous machine needs to be converted to gate signals of the inverter. This task is performed by the proposed FPGA-based Modulator as described in next subsection.

5.4.2 FPGA-Based Modulator

The pulse pattern produced by a SHE-like modulation requires a specific modulator, since it is based on programmed switching angles. This modulator must be capable of changing amplitude and frequency in real time. Conventional PWM modules present in commercial Digital Signal Controllers (DSCs) are not suitable to perform such task, since just two or three comparators are available per carrier. Besides that, SHE-like commutation patterns are not equally spaced in time, and also not synchronized among phases.

An interesting solution can be achieved using a Field-Programmable Gate Array (FPGA) based modulator, mainly due to the FPGA's flexibility. In this approach, two *up-down* counters are used, as shown in Fig. 5.3. Counter 2 is at the fundamental grid frequency and it is used to define the polarity of the output voltage, generating the Signal A. On the other hand, Counter 1 is at the double of the fundamental grid frequency. The switching angles α, obtained by solving the optimization equations from Sect. 5.3 and shown in Fig. 5.9, are compared to Counter 1 to generate the signals B to H. These signals are then used in a combinational logic to produce the gate signals for the inverter (S_1 and S_2). Figure 5.3 shows the internal signals from the modulator. The drive signals for the other phases of the inverter can be obtained creating other two groups of Counter 1 and Counter 2 and shifting them by 120°.

The FPGA-modulator receives two information from the DSC, i.e., amplitude of the output voltage and frequency. The desired output voltage amplitude signal is used to calculate the modulation index of the inverter, which is defined as

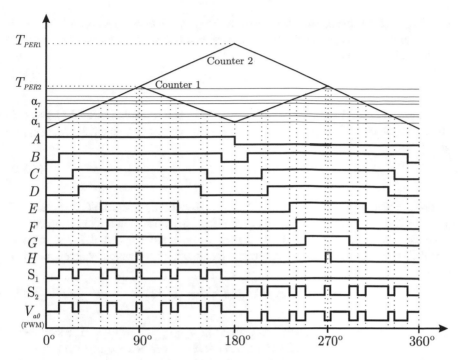

Fig. 5.3 Drive signals generation for the three-level NPC inverter

$$m_i = \frac{\Psi_{fv}\omega}{V_{dc}/2}. \tag{5.29}$$

where V_{dc} is the voltage of the DC link. The modulation index belongs to a range from 0 to a theoretical maximum of $4/\pi \approx 1.27$. At the DSC, the m_i value is multiplied by 100 and rounded to a 7 bits binary number and then sent to the FPGA through a parallel communication. At the FPGA, m_i is used as an index in the look-up table with the information of the α angles.

On the other hand, the information about the virtual synchronous machine frequency is sent to the modulator as a variable clock frequency. This clock is generated using the conventional PWM modules, or counter timer, available in most common DSCs. This PWM module usually works at a high frequency (f_{DSC}). By setting a maximum count value to T_{PER}, i.e., the maximum value of the counter, and the duty cycle to 50%, this module can work as a variable frequency clock generator, as shown in Eq. 5.30.

$$T_{PER} = \text{round}\left(N_b \cdot \frac{\omega_{gn}}{\omega}\right)$$

$$COMP = \frac{T_{PER}}{2}. \tag{5.30}$$

where, $COMP$ is the duty cycle comparator register set to 50%, N_b is the base clock divider and depends on the chosen frequency resolution of the modulator (Δf_{min}), as shown in Eq. 5.31.

$$\Delta f_{min} = \frac{f_{gn}}{N_b} \tag{5.31}$$

Considering a grid frequency of $f_{gn} = 60$ Hz and a resolution of $\Delta f_{min} = 0.1$ Hz, a base clock divider of $N_b = 600$ is obtained. The clock frequency (f_{CLOCK}) is given by

$$f_{CLOCK} = \frac{f_{DSC}}{T_{PER}} \tag{5.32}$$

For a VSM at the grid frequency, $T_{PER} = N_b$ with a typical DSC frequency of $f_{DSC} = 100$ MHz and $f_{CLOCK} = 166.66$ kHz. This clock signal will increment all Counters 1–2 at the FPGA, whose maximum values are T_{PER1} and T_{PER2} respectively. These values are used to define the frequency of the output voltage, defined as

$$T_{PER1} = round\left(\frac{f_{DSC}}{4 f_{gn} N_b}\right), \tag{5.33}$$

The maximum value of counter 2 is defined as $T_{PER2} = 2 \cdot T_{PER1}$. The switching angles stored in the FPGA are converted to counter values as $\alpha_{FPGA} = round(\alpha_{rad} \cdot T_{PER1})$. If the VSM frequency ω increases, then f_{CLOCK} increases proportionally; as a consequence, Counters 1 and 2 are going to be incremented in a faster way, as shown in Fig. 5.4.

The modulator is also responsible for calculating the electrical angle (θ) of the VSM, which is defined as the integral of the frequency:

$$\theta = \frac{\omega}{s} \tag{5.34}$$

In the FPGA, the electrical angle (θ) is implemented as a Counter 3, which is an up-count only counter. This Counter has the same period of Counter 2 and is synchronized with phase A. Therefore, the maximum value of Counter 3 is $2 \cdot T_{PER2}$ which, in this case, is a 13-bit long binary number sent back to the DSC. The electrical angle is sent from the FPGA to the DSC using a 4-bit parallel communication plus a bit for the clock signal, as shown in Fig. 5.5.

Figure 5.5 shows the complete block diagram of the internal structure of the FPGA-based modulator, which is implemented in VHDL (VHSIC Hardware Description Language). The Clock signal is received from the DSC and used to drive Counters 1–2 of each phase and Counter 3 is used for the electrical angle of the VSM. The modulation index is used to select the nearest switching angle (α_{1-7}) at the look-up table. The chosen angles α are buffered and transferred to the comparator register at the next update cycle. The update rate of the switching angles is the same as the discretization period (T_s) used for the control strategy.

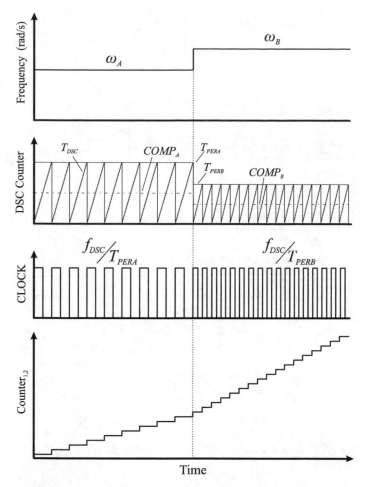

Fig. 5.4 Clock generation at the DSC to drive the FPGA counters. Example when the frequency changes from ω_A to ω_B, with $\omega_B > \omega_A$

The following Boolean logic can be employed to derive the drive signals for the inverter power switches:

$$
\begin{aligned}
S_1 &= A \cdot \left(B \cdot \overline{C} + D \cdot \overline{E} + F \cdot \overline{G} + H \right) \\
S_2 &= \overline{A} \cdot \left(B \cdot \overline{C} + D \cdot \overline{E} + F \cdot \overline{G} + H \right).
\end{aligned}
\tag{5.35}
$$

Appendix B shows the VHDL code to implement the main part of the modulator proposed above for phase A.

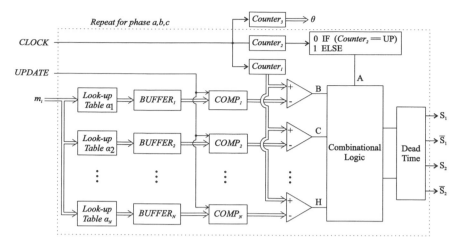

Fig. 5.5 Drive signals generation for the three-level NPC inverter

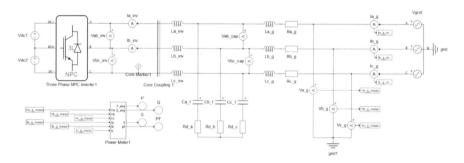

Fig. 5.6 Typhoon Model composed of NPC inverter, LCL filter, three-phase grid, current e voltage sensor, power meter

5.5 Hardware-in-the-Loop Results

Hardware-in-the-Loop results were obtained using a Typhoon HIL 402. The model of the converter has been developed in the Typhoon HIL software using a Three-Phase Neutral-Point Clamped converter (NPC) connected to the grid by means of an LCL filter, as shown in Fig. 5.6.

The implementation of the control strategy is performed in a Texas Instruments DSC TMS320F28379D. This DSC is a 32-bit floating point Digital Signal Controller with a processor clock of 200 Mhz and a peripheral clock of $f_{DSC} = 100$ Mhz. The VSM is discretized using the Euler method with a time step of $T_s = 1/2760$ s.

The voltages and currents from the emulated system in Typhoon Hil are converted to analog signals with a ± 10 V range. These analog signals are scaled down with operational amplifiers to match the range of 0 to 3 V, that is the input of the ADC channels of the DSC.

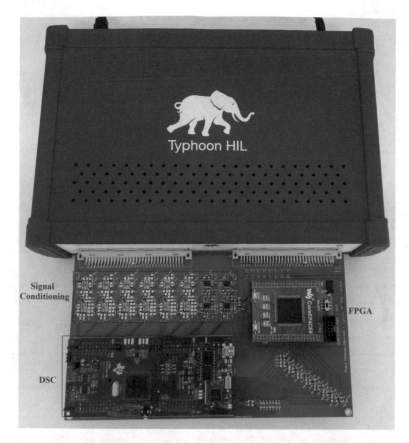

Fig. 5.7 Hardware in the loop setup

Digital signals are sent from the DSC to the ALTERA FPGA Cyclone IV EP4CE6E22C8N, where the modulator and the dead-time for the gate signals are implemented in VHDL language. The modulator also returns the electrical angle of the virtual synchronous machine. Figure 5.7 shows the setup employed to obtain the results and Fig. 5.8 illustrates the implementation of each stage in their respective device. In summary, the grid voltage (v_g) and the converter output current (i_{inv}) are measured by the TMS320F28379D DSC, that communicates with the CoreEP4CE6 FPGA sending frequency (ω) and modulation index (m_i) and receiving the angle (θ). Moreover, the computed switching signals ($S_1...S_6$) are sent to the converter that is emulated in the Typhoon HIL 402.

Table 5.2 shows the main parameters of the HIL test. It is important to mention that the parameters of the LCL filter have to be known a priori, since the transfer function of the filter is part of the algorithm that searches the switching angles. Therefore, in this chapter, the following steps are presented for the LCL filter design: (i) the capacitor is derived from the reactive power absorbed at rated conditions, where the

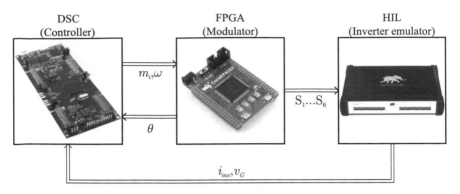

Fig. 5.8 Block diagram of implemented control system

Table 5.2 Main simulation parameters

Parameter	Value	Parameter	Value
L_1	4 mH	L_2	400 μH
C	170 μF	R_d	0,1 Ω
L_g	100 μH	R_g	0,4 Ω
V_{dc}	4 kV	S	700 kVA
I_{nom}	200 A	f_{res}	579 Hz
J	0.5	D_p	200
K	80000		

capacitance generally is around 5% of the base value (Liserre et al. 2005). (ii) The converter-side inductor L_1 is considered ten times greater than the grid-side inductor L_2. This value defines the ratio of harmonic current attenuation between both sides, i.e., converter and grid. (iii) The resonance frequency is lower than the equivalent switching frequency and it must not coincide with the harmonic components generated by the converter. In this chapter, the equivalent switching frequency is 840 Hz and the resonance frequency of the LCL filter is 579 Hz. Furthermore, the LCL filter presents a resonance that can be damped for a stable grid-connected operation. The resonance can be damped via control feedback whenever there is enough bandwidth. However, the virtual synchronous machine acts mainly at nominal frequency and cannot properly damp the LCL resonance. Therefore, a resistor is added in series with the capacitor as passive damping.

Figure 5.9 shows the switching angles for the proposed OP-PWM and for the convectional SHE-PWM. Note that the solution of the proposed strategy results in discontinuous switching angles as a function of the modulation index.

Based on the switching angles from Fig. 5.9, two simulations are carried out on MATLAB with both SHE-PWM and OP-PWM. Figure 5.10 shows the behavior of the output grid currents.

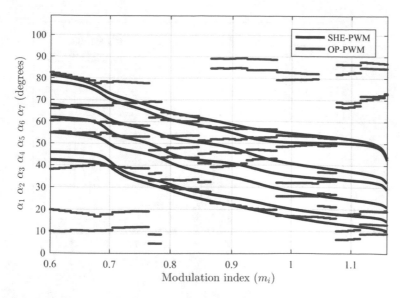

Fig. 5.9 Switching angles for the conventional SHE and the proposed OP modulation strategy

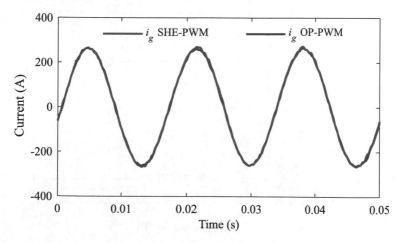

Fig. 5.10 Output current using conventional SHE modulation versus proposed OP-PWM

The spectrum of the current for the proposed OP-PWM and the conventional SHE can be seen in Fig. 5.11. It is possible the note from Fig. 5.11a that the harmonic components 5^a, 7^a, 11^a, 13^a, 17^a and 19^a are eliminated when the conventional SHE is applied. However, the harmonic components 23^a, 25^a e 29^a exceed the limits established by the standard. This issue can be solved by the increase of the number of angles or the design of the output filter, although both solutions can represent an increase of costs or volume/weight to the system. On the other hand, as shown in

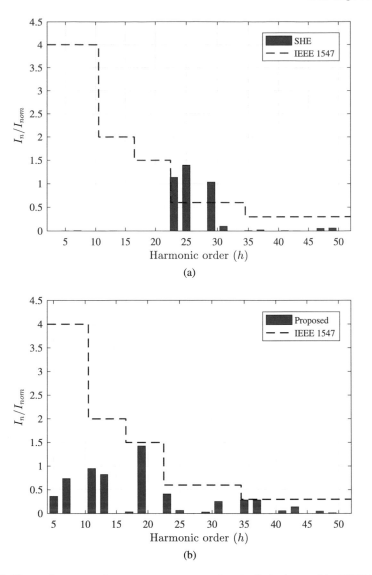

Fig. 5.11 Harmonic spectra of output current compared to the limits imposed by IEEE 1547. **a** Conventional SHE-PWM. **b** proposed OP-PWM strategy

Fig. 5.11b, the proposed modulation strategy keeps all harmonic components under the limits established by the standard.

Furthermore, Fig. 5.12 shows that the proposed strategy keeps the THD of the output current within the limits imposed by the standard for the modulation indexes from 0.6 to 1.16. Additionally, Fig. 5.13 shows the behavior of the harmonic content

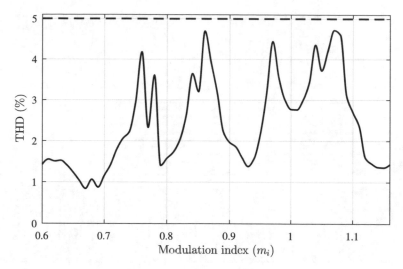

Fig. 5.12 THD of output current and limits imposed by IEEE 1547

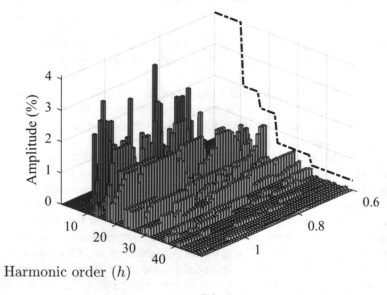

Fig. 5.13 3D graphic of the harmonic spectrum of output for the proposed strategy

of the output current until the 50^a harmonic order, which are also within the limits imposed by the standard for the modulation indexes from 0.6 to 1.16.

Figure 5.14 shows a step in the active power reference from $P_{\text{ref}} = 300$ kW to $P_{\text{ref}} = 750$ kW. It can be seen that the system presents a good transient response with

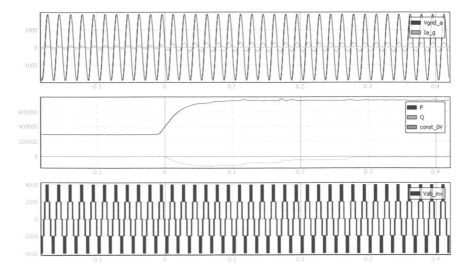

Fig. 5.14 Step on the active power reference from $P_{ref} = 300$ kW to $P_{ref} = 750$ kW

a settling time of less than 0.1 s. This response can be adjusted through the inertia of the virtual synchronous machine. This result was taken from the Typhoon HIL software.

Figure 5.15 shows the internal variables of the DSC for the same power step shown in Fig. 5.14. An increase in the frequency when the active power increases can be seen. The modulation index also increases to compensate the voltage drop over the LCL filter due to the current increase.

Data from the DSC memory is gathered by saving all needed variables in a specific memory region, and it is imported with the "Save Memory functionality" of the Code Composer Studio. This function is present in the most common Eclipse-based IDEs. "Save Memory" outputs a single-column file with all the content of a given memory region. This file is then processed with MATLAB to generate Fig. 5.15.

Figure 5.16 shows the steady-state current at the inverter-side inductor (i_{inv}) and at the grid-side inductor (i_g). The amplitude difference is due to the reactive power of the capacitor. A higher current ripple can be seen at the inverter-side inductor. On the other hand, at the grid-side inductor, a very small ripple is perceived due to the third-order harmonic attenuation of the LCL filter. The THD of the grid current is kept below 5%, which meets the IEEE 1547 standard, as expected.

Figure 5.17 shows a FFT analysis of the line current at the grid-side inductor (i_g), while Figure 5.18 shows the inverter output line-to-line voltage with the OP-PWM strategy.

The proposed algorithm finds valid solutions for modulation indexes lower than 1.16. Conversely, the range $m_i = 1.16$ to 1.27 comprises the overmodulation region where the loss of degrees of freedom generates additional harmonic components that deteriorate the quality of the output voltages. Therefore, the proposed algorithm can

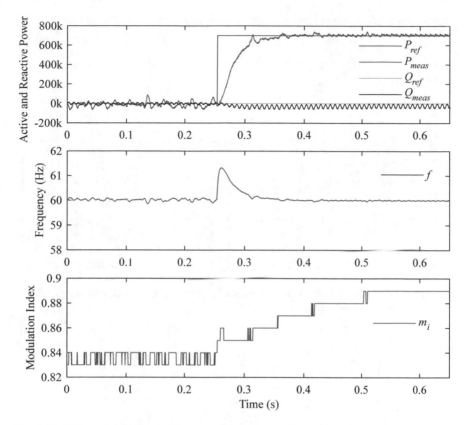

Fig. 5.15 Main variables from DSC memory equivalent to the step on active power reference in Fig. 5.14

not find a solution that keeps the THD and individual harmonics under the limits of the IEEE 1547 standard for $m_i \geq 1.16$. As the grid voltage generally does not experience large variations in its amplitude and frequency at nominal conditions, the converter is designed to operate with modulation indexes lower than 1.16. Supposing that a modulation index greater than 1.16 is required during the operation of the grid-connected converter, then there will be a limitation of the control action. The converter still synthesizes sinusoidal output voltages with a limited maximum value, however, this fact may result in active and reactive powers different from the reference.

On the other hand, when it comes to a five-level multilevel converter, low modulation indexes lead to a reduction of the number of output voltage levels from 5 to 3, increasing the output current distortion. Therefore, the operation of a grid-connected converter at low modulation indexes is avoided. It is worth mentioning that low modulation indexes generally are useful for applications such as motor drives operating at the variable voltage/frequency; therefore this topic is out of the scope of this chapter.

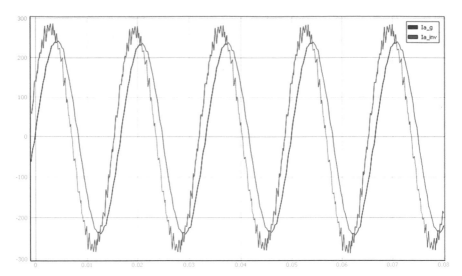

Fig. 5.16 Line current at the inverter-side inductor (i_{inv}) and at the grid-side inductor (i_g). Image from the Typhoon HIL software

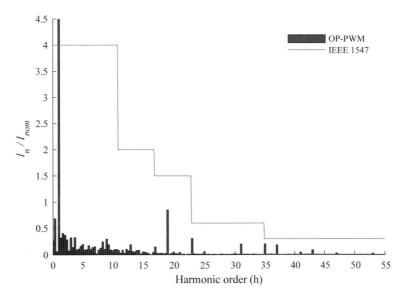

Fig. 5.17 FFT analysis of the line current at the grid-side inductor (i_g). Data from the Typhoon HIL software, plot, and FFT performed on MATLAB

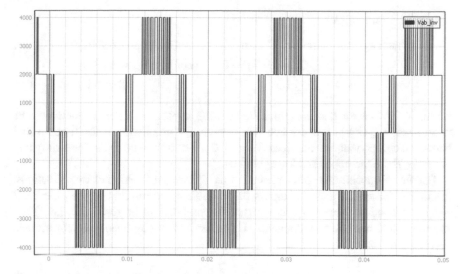

Fig. 5.18 Inverter line-to-line output voltage. Image from the Typhoon HIL software

5.6 Conclusion

This chapter proposed an optimal programmed PWM strategy coordinated with the virtual synchronous machine concept for grid-connected multilevel converters. The switching angles were obtained offline by an optimization algorithm that minimized the THD and constrained the individual harmonics in compliance with the harmonic limits of the IEEE 1547 standard. The proposed technique was applied to a three-phase grid-connected NPC converter; notwithstanding the principles described in this chapter can be expanded to other topologies with any number of levels. Moreover, the virtual synchronous machine concept supported the control and grid-connection, as it incorporates the angle for synchronization and dismisses additional current or voltage control loops. HIL results were obtained with a setup comprising Typhoon HIL 402, a TMS320F28379D DSC and an EP4CE6E22C8N FPGA. These results demonstrated the good performance of the proposed modulation technique compared to the conventional selective harmonic elimination strategy.

Appendix A

The search of the switching angles is performed by means of the software MATLAB. The command *fmincon* finds a constrained minimum of a function of several variables. The *fmincon* attempts to solve problems of the form:

$$\text{min} \quad f(X)$$

$$
\begin{array}{lll}
\text{subject to} & AX \le B, \; A_{eq}X = B_{eq} & \text{(linear constraints)} \\
& C(X) \le 0, \; Ceq(X) = 0 & \text{(nonlinear constraints)} \\
& LB \le X \le UB & \text{(bounds)}
\end{array}
\qquad (5.36)
$$

where X is a vector that contain the switching angles, $f(X)$ is a function to be minimized, $C_{eq}(X)$ is a restriction that ensures the amplitude of the fundamental harmonic component, the matrices A_{eq} and B_{eq} determine the linear inequality and LB and UB are the upper and lower bounds.

For the proposed algorithm presented as follows, it is defined two functions: the first is *fminimize* that encompass the variable to be minimized, i.e., the THD and the second is *frestric* that comprises the variables to be restricted, i.e., THD and individual harmonic components. These harmonic components are constrained according to Table 5.1.

```
input  : fminimize, frestric;
output: alpha;

for k ← 1 to n_p do
    alpha ← fmincon( fminimize, init, frestrict, LB, UB);
    fminimize ← Eq. (5.18);
    frestrict ← Eqs. (5.12), (5.18);
end
```

Algorithm 5.1: Search of the switching angles

The vector *init* contains the initialization of the seven switching angles. Whenever one solution is achieved, the initialized angles are refreshed for the new search. If the algorithm cannot find a solution next to the discontinuities of the angles, therefore the initialization angles can be randomized. The lower bound (LB) of the solutions is zero and the upper bound (UB) is $\pi/2$.

Appendix B

Figure 5.19 shows the VHDL code to implement the main part of the proposed modulator for the phase A. A VHDL code usually is composed of *processes*. These *processes* are activated by a sensitivity list. There are two signals at the sensitivity list: *reset* and *clock*. Reset is used to asynchronously put all variables to default states. On the other hand, an uprising edge clock is used to run the modulator itself.

There are other *processes* to implement all the functions of the modulator, e.g., dead-time generation, receiving/sending data from/to the DSC, but they are not shown here for the sake of simplicity.

```vhdl
1  PHASE_A:PROCESS (reset,clock) -- Modulation process for phase A
2
3     VARIABLE count1_A  : integer RANGE 0 to TPER1 := 0;   -- Counter 1
4     VARIABLE count2_A  : integer RANGE 0 to TPER2 := 0;   -- Counter 2
5     VARIABLE dir1_A    : BIT;                    -- Direction of Counter1-A
6     VARIABLE dir2_A    : BIT;                    -- Direction of Counter2-A
7     VARIABLE aux_a0    : BIT;                    -- Auxiliary Variable
8     VARIABLE Signal_A  : BIT_VECTOR(6 DOWNTO 0); -- Output signals for the comparators
9
10 BEGIN
11    IF (reset = '1') THEN -- Asynchronous Reset
12        count1_A := 0;
13        count2_A := 0;
14        dir1_A := '1';
15        dir2_A := '1';
16
17        ELSIF (clock = '1' AND clock'EVENT) THEN -- Run this process at the uprising clock event
18
19           --Increase/decrease Counter 1 based on its direction ------------------------------------
20           IF(dir1_A = '1')THEN  count1_A := count1_A + 1; -- Increase Counter1-A if phase A is counting up
21                      ELSE  count1_A := count1_A - 1; -- Decrease Counter1-A if phase A is counting down
22           END IF;
23
24           --Increase/decrease Counter 2 based on its direction ------------------------------------
25           IF(dir2_A = '1')THEN
26                    count2_A := count2_A + 1;  -- Increase Counter2-A if phase A is counting up
27                    count3 <= count2_A;        -- Sample the Converter angle to be sent to the DSC
28           ELSE
29                    count2_A := count2_A - 1;  -- Decrease Counter2-A if phase A is counting down
30                    count3 <= V360deg - count2_A; -- Sample the Converter angle to be sent to the DSC
31           END IF;
32
33           --Change direction of counter 1 based on top/bottom values---------------------------
34           IF(count1_A>=V90deg) THEN dir1_A := '0'; END IF;  -- Set direction down of Counter1-A at the top
35           IF(count1_A<=0)      THEN dir1_A := '1'; END IF;  -- Set direction up of Counter1-A at the bottom
36
37           --Change direction of counter 2 based on top/bottom values---------------------------
38           IF(count2_A>=V180deg) THEN dir2_A := '0'; END IF;  -- Set direction down of Counter2-A at the top
39           IF(count2_A<=0)      THEN
40                    dir2_A := '1';            -- Set direction up of Counter2-A at the bottom
41                    count1_A:=0;              -- Force synchronization between counters 1A and 2A
42           END IF;
43
44           -- Generation of signals B-H for phase A
45           IF(count1_A>=alpha1_COMP) THEN Signal_A(0):='1'; ELSE Signal_A(0):='0'; END IF;  --B
46           IF(count1_A>=alpha2_COMP) THEN Signal_A(1):='1'; ELSE Signal_A(1):='0'; END IF;  --C
47           IF(count1_A>=alpha3_COMP) THEN Signal_A(2):='1'; ELSE Signal_A(2):='0'; END IF;  --D
48           IF(count1_A>=alpha4_COMP) THEN Signal_A(3):='1'; ELSE Signal_A(3):='0'; END IF;  --E
49           IF(count1_A>=alpha5_COMP) THEN Signal_A(4):='1'; ELSE Signal_A(4):='0'; END IF;  --F
50           IF(count1_A>=alpha6_COMP) THEN Signal_A(5):='1'; ELSE Signal_A(5):='0'; END IF;  --G
51           IF(count1_A>=alpha7_COMP) THEN Signal_A(6):='1'; ELSE Signal_A(6):='0'; END IF;  --H
52
53           -- Perform boolean logic
54           aux_a0:=(Signal_A(0)AND(NOT Signal_A(1))) OR (Signal_A(2)AND(NOT Signal_A(3)))
55                           OR (Signal_A(4)AND(NOT Signal_A(5))) OR Signal_A(6);
56
57           -- Conditionals below are used to define the polarity of the output signals based
58           -- on signal A (direction of Counter2)
59           IF (dir2_A = '1') THEN
60                    aux_a1 := NOT (aux_a0);
61                    aux_a2 := '1';
62           ELSE
63                    aux_a1 := '0';
64                    aux_a2 :=  (aux_a0);
65           END IF;
66        END IF;
67
68        S1aux <= aux_a1; -- Send the two signals to Dead-time generation process
69        S2aux <= aux_a2;
70
71 END PROCESS PHASE_A;
```

Fig. 5.19   VHDL code to implement the proposed modulation for phase A

References

Agelidis VG, Balouktsis A, Balouktsis I, Cossar C (2006) Multiple sets of solutions for harmonic elimination PWM bipolar waveforms: analysis and experimental verification. IEEE Trans Power Electron 21(2):415–421. https://doi.org/10.1109/tpel.2005.869752

Beck H-P, Hesse R (2007) Virtual synchronous machine. In: 9th international conference on electrical power quality and utilisation, pp 1–6. https://doi.org/10.1109/epqu.2007.4424220

Chabni F, Taleb R, Helaimi M (2017) ANN-based SHEPWM using a harmony search on a new multilevel inverter topology. Turkish J Electr Eng Comput Sci 25:4867–4879. https://doi.org/10.3906/elk-1703-122

Dahidah MSA, Konstantinou G, Agelidis VG (2015) A review of multilevel selective harmonic elimination PWM: formulations, solving algorithms, implementation and applications. IEEE Trans Power Electron 30(8):4091–4106. https://doi.org/10.1109/tpel.2014.2355226

Enjeti PN, Ziogas PD, Lindsay JF (1990) Programmed PWM techniques to eliminate harmonics: a critical evaluation. IEEE Trans Ind Appl 26(2):302–316. https://doi.org/10.1109/28.54257

Franquelo LG, Napoles J, Guisado RCP, Leon JI, Aguirre MA (2007) A flexible selective harmonic mitigation technique to meet grid codes in three-level PWM converters. IEEE Trans Indust Electron 54(6):3022–3029. https://doi.org/10.1109/tie.2007.907045

Grigoletto FB, Pinheiro H (2009) A space vector PWM modulation scheme for back-to-back three-level diode-clamped converters. In: Brazilian power electronics conference, pp 1058–1065. https://doi.org/10.1109/COBEP.2009.5347717

Grigoletto FB, Pinheiro H (2011) Generalised pulse width modulation approach for DC capacitor voltage balancing in diode-clamped multilevel converters. IET Power Electron 4(1):89–100. https://doi.org/10.1049/iet-pel.2009.0214

Holmes DG, Lipo TA (2003) Pulse width modulation for power converters: principles and practice, Wiley-IEEE Press, ISBN: 978-0-471-20814-3

IEEE standard for interconnection and interoperability of distributed energy resources with associated electric power systems interfaces (2018), IEEE Std 1547-2018 (Revision of IEEE Std 1547-2003), 19(5):1–138. https://doi.org/10.1109/IEEESTD.2018.8332112

Jayakumar V, Chokkalingam B, Munda JL (2021) A comprehensive review on space vector modulation techniques for neutral point clamped multi-level inverters. IEEE Access 9:112104–112144. https://doi.org/10.1109/ACCESS.2021.3100346

Jiang Y, Li X, Qin C, Xing X, Chen Z (2022) Improved particle swarm optimization based selective harmonic elimination and neutral point balance control for three-level inverter in low-voltage ride-through operation. IEEE Trans Indust Inf 18(1):642–652. https://doi.org/10.1109/TII.2021.3062625

Leon JI, Kouro S, Franquelo LG, Rodriguez J, Wu B (2016) The essential role and the continuous evolution of modulation techniques for voltage-source inverters in the past. Present, Future Power Electron, IEEE Trans Indust Electron 63(5):2688–2701. https://doi.org/10.1109/TIE.2016.2519321

Leon JI, Vazquez S, Franquelo LG (2017) Multilevel converters: control and modulation techniques for their operation and industrial applications. Proc IEEE 105(11):2066–2081. https://doi.org/10.1109/JPROC.2017.2726583

Li Y, Zhang X-P, Li N (2022) An improved hybrid PSO-TS algorithm for solving nonlinear equations of SHEPWM in multilevel inverters. IEEE Access 10:48112–48125. https://doi.org/10.1109/ACCESS.2022.3170442

Liserre M, Blaabjerg F, Hansen S (2005) Design and control of an LCL-filter-based three-phase active rectifier. IEEE Trans Ind Appl 41(5):1281–1291. https://doi.org/10.1109/TIA.2005.853373

McGrath BP, Holmes DG (2022) An analytical technique for the determination of spectral components of multilevel carrier based PWM methods. IEEE Trans Indust Electron 49(4):847–857. https://doi.org/10.1109/TIE.2002.801071

Memon MA, Mekhilef S, Mubin M, Aami M (2018) Selective harmonic elimination in inverters using bio-inspired intelligent algorithms for renewable energy conversion applications: a review. Renew Sustain Energy Rev 82(3):2235–2253. https://doi.org/10.1016/j.rser.2017.08.068

Napoles J, Leon JI, Portillo R, Franquelo LG, Aguirre MA (2010) Selective harmonic mitigation technique for high-power converters. IEEE Trans Indust Electron 57(7):2315–2323. https://doi.org/10.1109/tie.2009.2026759

Napoles J, Portillo R, Leon JI, Aguirre MA, Franquelo LG (2008) Implementation of a closed loop SHMPWM technique for three level converters. In: 2008 34th annual conference of IEEE industrial electronics. https://doi.org/10.1109/iecon.2008.4758482

Patel HS, Hoft RG (1973) Generalized techniques of harmonic elimination and voltage control in thyristor inverters: part i-harmonic elimination. IEEE Trans Ind Appl 3:310–317

Qing-Chang Z, Phi-Long N, Zhenyu M, Wanxing S (2014) Self-synchronized synchronverters: inverters without a dedicated synchronization unit. IEEE Trans Power Electron 29(2):617–630. https://doi.org/10.1109/tpel.2013.2258684

Sadoughi M, Pourdadashnia A, Farhadi-Kangarlu M, Galvani S (2022) PSO-optimized SHE-PWM technique in a cascaded H-bridge multilevel inverter for variable output voltage applications. IEEE Trans Power Electron 37(7):8065–8075. https://doi.org/10.1109/TPEL.2022.3146825

Schuetz DA, Grigoletto FB, Carnielutti F, Pinheiro H (2021) Discontinuous space vector modulation for three-phase five-levels packed-u-cell converter. IEEE Trans Power Electron 36(12):14353–1436. https://doi.org/10.1109/TPEL.2021.3086407

Tibola JR, Pinheiro H, de Camargo RF (2011) Closed loop selective harmonic elimination applied to a grid connected PWM converter with LCL filter. In: XI Brazilian power electronics conference. https://doi.org/10.1109/cobep.2011.6085282

Turnbull FG (1964) Selected harmonic reduction in static D-C - A-C inverters. IEEE Trans Commun Electron 83(73):374–378. https://doi.org/10.1109/tcome.1964.6541241

Vijeh M, Rezanejad M, Samadaei E, Bertilsson K (2019) A general review of multilevel inverters based on main submodules: structural point of view. IEEE Trans Power Electron 34(10):9479–9502. https://doi.org/10.1109/TPEL.2018.2890649

Zhao H, Jin T, Wang S, Sun L (2016) A real-time selective harmonic elimination based on a transient-free inner closed-loop control for cascaded multilevel inverters. IEEE Trans Power Electron 31(2):1000–1014. https://doi.org/10.1109/tpel.2015.2413898

Zhong Q-C, Weiss G (2011) Synchronverters: inverters that mimic synchronous generators. IEEE Trans Indust Electron 58(4):1259–1267. https://doi.org/10.1109/tie.2010.2048839

Chapter 6
Selective Harmonic Compensation in Active Power Filter Using Nonlinear Predictive Current Control Method

Sandeep Ojha⊙ **and Rajesh Gupta**⊙

Abstract In this chapter, a non-linear predictive current control (PCC) scheme is proposed based on the inner current loop and outer voltage loop using selective harmonic compensation for single-phase shunt active power filters (SAPFs). This study designs the controller parameters utilizing conventional non-linear techniques, namely high-frequency rejection transfer functions (combination of band-pass filter and harmonic compensator transfer function). The proposed control algorithm is executed and a comparative study between the conventional and proposed technique in APFs has been presented. The novelty of the proposed PCC technique is the selective harmonic elimination from the grid current which comes from non-linear loads. In this chapter, the third harmonic component is eliminated from the grid current. The proposed technique is simulated on a virtual HIL environment and real-time simulation validation is done by using FPGA based Typhoon HIL 402 kit.

Keywords Active Power Filter (APF) · Predictive Current Control (PCC) · Selective Current Harmonic (SCH) · Total Harmonic Distortion (THD)

6.1 Introduction

In recent years, the distorting loads have caused an increasing amount of harmonic currents to enter the grid (Marcos-Pastor et al. 2020). Harmonic distortion currents can be minimized by utilizing power line conditioner techniques. Active power filters (APFs) prove to be the popular and efficient power line conditioners in the field of power electronics (Hsu and Wu 1996). The APFs are effective applications not only for the harmonic current compensation of grid current which is produced by the distorting loads, but also for the reactive power compensation, unbalanced, non-linear loads, etc. (Ojha and Gupta 2020). During the past two decades, the design and

S. Ojha (✉) · R. Gupta
Electrical Engineering Department, Motilal Nehru National Institute of Technology Allahabad, Prayagraj 211004, Uttar Pradesh, India
e-mail: ojha.sandeep89@gmail.com

© The Author(s), under exclusive license to Springer Nature Singapore Pte Ltd. 2023   145
S. M. Tripathi and F. M. Gonzalez-Longatt (eds.), *Real-Time Simulation and Hardware-in-the-Loop Testing Using Typhoon HIL*, Transactions on Computer Systems and Networks, https://doi.org/10.1007/978-981-99-0224-8_6

control of APFs have been rigorously investigated, especially for the compensation of harmonics produced by the distorting loads. Due to the direct control of the harmonics, the selective current harmonics (SCH) elimination method is useful, especially in low switching frequency applications (Makhlouf et al. 2016; Liang et al. 2016; Cid-Pastor et al. 2013). Generally, the three-phase shunt filters find application in the systems for large capacity non-linear loads. On the other hand, single-phase APFs are useful for small-scale applications with single wire earth return power systems as shown in Fig. 6.1.

The SCH elimination is a pre-programmed modulation technique that eliminates lower-order harmonics in the reference current (Ojha and Pandey 2016). The SCH elimination has some distinct features in comparison to other control techniques; these include (i) low converter switching frequency, (ii) possibility of over-modulation, which increases the utilization of the DC-link voltage, (iii) elimination of lower-order harmonics, and (iv) reduction in the size of the DC-link filter components. The SCH elimination scheme uses trigonometric equations with the variety of angles obtained after applying the Discrete Fourier Transform (DFT)-based filtering of the output waveforms. The most common methods used to solve these equations are numerical methods and optimization-based methods.

In the proposed technique, a Diode clamped 2-level inverter with a DC-link capacitor is employed. The performance of the proposed predictive current controller is superior to classical selective harmonic elimination (SHE) owing to the inclusion of the band-pass filter with a harmonic compensator which eliminates the selective harmonics present in the system without using the physical filter in the hardware

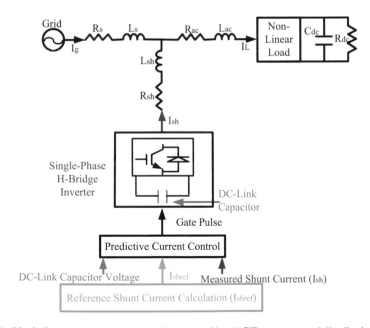

Fig. 6.1 Block diagram of single-phase active power filter (APF) compensated distribution system

model. It reduces the cost and bulkiness of the system. The advantages of model predictive control (MPC) are its robustness and stability (due to the presence of zero in transfer function) and the disadvantages of MPC (the conventional and the proposed ones), when compared to SHE, is that it contains variable switching frequency which increases switching losses.

This chapter presents an SCH elimination based on the PCC algorithm (Hsu and Wu 1996) for the APF. It uses the reference shunt current (I_{shref}) and a PCC algorithm to achieve tracking in presence of the periodic disturbances. The disturbance rejection transfer functions are used to eliminate the selected frequencies (Ojha et al. 2017). However, the proposed method gives a more efficient algorithm in terms of the complexity, sensitivity, and quantization effect with an algorithmic configuration that is tested through the Typhoon-HIL implementation.

This chapter is divided into five sections: mathematical analysis of PCC for a single-phase inverter, calculation of shunt reference current having selective harmonic (third harmonic in this chapter), simulation on Typhoon-HIL simulation environment, real-time simulation validation by using FPGA-based Typhoon HIL 402 kit for APF, and finally the conclusions.

6.2 Mathematical Analysis of PCC

The mathematical analysis of the PCC is described in this section. The controller logic uses the past and present state of the variable and predicts the future state of the variable. In the PCC logic, the variable is the shunt APF current. The predicted current (at an instant k + 1) is then compared with the reference current (at a particular instant k) and produces an error, which is attempted to be eliminated using the proposed criteria known as the minimization of the cost function (Bodetto et al. 2014; Ojha and Pandey 2015).

The PCC control algorithm in the form of a flow chart is shown in Fig. 6.2. The controller senses the initial value of the shunt current, DC-link voltage, stored reference shunt current, and switching state at instant k. Now, the controller calculates or predicts the future value of the shunt current using the voltage vector state at instant k + 1. Then, the controller iterates the program for four times due to the availability of four switching vector states, and using (6.3), it will find the best switching states. If the estimated cost function is less than the optimal value of the cost function, it will set the estimated cost function as the optimal cost function. If this condition is satisfied, it will generate gate pulse, otherwise it will return to the for loop block for the next iteration.

The single-phase inverter dynamic equation can be written as

$$\frac{di_{sh}}{dt} = -\frac{R_{sh}i_{sh}}{L_{sh}} + \frac{uV_{dc}}{L_{sh}} \quad (6.1)$$

where i_{sh} is shunt current which varies from $i_{sh}(kT_S) \rightarrow i_{sh}(kT_S + T_S)$ (i.e., from particular instant kT_S to very next instant $(kT_S + T_S)$, T_S is sample time, dt is the

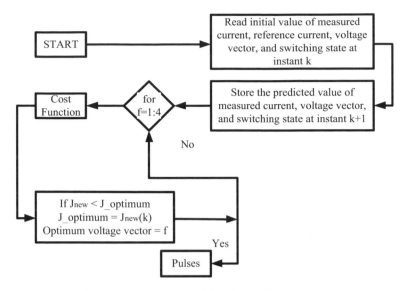

Fig. 6.2 Flow chart of predictive current control (PCC) algorithm

continuous time in the small differential form which varies from kT_S to $kT_S + T_S$. in dicrete time domain, R_{sh} is the shunt resistor, V_{dc} is the DC-link voltage, and L_{sh} is the shunt inductor. By solving the differential Eq. (6.1), the following can be written as:

$$i_{sh}^P(kT_S + T_S) = \frac{u V_{dc}(kT_S)}{R_{sh}}\left(1 - e^{\frac{-R_{sh}T_S}{L_{sh}}}\right) + e^{\frac{-R_{sh}T_S}{L_{sh}}} i_{sh}(kT_S) \qquad (6.2)$$

where i_{sh}^P is the predicted shunt current. The property of the controller depends on the sample time but its effect is negligible when we consider the measured shunt current one step ahead in the system.

The cost function ($J_{new}(kT_S)$) can be defined as

$$J_{new}(kT_S) = \left\{i_{shref}(kT_S) - i_{sh}^P(kT_S + T_S)\right\} \qquad (6.3)$$

6.3 Calculation of Reference Current

6.3.1 Conventional Method

Let I_L be the load current or compensating current as shown in Fig. 6.1. The components of the load current are (a) real (I_{PL}), (b) reactive (I_{QL}), and (c) harmonic components (I_{hL}), or (a) grid current (I_g) and (b) shunt current (I_{sh}) as defined below.

Outer Voltage Loop   Inner Current Loop

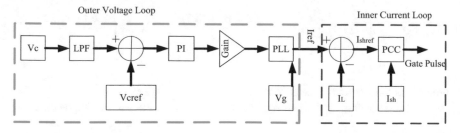

Fig. 6.3 General diagram of the current mode voltage control method for single-phase APF using conventional PCC algorithm

$$I_L = I_{PL} + I_{QL} + I_{hL} = I_g + I_{sh} \qquad (6.4)$$

Figure 6.3 shows a general diagram of the current mode voltage control technique. The control is based on an outer voltage loop and an inner current loop. The current is controlled by using PCC techniques. Where the reference current is generated with the help of the outer voltage loop.

The reactive and harmonic components of the grid current can be eliminated by introducing the shunt current in the system (Liu et al. 2019). It implies that the shunt current consists of two components of the load current, namely reactive and harmonic, which have equal magnitude and are out of phase with each other to convert the grid current to have a purely real component. To provide complete compensation of reactive and harmonics components, the compensation current can be evaluated by (6.5)

$$I_{shref} = I_L - \left(\frac{\sqrt{2}(P_{Lac} + P_{Loss})}{V_{sRMS}} \right) sin(\omega_o t) \qquad (6.5)$$

where P_{Lac} is the load instantaneous power, V_{sRMS} is the RMS value of grid voltage, and P_{Loss} is the power loss of the DC-link capacitor. The power loss P_{Loss} can be defined as follows:

$$P_{Loss} = K_{pdc}(V_{dcref} - V_{dcfilter}) + K_{idc} \int (V_{dcref} - V_{dcfilter})dt \qquad (6.6)$$

6.3.2 Proposed Method

For harmonic detection, a digital analysis filter bank can be used (Ojha and Pandey 2016; Ojha et al. 2017). A digital filter bank is based on the Fourier series, also called DFT (Hsu and Wu 1996). So, the distorted load current can be written as

$$i_L(kT_S) = I_o(kT_S) + \sum_{h=1}^{N}\left(A_h\cos\left(\frac{2*pi*hkT_S}{N}\right) + B_h\sin\left(\frac{2*pi*hkT_S}{N}\right)\right)$$

$$(6.7)$$

where h is the harmonic coefficient, $N = \frac{switching\ frequency}{fundamental\ frequency}$.

The time-varying coefficient $A_h(kT_S)$ and $B_h(kT_S)$ value can be calculated as

$$\left.\begin{array}{l} A_h(kT_S) = A_h(kT_S - T_S) + \dfrac{2}{N}(i_L(kT_S) - i_L(kT_S - N))\sin\left(\dfrac{2*pi*h*kT_S}{N}\right) \\[4mm] B_h(kT_S)_h(kT_S - T_S) + \dfrac{2}{N}(i_L(kT_S) - i_L(kT_S - N))\cos\left(\dfrac{2*pi*h*kT_S}{N}\right) \end{array}\right\}$$

$$(6.8)$$

Figure 6.4 Shows how the SCH elimination technique has been implemented to generate I^*_{shref}. This I^*_{shref} differs from conventional I_{shref}.

The transfer function of band-pass filter (TF1) can be written as

$$TF1 = \frac{2*\xi w_1*f_1*s}{s^2 + 2*\xi w_1*s + w_1^2}$$

$$(6.9)$$

where ξ is damping factor, w_1 is the fundamental angular frequency.

The transfer function of the harmonic compensator (TF2) can be written as

$$TF2 = \sum_{h=3}^{N}\frac{2*\xi w_1*f_h*h*s}{s^2 + 2*\xi w_1*h*s + (h*w_1)^2}$$

$$(6.10)$$

Now, the modified I_{shref} can be written as

$$I^*_{shref} = \left(\left(I_{ref} - I_L\right)*TF1\right) - (I_L*TF2)$$

$$(6.11)$$

Fig. 6.4 Implementation of SCH elimination method in I^*_{shref}

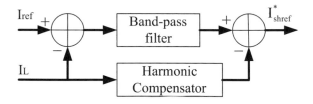

6.4 Performance Evaluation

The simulations are performed in the Typhoon-HIL simulation environment and real-time simulation results are obtained for the SCH for single-phase SAPFs, as shown in Fig. 6.1. The emulated system has been accomplished in real time, using Typhoon HIL 402, as shown in Fig. 6.5, where the circuit parameters are given in Table 6.1.

6.4.1 Virtual HIL Simulation Results

6.4.1.1 Conventional Method

Figure 6.6 shows the grid voltage, grid current, shunt current, reference shunt current, DC-link voltage, and load current waveform when conventional PCC is implemented using (6.5) in (6.3). Figure 6.7 shows the Fast Fourier transform (FFT) of grid current, where it shows the frequency spectra to 19th harmonic order. Except fundamental frequency the grid current has magnitude of 3rd, 5th etc. harmonic component, the numerical value is mentioned in Table 6.2, while using conventional method, the THD of grid current is 6.34%.

6.4.1.2 Proposed Method

Figure 6.8 shows the grid voltage, grid current, shunt current, reference shunt current, DC-link voltage, and load current waveform when the proposed PCC technique is implemented using (6.11) in (6.3). Figure 6.9 shows the Fast Fourier Transform (FFT) of grid current, where it shows the frequency spectra till 19th harmonic order. Except for fundamental frequency, the grid current has a magnitude of 5th etc harmonic

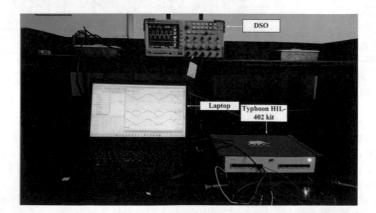

Fig. 6.5 Typhoon HIL 402 interfaced with console PC through USB cable

Table 6.1 Simulation Parameters

Parameters	Value
Grid voltage (V_g)	220 V
Fundamental frequency	50 Hz
Grid feeder line impedance Inductance (L_s) and resistance (R_s)	3.78 mH, 0.5 Ω
Shunt line impedance Inductance (L_{sh}) and resistance (R_{sh})	3.7 mH, 1 Ω
DC-link voltage and initial capacitor	500 V, and 4400 μF
Non-linear load with smoothing reactor L_{ac}, and DC side resistor and capacitor R_{DC}, and C_{DC}	3.46 mH, 25 Ω, and 150 μF
Damping factor (ξ)	0.02
Sampling frequency	10 kHz

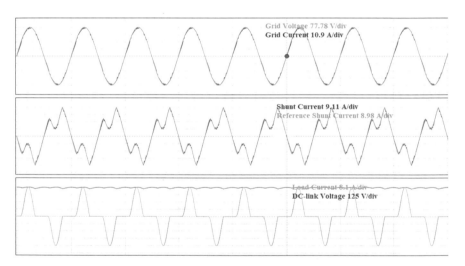

Fig. 6.6 Grid voltage, grid current, shunt current, reference shunt current, DC-link voltage, and load current when conventional PCC technique is implemented

component. It can be noted that load current has 3rd harmonic component of very low magnitude, while using the proposed method the load current has very low magnitude of 3rd harmonic component. The numerical value is mentioned in Table 6.2. The THD of grid current is 2.34%.

The measured values of grid current and load current with conventional and proposed PCC techniques implemented are shown in Table 6.2.

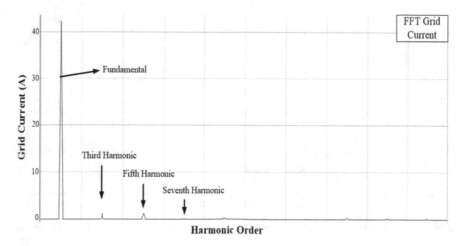

Fig. 6.7 FFT of grid current when conventional PCC technique is implemented

Table 6.2 Measured values of grid current and load current with and without SCH elimination method

Mode/Frequency (Hz)		50	150	250	350	450	550	650
Conventional PCC	Grid current (A)	43.9	0.8	1	0.04	0.08	0.01	0.01
	Load current (A)	39.8	24	7.1	1.2	1.1	0.6	0.43
Proposed PCC	Grid current (A)	43.8	0.22	1	0.04	0.08	0.01	0.01
	Load current (A)	39.7	7.68	7.1	1.2	1.1	0.6	0.43

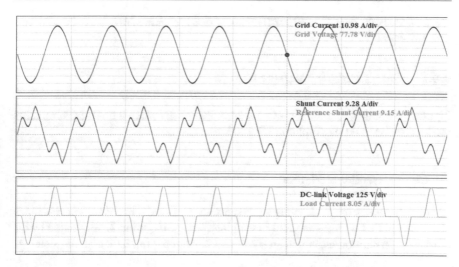

Fig. 6.8 Grid voltage, grid current, shunt current, reference shunt current, DC-link voltage, and load current with the implementation of the proposed PCC technique by eliminating 3rd harmonic

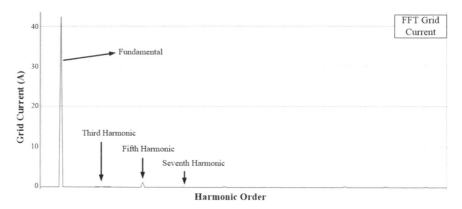

Fig. 6.9 FFT of grid current implementing proposed PCC technique by eliminating 3rd harmonic

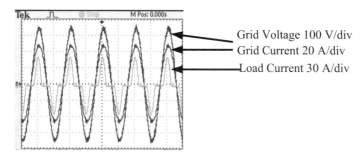

Fig. 6.10 Grid voltage, grid current, and load current waveform when conventional PCC techniques are implemented

6.4.2 Real-Time Simulation Results

Figures 6.10 and 6.11 show real-time simulation results when the virtual simulation model is dumped on the Typhoon HIL 402 kit.

6.4.2.1 Conventional Method

Figure 6.10 shows the grid voltage, grid current, and load current waveforms when conventional PCC is implemented in the plant. The grid current is dominated by the third and fifth harmonics, which are 1.8% and 2.2% of the fundamental, respectively.

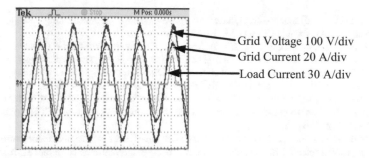

Fig. 6.11 Grid voltage, grid current, and load current waveforms when the proposed PCC technique is implemented by eliminating the 3rd harmonic

6.4.2.2   Proposed Method

Figure 6.11 shows the grid voltage, grid current and load current waveforms respectively when proposed PCC technique is implemented here 3rd harmonic eliminate in the grid current. The grid current is dominated by the third and fifth harmonics, which are 0.5% and 2.2% of the fundamental, respectively. The proposed method has a reduction of 72.2% of the third harmonic in the grid current than the conventional methods.

6.5   Conclusion

In this chapter, conventional predictive current control (PCC) scheme and proposed PCC technique based on a selective harmonic (3rd harmonic) compensator for single-phase SAPFs have been presented. In view of the performance of the SAPFs confirmed by the results of virtual simulation and real-time simulation, the proposed algorithm is highly suitable for SCH elimination. In the proposed method, 72.2% of the third harmonic has been eliminated from the grid current as compared to the conventional methods. This chapter designs the control parameters utilizing conventional non-linear techniques, namely, high-frequency rejection transfer functions. And control circuit should be synchronized with the power line using the PLL circuit.

References

Bodetto M, El Aroudi A, Cid-Pastor A, Martinez-Salamero L (2014) High performance hysteresis modulation technique for high-order PFC circuits. Electron Lett 50(2):113–114
Cid-Pastor A, Martinez-Salamero L, El Aroudi A, Giral R, Calvente J, Leyva R (2013) Synthesis of loss-free resistors based on sliding-mode control and its applications in power processing. Control Eng Pract 21(5):689–699

Hsu CY, Wu HY (1996) A new single-phase active power filter with reduced energy-storage capacity. IEE Proc Elect Power Appl 143(1):25–30

Liang Q, Hui Z, Kaipei L, Liu Q (2016) An improved modulation of the selective harmonic elimination controlling. 2006 international conference on power system technology, pp 1–5

Liu X, Li X, Zhou Q, Xu J (2019) Flicker-free single switch multi-string LED driver with high power factor and current balancing. IEEE Trans Power Electron 34(7):6747–6759

Marcos-Pastor A, Vidal-Idiarte E, Cid-Pastor A, Martínez-Salamero L (2020) Minimum DC-link capacitance for single-phase applications with power factor correction. IEEE Trans Industr Electron 67(6):5204–5208

Makhlouf B, Bouchhida O, Nibouche M (2016) Extension of real time harmonic elimination theory to the traditional selective harmonic elimination. 2016 8th international conference on modelling, identification and control (ICMIC), pp 597–602

Ojha S, Gupta R (2020) Performance comparison of sampled hysteresis and predictive control methods for tracking current in APF. 2020 IEEE 17th India council international conference (INDICON), pp 1–6

Ojha S, Pandey AK (2015) Comparative analysis of voltage source inverter using sinusoidal pulse width modulation and third harmonic injection method for different levels and loads. IJAER 10(20):41451–41457

Ojha S, Pandey AK (2016) Close loop V/F control of voltage source inverter using sinusoidal PWM. Third harmonic injection PWM and space cector PWM method for induction motor. IJPEDS 7(1)

Ojha S, Sharma C, Pandey AK (2017) Comparative analysis of closed loop three level voltage source inverter using sinusoidal Pulase Width Modulation and Third Harmonic Injection method for different loads. Second international conference on electrical, computer and communication technologies (ICECCT) Year: pp 1–6

Chapter 7
Development of Electric Vehicles Applications Using AURIX™ Microcontroller and Typhoon HIL

Ivan Todorović◉, Ivana Isakov◉, and Marko Gecic

Abstract Nowadays, a lot of countries have announced a zero-emission vehicle target or the phase-out of internal combustion engine (ICE) vehicles by 2050 to meet the 2015 Paris Climate Change Agreement targets. Electromobility is becoming a reality. The development of energy-efficient xEV applications such as main and auxiliary inverters, on-board chargers (OBC), DC-DC converters, and battery management systems (BMS) is very important. In this chapter, the setup for the development and controller testing of xEV applications is proposed. The setup consists of mass-market evaluation boards with Infineon 32-Bit Single-Chip AURIX™ TriCore™-based Microcontroller AURIX™ TC275, interface board, and Typhoon HIL 602+ devices. Based on the given flexibility of the Typhoon HIL 602+ emulator and the features of selected microcontrollers, the development and controller testing of different topologies are possible. Potential examples are described and test results of field-oriented control (FOC) of permanent magnet synchronous motor (PMSM) are presented.

Keywords EV · AURIX™ · Typhoon HIL · PMSM · FOC

Note: xEV denotes any kind of vehicle that utilizes electric motor traction (electric vehicle).

I. Todorović (✉) · I. Isakov
Faculty of Technical Sciences, University of Novi Sad, Trg Dositeja Obradovića 6, 21000 Novi Sad, Serbia
e-mail: ivan.todorovic@uns.ac.rs

I. Isakov
e-mail: ivana.isakov@uns.ac.rs

M. Gecic
Infineon Technologies AG, Am Campeon 1-15, 85579 Neubiberg, Germany
e-mail: Marko.Gecic@infineon.com

© The Author(s), under exclusive license to Springer Nature Singapore Pte Ltd. 2023   157
S. M. Tripathi and F. M. Gonzalez-Longatt (eds.), *Real-Time Simulation and Hardware-in-the-Loop Testing Using Typhoon HIL*, Transactions on Computer Systems and Networks, https://doi.org/10.1007/978-981-99-0224-8_7

7.1 Introduction

Fossil fuels usage and greenhouse gas emissions reductions are adopted as pivotal objectives in the domain of the development and application of emerging technologies. This has become a necessity as several ecological and societal problems became impossible to ignore. Since the transportation sector is one of the largest contributors to CO_2 emissions and one of the largest consumers of non-renewable fuels, this sector will be transformed in the following years (World Energy Outlook 2020). Primarily, this transformation will come in form of transportation electrification and xEVs' proliferation. xEVs should decrease fossil fuel usage and greenhouse gas emissions, but they will also reduce the residential area air and noise pollution, which are associated with a significant number of health issues. Moreover, these vehicles bring several advantageous features of direct interest to users. They are constructed to be safe and reliable and have lower maintenance costs than conventional vehicles. They can be charged at home and their usage is significantly cheaper since the electricity price is comparatively significantly lower than the price of fossil fuels (Report on EV 2020; Electric Cars 2021). Traditional vehicles in turn used internal combustion engines for traction, which led to rather poor reliability and high maintenance costs, while the fuel cost significantly increased usage expenses.

The electromobility development is still in progress and the adoption of xEV technologies began with the introduction of hybrid electric vehicles (HEV), as an interim solution. On the other hand, the number of vehicles that rely only on electric traction is increasing. Both HEVs and EVs represent highly complex systems since the essential requirements, such as reliability and safety need to be fulfilled. Therefore, all pertaining subsystems have to be carefully developed and tested.

Among others, in recent years, the vigorous development of various power electronics-based devices and systems has been undertaken. Various charger topologies have been investigated, both high voltage and low voltage DC-DC converters were developed, and the inverter design was improved, as well as the design of the other auxiliary converters. Additionally, the battery management system, as one of the most important control elements, has gained considerable attention. It directly affects the efficiency, reliability, and safety of xEVs, and thus researchers are putting a lot of effort into the enhancement of battery management strategies. Another relevant xEVs' component is the powertrain domain controller that controls centralized powertrain functions and governs the inter-communication. All these systems require careful planning and testing methods. Several development strategies can be undertaken, but the most common method in the praxis is the utilization of the V-model or its derivatives (Anselma and Belingardi 2019).

The V-model is a representation of the xEV's development lifecycle, i.e., it defines the activities to be performed and the results to be produced. Firstly, the system's specifications have to be provided and these include:

- Concept of operations,
- Requirements and architecture,

- Detailed design.

Secondly, a project test and integration procedures are conducted, which incorporate:

- Integration, test, and verification,
- System verification and validation,
- Operation and maintenance.

The use of the V-model ensures project risk minimization, cost reduction, and secures reliable development process outcomes.

This chapter provides a detailed insight into how two V-model stages—"Integration, test, and verification" and "System verification and validation" are conducted. It will be shown how these stages can be accelerated and qualitatively improved using C-HIL (controller hardware-in-the-loop) approach.

The usage of C-HIL setups during xEV development lifecycles is becoming an industry standard and the C-HIL environment addressed in the chapter is based on sophisticated real-time simulators and digital signal processors that are widely used in the automotive industry in general.

Hardware-wise, the C-HIL development, testing, and verification environment consists of:

- Real-time simulator,
- Digital signal processor or microcontroller,
- Interface board.

A real-time simulator, often denoted as an emulator, is used to emulate the power stage of the system. Essentially, it replaces and mimics the vehicle's power stage operation. The microcontroller is used to implement desired control structures. The interface board adjusts the signals that are exchanged between the real-time simulator and the microcontroller.

Considering Fig. 7.1, the control code downloaded to the Infineon 32-Bit Single-Chip AURIX™ TriCore™-based Microcontroller (represented with a circle) is used to govern different domains of the power stage (represented with rectangles) emulated in the real-time simulator.

Considering the software tools used, the power stage model was implemented using the Typhoon HIL Control Center software package, while the control code was synthesized using AURIX™ Development Studio. Easy controller peripheral configuration, initialization, and usage were achieved using the Infineon Low-Level Drivers (iLLD) software package. Software tools OneEye (for controller variables observation) and Typhoon HIL SCADA (for power stage variables observation) provided the graphical representation of the results. Both OneEye and Typhoon HIL SCADA enable users to create GUIs, communicate with different controllers and emulators, respectively, and easily depict gathered data.

The more precise the power stage model is, the more accurate the obtained results are. Additionally, if more complex systems are to be emulated, several

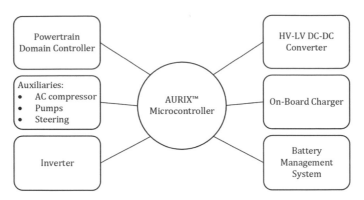

Fig. 7.1 AURIX™ for xEV applications

emulator units can be used. The power stage is defined on the PC, using dedicated software, compiled and downloaded on an emulator. Emulator- and controller-
relevant signals are routed through the interface board. The emulator sends analog
and digital signals that correspond to the signals gathered from real sensors in the
physical system. Then, these signals are adjusted using operational amplifier circuits
and forwarded to the real controller with the control code. Finally, according to the
received information, the controller takes necessary actions and produces gating and
other control signals that are sent to the inputs of the emulator.

Alternatively, the testing environment can be constituted using different tools and
devices. Firstly, other microcontrollers can be used (Texas Instruments C2000 series
(Texas Instruments C2000 real-time microcontrollers n.d.), STMicroelectronics
microcontrollers (STMicroelectronics Microcontrollers Microprocessors n.d.), etc.),
but the AURIX™ microcontrollers are dedicated to electric vehicle applications.
Infineon provides users with code examples, drivers, and software routines that are
compatible with and standardized for various electric vehicle applications. Hence,
the users can quickly start developing complex control structures and pertaining
safety and compliance software modules.

Moreover, instead of the C-HIL approach (based on real-time simulators), other
configurations can be used. Firstly, software in the loop (SIL) setup can be used.
Software tools such as MATLAB, PSIM, PLECS, and others can be used to build
digital twins of electric vehicle infrastructure and control structures. SIL enables
users to build testing setups of almost arbitrary complexity and size and analyze
them with the desired level of detail. Still, offline simulations, on which SIL is
based, are executed significantly slower than in real-time, hence extending testing
and development time (MathWorks n.d.).

Next, the processor in the loop (PIL) setups can be used (Infineon n.d.). They
enable the processor of a specific controller and the implemented code to be tested and
analyzed in great detail. The disadvantage of PIL setups is that the testing procedure
is also slower than in real-time since the execution of the implemented code is slowed
down in order to synchronize code execution with the model execution (the model

is implemented using offline simulators). SIL and PIL are usually employed at the beginning of the development lifecycles to conduct preliminary tests.

In recent years, the controller model in the loop (CMIL) approach is becoming increasingly popular. The model of the complete microcontroller is implemented and executed using dedicated computer software and it is connected via specific software routines to offline simulators and the model is implemented using those simulators. Using this approach, comprehensive examination of the control code behavior, microcontroller resources utilization, etc., as if the real controller was tested. Again, CMIL is based on offline simulators run on personal (general-purpose) computers, and testing procedures are unnecessarily prolonged in comparison to the C-HIL testing procedures. Also, only several microcontroller vendors are providing the users with corresponding microcontroller models, and only for flagship controllers (Infineon n.d.).

Lastly, prototypes, full-scale and scaled-down versions, are platforms for the most comprehensive testing of EV systems. The disadvantage of the usage of prototypes is that they do pose a threat to personnel and equipment safety and if damaged or destroyed by misuse or otherwise, expensive and time-consuming repairs must take place. Generally, if employed, prototypes are used at the end of the development process.

In the following paragraphs, the data describing the microcontroller, emulator, and pertaining software tools, i.e., specificities of the environment for the development of electric vehicle applications, will be provided. Also, for which applications the environment can be used will be proposed, where implementation of field-oriented control of permanent magnet synchronous motor will be explained in more detail.

7.2 Typhoon HIL 602+

Typhoon HIL 602+ real-time simulator is a powerful tool for the development, automated testing, and optimization of a wide range of power electronic devices used in grid-connected applications, microgrids, and automotive and electric propulsion drives. In view of the available I/O ports, the emulator contains 16 analog input channels with 16-bit resolution (voltage range ± 10 V), 32 analog outputs (with the same precision and same voltage range), 32 digital inputs (input voltage range is $[-15$ V, 15 V]), and the same number of digital outputs (output voltage range is [0 V, 5 V]). Considering the communication features, Ethernet, CAN, RS232, USB 2.0, and high-speed, a serial link can be used (Typhoon HIL General Specification n.d.).

7.2.1 Typhoon HIL Schematic Editor

Typhoon HIL Schematic Editor, shown in Fig. 7.2, is a software tool used for the graphical representation of the power stage models to be emulated. It contains a

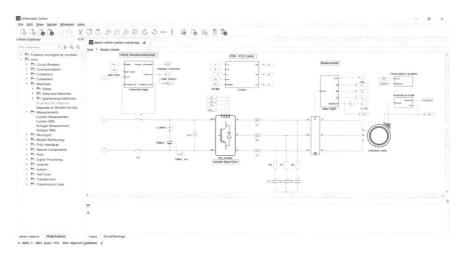

Fig. 7.2 Typhoon HIL schematic editor

rich library with different components used in a wide range of applications—from simple DC-DC converters to electric machines and microgrid building blocks. All components are parametrizable and can be dragged and dropped into the model schematic.

7.2.2 Typhoon HIL SCADA

Typhoon HIL SCADA, represented in Fig. 7.3, is another tool from the Typhoon HIL software toolchain used for the visualization of a broad spectrum of real-time simulations. It contains a vast number of user-friendly widgets that can be simply configured and enables easy and convenient monitoring of all significant signals. The drag and drop principle is used to design SCADA panels and different advanced functions such as signal FFT are also available.

7.3 AURIX™ Microcontroller

The Infineon 32-Bit Single-Chip AURIX™ TriCore™-based microcontroller can be used for a wide range of automotive and industrial applications. The AURIX™ TriCore™ unites the elements of a RISC processor core, a microcontroller, and a DSP in one single MCU. The main highlights are shown in Table 7.1.

The AURIX™ provides an already built-in hardware security module, which makes it a perfect match for xEV applications. Besides other Infineon automotive microcontroller families, there are several AURIX™ families, as shown in Fig. 7.4.

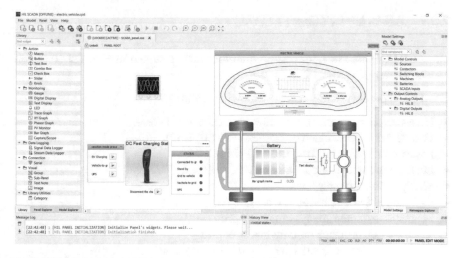

Fig. 7.3 Typhoon HIL SCADA

Table 7.1 AURIX™ TriCore™ elements

Microcontroller	RISC processor	DSP
Fast context switch and interrupt response	32-bit load/store Harvard architecture	Sustainable single-cycle dual MAC
Integrated Peripheral support	Super-scalar execution and uniform register set	DSP addressing modes and Zero overhead modes
Powerful bit manipulation unit and comparison Instructions	Memory Protection Unit (MPU) and C/C++ and RTOS support	Saturation, Rounding and Q-Math (fraction format)

Those AURIX™ families come with a different range of memories, frequencies, temperatures, packaging options, and a different number of peripherals.

In this chapter, we focus on AURIX™ TC275 and the low-cost evaluation board AURIX™ TC275 LiteKit (KIT_AURIX_TC275_LITE).

Fig. 7.4 Evolution of Infineon automotive microcontrollers

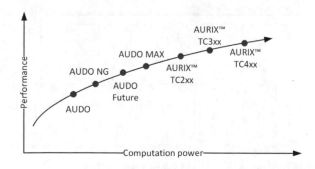

7.3.1 AURIX™ TC2xx (TC275)

The AURIX™ TC2xx family system architecture is shown in Fig. 7.5 (Infineon-TC27x_D-step-UM-v02_02-EN n.d.).

The TC27x product family has the following features:

- High-performance microcontroller with three CPU cores,
- Two 32-bit super-scalar TriCore™ CPUs (TC1.6P),
- Power-efficient scalar TriCore™ CPU (TC1.6E),
- Lock stepped shadow cores for one TC1.6P and TC1.6E,
- Multiple on-chip memories,
- 64-Channel DMA controller with safe data transfer,
- Sophisticated interrupt system (ECC protected),
- The high-performance on-chip bus structure,
- Optional hardware security module (HSM) on some variants,
- The safety management unit (SMU) handling safety monitor alarms,
- Memory test unit with ECC, Memory Initialization, and MBIST functions (MTU),
- Hardware I/O Monitor (IOM) for checking digital I/O,

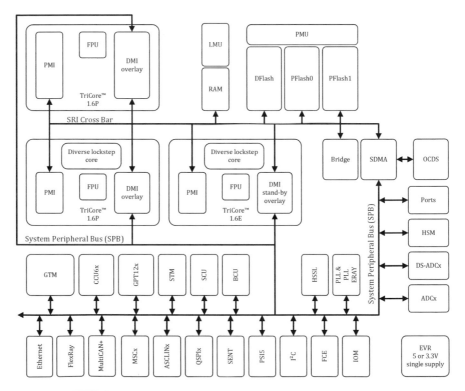

Fig. 7.5 AURIX™ TC2xx family system architecture

- Versatile on-chip peripheral units,
- Versatile successive approximation ADC (VADC),
- Delta-sigma ADC (DSADC),
- Digital programmable I/O ports,
- On-chip debug support for OCDS Level 1 (CPUs, DMA, On-Chip Buses),
- Multi-core debugging, real-time tracing, and calibration,
- Four/five-wire JTAG (IEEE 1149.1) or DAP (device access port) interface,
- Power management system and on-chip regulators,
- Clock generation unit with system PLL and flex-ray PLL,
- Embedded voltage regulator.

The AURIX™ TC275 has two modules usually used for the PWM generation. Those modules are Capture/Compare Unit 6 (CCU6) and General Timer Module (GTM) (Infineon-TC27x_D-step-UM-v02_02-EN n.d.).

The CCU6 is made up of a Timer T12 block and a Timer T13 block:

- Timer T12 has three capture/compare channels that can independently generate PWM signals or accept capture triggers. Timer T12 has a 16-bit resolution,
- Timer T13 block has one compare channel and 16-bit resolution.

The GTM has two submodules that can be used for PWM generation:

- Timer Output Module (TOM): There are 3 TOM submodules and each submodule has 16 channels. Those channels are separated into two groups. Each channel has its own 16-bit counter,
- ARU-connected Timer Output Module (ATOM): There are 5 ATOM submodules and each submodule has 8 channels. Each channel has its own 24-bit counter.

In electrical drive applications, generation of the center-aligned and edge-aligned three-phase PWM is supported (six outputs, individual signals for high-side and low-side switches) by both mentioned modules.

If three Hall sensors are used to determine the angular rotor position in electrical drive applications, the processing of signals can be done with GTM or CCU6 modules. The GTM has a submodule called Sensor Pattern Evaluation (SPE) which can control the outputs of a dedicated connected TOM submodule if a defined input pattern is detected. Similarly, CCU6 has a multi-channel mode, which is introduced to provide efficient means for switching pattern generation.

If an incremental encoder is used to determine the angular rotor position in electrical drive applications, the processing of signals can be done with GTM or General-Purpose Timer (GPT12) modules. The GTM has a submodule called Timer Input Module (TIM), which can be used for encoder support. The GPT12 has two timer blocks, GPT1 and GPT2. The GPT1 has main and two auxiliary timers. Those timers have 4 operating modes. One of the modes is Incremental Interface mode and it can be used for incremental encoder support.

Table 7.2 AURIX™ TC27x series feature set

TriCore™ 1.6P	# Cores/Checker	2/1
	Frequency	200 MHz
TriCore™ 1.6E	# Cores/Checker	1/1
	Frequency	200 MHz
	Program Flash	4 MB
	EEProm @ w/e cycles	64 KB @ 500 k
SRAM	Total (DMI, PMI, LMU)	472 KB
DMA	Channels	64
ADC	Modules 12bit/DS	8/6
	Channels 12bit/DS	60/6 diff
Timer	GTM Input/Output	32/88 channels
	CCU/GPT modules	2/1
Interfaces	FlexRay (#/ch.)	1/2
	CAN FD (nodes/obj)	4/256
	QSPI/ASCLIN/I2C	4/4/1
	SENT/PSI5/PSI5S	10/3/1
	HSCT/MSC/EBU	1/2 diff LVDS/–
	Other	Ethernet
Safety	SIL Level	ASIL-D
Security	HSM	Optional
Power	EVR	Yes

In AURIX™ TC275, there are two modules dedicated to the conversion of analog input values (voltages) to discrete digital values. Those modules are the Versatile Analog-to-Digital Converter (VADC) and the Delta-Sigma Analog-to-Digital Converter (DSADC).

If a resolver is used to determine the angular rotor position in electrical drive applications, the carrier generation and processing of resolver signals can be done with DSADC.

Table 7.2 summarizes the feature set of the AURIX™ TC27x series. More details can be found in the product-specific user manual (World Energy Outlook 2022) and datasheet (Infineon-TC27xDC-DataSheet-v01_00-EN n.d.).

7.3.2 AURIX™ TC275 LiteKit

The AURIX™ TC275 LiteKit, shown in Fig. 7.6, is equipped with a 32-Bit Single-Chip AURIX™ TriCore™-based Microcontroller Aurix™ TC275.

Summary of features (Infineon-AURIX_TC275_Lite_Kit-UserManual n.d.):

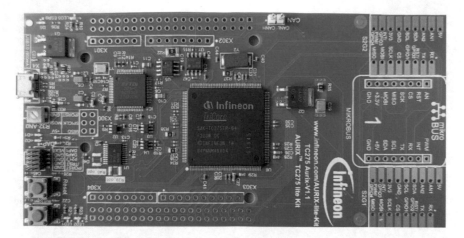

Fig. 7.6 AURIX™ TC275 Lite Kit

- Arduino connector,
- Arduino ICSP connector,
- Voltage regulator 5 V to 3.3 V,
- Optional 0 Ω resistors (R39_opt/R40_opt in 1210 imperial),
- Arduino connector (Digital),
- 20 MHz Crystal for AURIX™ and 12 MHz Crystal for OCDS,
- Mikrobus connector for WIFI/BLE,
- Infineon CAN transceiver TLE9251VSJ & CAN connector,
- Pin connector X2,
- Power LED (D5),
- LEDs D1/D2 for ADBUS7/4 and LED3 for ESR0 Signal (low-active),
- Arduino pin connector (POWER & ANALOG IN),
- Potentiometer (10 kOhm) and solderable 0 Ω resistor (R33 in 0805 imperial),
- Micro USB (USB3.0 recommended),
- 10-pin DAP connector,
- Reset button,
- 2 × Shield2GO connector for Infineon Maker Shields,
- EEPROM 1Kbit.

7.3.3   AURIX™ Development Studio

The AURIX™ Development Studio, shown in Fig. 7.7, is a free-of-charge Integrated Development Environment (IDE) for the TriCore™-based AURIX™ microcontroller family. It is a comprehensive development environment, including Eclipse IDE, C-compiler multicore debugger, and Infineon Low-Level Driver (iLLD). There are no time or project size restrictions. It enables editing, compiling, and debugging

Fig. 7.7 The AURIX™ development studio

of application code. The debugger and compiler are intended to be used in non-productive projects.

The AURIX™ Development Studio also comes with examples related to:

- Communication protocols (ASCLIN module handling for LIN, UART, SPI, and CAN communication),
- Data handling (e.g., flexible CRC engine control, input–output monitor control, bus register protection, memory protection control, data and program flash programming, direct memory access control, memory test),
- Timers (e.g., signal capturing, clock system, PWM signal generation),
- Analog signals (e.g., analog-to-digital converter control, delta-sigma analog-to-digital converter control),
- General-purpose input–output control, CPU management (e.g., code execution from SRAM, performance register usage, power management system control, multicore, trap recognition, watchdog handling),
- Alarms (e.g., device reset type trigger and detection, interrupt handling, safety management unit control).

Those examples can be easily imported to AURIX™ Development Studio. Figure 7.8 partially shows the list of examples available for the AURIX™ TC275 Lite Kit.

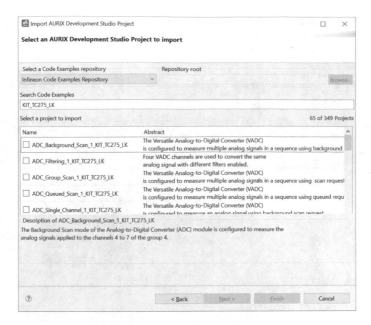

Fig. 7.8 Infineon code examples, import to the AURIX™ development studio

7.3.4 Infineon Low-Level Drivers (iLLD)

The Infineon Low-Level Driver library (iLLD) provides access and configuration functions to the integrated peripherals of the Infineon AURIX™ family of microcontrollers. iLLD is intended to be used as non-productive software. Together with SFR header files, iLLDs are a fundamental part of the infrastructure for tests and applications.

iLLDs consist of the following components:

- Standard layer with (verbose) access functions to SFRs,
- The interface layer has configuration and handling functions, which will implement the use cases. Multiple interface layers could be available if the peripheral is used for different purposes,
- The infrastructure layer consists of SFR headers and application startup software, which are also used by the MCAL software,
- The implementation layer consists of device configuration,
- Pin maps configuration to configure I/O pins,
- The service software layer provides a generic runtime API for a common application use case. It abstracts the underlying interface driver.

As shown in Fig. 7.9, the documentation is generated using Doxygen. The documentation can be downloaded from the Infineon website (Infineon Low Level Driver documentation n.d.).

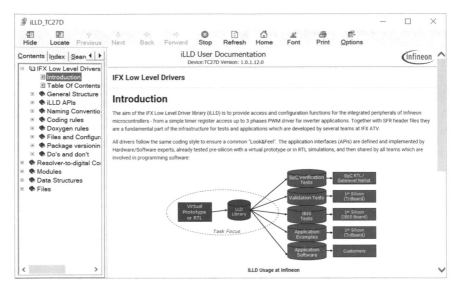

Fig. 7.9 iLLD user documentation

7.3.5 OneEye

OneEye is a free of charge Windows application that enables the users to create their own Graphical User Interface (GUI) and to interact with an Electronic Control Unit (ECU) during runtime. Different kinds of widgets like graphs, oscilloscopes, sliders, numeric fields, and pictures enable the user to visualize the state of the ECU and control the ECU. OneEye communicates with the target over the selected communication interface (such as CAN, COM, Ethernet, etc.) using text or binary protocols.

OneEye can be downloaded from the Infineon toolbox (Infineon Toolbox n.d.; softwaretools.infineon.com n.d.). The latest version supports the Infineon device access server. This feature enables the user to have direct read and write access to C variables.

Figure 7.10 shows OneEye in edit mode. Users can enable Tool Box, Property Box, Browser, or Debug Box. Those boxes and their usages are described in OneEye help.

7.4 Setup

The setup, shown in Fig. 7.11, for the development of xEV applications (but not limited to) consists of:

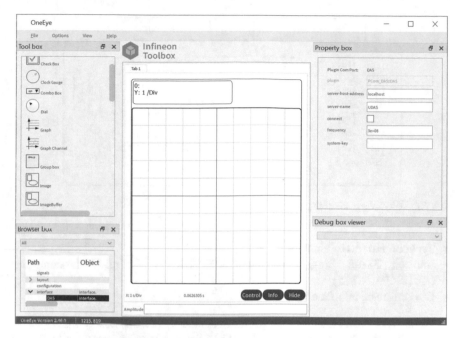

Fig. 7.10 OneEye configuration example

Fig. 7.11 Setup for development of xEV applications (University of Novi Sad, Serbia)

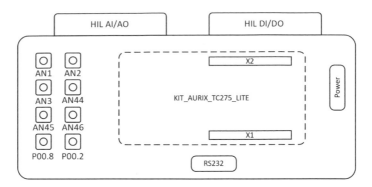

Fig. 7.12 Block diagram of the interface board

- Typhoon HIL 602+ emulator,
- Infineon 32-Bit Single-Chip AURIX™ TriCore™-based Microcontroller (AURIX™ TC275),
- Custom-made interface board.

Figure 7.12 shows the block diagram of the interface board, which is designed to enable a seamless interface between AURIX™ development kits, such as AURIX™ TC275 LiteKit or TC375 LiteKit, and any of the HIL4 and HIL6 series devices.

Features:

- Headers for one AURIX™ TC275 LiteKit (or AURIX™ TC375 LiteKit),
- All HIL and development board signals available through sensing terminals,
- 5 V and 3.3 V power supply jumpers (HIL or development board) with LED indication,
- 16 HIL digital inputs,
- 16 HIL digital outputs (level shifted to 3.3 V),
- 16 HIL analog outputs (clamped to 3.3 V),
- A total of 8 BNC terminal posts are provided (e.g., for easy oscilloscope connection for analog signal monitoring).

The following figure shows the pinout of AURIX™ TC275 LiteKit connectors X1 and X2 connected to HIL digital inputs/outputs and HIL analog outputs (Fig. 7.13).

7.5 Potential Application

The development of energy-efficient xEV applications such as main and auxiliary inverters, on-board chargers (OBC), DC-DC converters, and battery management systems (BMS) is very important. The proposed setup can be used for fast prototyping and testing. The plant models might be emulated using the HIL, while the control

X1

	GND	1	2	VEXT	
DI16	P33.11	3	4	P33.12	DI15
DI14	P33.13	5	6	P33.14	DO12
DI13	P23.1	7	8	P23.0	DI12
DO11	P23.3	9	10	P23.2	DI11
DI10	P23.5	11	12	P23.4	DI9
DO10	P22.1	13	14	P22.0	DI8
DI7	P21.0	15	16	P22.2	DI6
DO9	P21.2	17	18	P22.3	DI5
DI4	P21.4	19	20	P21.3	DO8
DO7	P20.10	21	22	P21.5	DO6
DI3	P20.0	23	24	P20.1	DO5
DI2	P20.3	25	26	/ESR1	
	/ESR0	27	28	P20.14	DO4
DO3	P15.5	29	30	/PORST	
DI1	P15.4	31	32	P11.12	DI19
DI18	P11.11	33	34	P11.10	DI20
DI17	P11.9	35	36	P11.6	DO2
DO1	P11.3	37	38	P11.2	DI21
	VDD_USB	39	40	GND	

X2

	GND	1	2	VDD_USB	
	P00.0	3	4	P00.1	
AO16	P00.2	5	6	P00.3	DO13
AO14	P00.6	7	8	P00.5	AO15
AO13	P00.8	9	10	P00.7	
DI22	P00.10	11	12	P00.9	AO12
DI23	P00.12	13	14	P00.11	DI24
	VAREF1	15	16	AN47	AO11
AO10	AN46	17	18	AN45	AO9
AO8	AN44	19	20	AN7	AO7
AO6	AN6	21	22	AN5	AO5
AO4	AN4	23	24	AN3	AO3
AO2	AN2	25	26	AN1	AO1
	AN0	27	28	P33.0	DI25
DI26	P33.1	29	30	P33.2	DI27
DI28	P33.3	31	32	P33.4	DO14
DI29	P33.5	33	34	P33.6	DI30
DI31	P33.7	35	36	P33.8	DI32
DO15	P33.9	37	38	P33.10	DO16
	VEXT	39	40	GND	

Fig. 7.13 Pinout of AURIX™ TC275 LiteKit connectors X1 and X2 which are connected to HIL IO connectors

algorithm might be implemented using a control platform based on the Infineon AURIX™ TC275 LiteKit. All the plant model signals can be monitored using HIL SCADA. As a part of the AURIX™ tools ecosystem, a graphical user interface OneEye can be used to send commands to the microcontroller and plot desired system variables using a graph or oscilloscope functionality. In this chapter, potential applications are presented (but the setup is not limited to them).

Figure 7.14 shows a block diagram of a setup that can be used for developing a scalar control of AC motors, such as permanent magnet synchronous motors (PMSM) and induction motors (IM). The plant model can be created using the Typhoon HIL Schematic Editor by taking:

- The voltage source from the Sources library,
- The three-phase inverter (switches and diodes are modeled as ideal switches) component from the Converter library,
- The PMSM or IM from the Machines library.

One of the HIL digital outputs can be used as a start or stop command toward the microcontroller.

Figure 7.15 shows a block diagram of a setup that can be used for developing a sensorless FOC of a PMSM or IM. In addition to the previously described plant model, in this example, emulated phase current and DC-link voltage signals can be sensed using the microcontroller's ADC unit. During the development of a sensorless algorithm, one of the ADC channels can be used to sense the emulated rotor position.

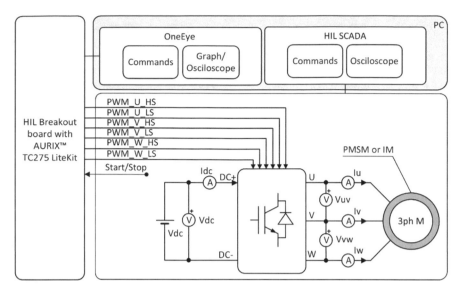

Fig. 7.14 Scalar control (V/f) of PMSM or IM

The estimated position can be compared with the emulated one and the output of the sensorless algorithm can be validated.

Figure 7.16 shows a block diagram of a setup that can be used for developing a dual motor drive. Two sensorless FOC of a PMSM or IM can be implemented in

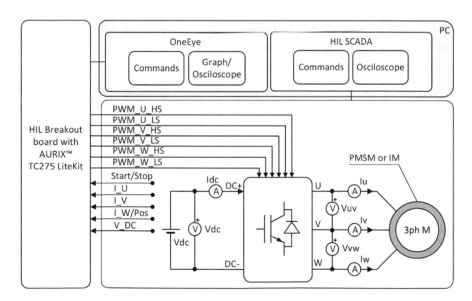

Fig. 7.15 Sensorless FOC of PMSM or IM

parallel. The emulated phase current and DC-link voltage signals can be sensed using the microcontroller ADC unit. During the development of a sensorless algorithm, one of the ADC channels, per drive, can be used to sense the emulated rotor position. The estimated position can be compared with emulated one and the output of the sensorless algorithm can be validated.

One of the HIL digital outputs, per drive, can be used as a start or stop command toward the microcontroller.

Figure 7.17 shows a block diagram of a setup that can be used for developing a sensored FOC of PMSM or IM. The emulated phase currents and DC-link voltage signals can be sensed using the microcontroller ADC unit. The DSADC unit can be used to generate resolver carrier signals and acquire resolver sine and cosine signals. In addition, during development, one ADC channel might be used to sense

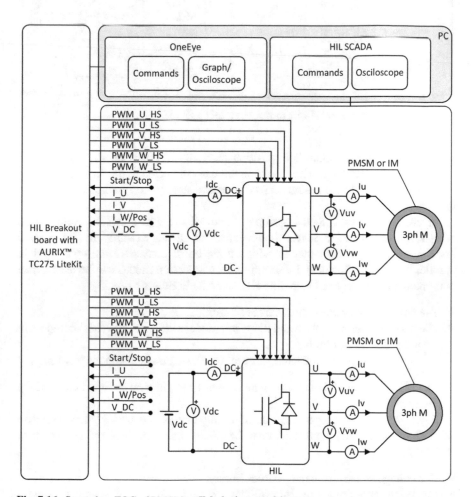

Fig. 7.16 Sensorless FOC of PMSM or IM, dual motor drive

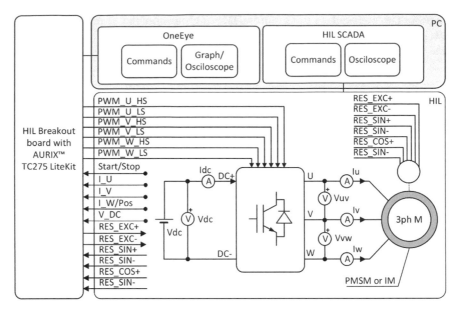

Fig. 7.17 FOC of PMSM or IM with resolver as position sensor

the emulated rotor position. The position recalculated based on signals from the DSADC can be compared with the emulated one connected to the ADC channel.

One of the HIL digital outputs can be used as a start or stop command toward the microcontroller.

Figure 7.18 shows a block diagram of a setup that can be used for developing an open-loop control of interleaved boost converter. In this example, the interleaved boost converter is a converter in which three boost converters are connected in parallel. Both, continuous and discontinuous conduction modes can be tested. The plant model can be created using the HIL Schematic Editor by taking:

- The voltage source from the Sources library,
- The boost converter (switch and diode are modeled as ideal switches) component from the Converter library,
- The inductor, resistor, and capacitor are from the Passive Components library.

One of the HIL digital outputs can be used as a start or stop command toward the microcontroller.

Figure 7.19 shows a block diagram of a setup that can be used for developing an AC/DC converter. The plant model can be created using the HIL schematic editor by taking:

- The voltage source from the Sources library,
- The single-phase diode rectifier (diodes are modelled as ideal switches), single-phase inverter (switches and diodes are modelled as ideal switches), and boost

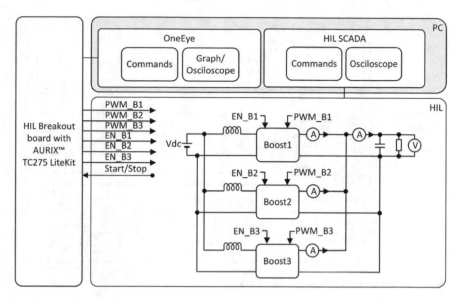

Fig. 7.18 Interleaved boost

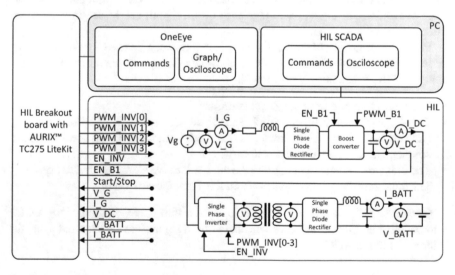

Fig. 7.19 AC/DC converter with PFC and DAB

converter (switch and diode are modelled as ideal switches) components from the Converter library,

- The inductor, resistor, and capacitor are from the Passive Components library.

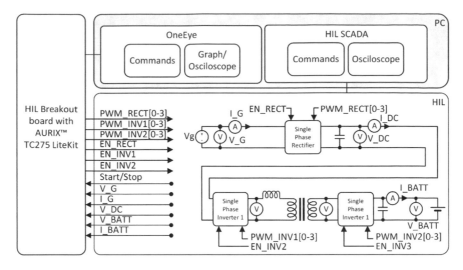

Fig. 7.20 Active front-end and dual-active bridge

One of the HIL digital outputs can be used as a start or stop command toward the microcontroller. The emulated voltage and current signals can be acquired using the microcontroller's ADC unit.

Figure 7.20 shows a block diagram of a setup that can be used for developing an active front-end and dual-active bridge converter. The plant model can be created using the HIL Schematic Editor by taking:

- The voltage source from the Sources library,
- The single phase rectifier (switch and diode are modelled as ideal switches), and single-phase inverters (switches and diodes are modelled as ideal switches) components from the Converter library,
- The inductor, resistor, and capacitor from the Passive Components library,
- The ideal transformer component from the Transformers library.

One of the HIL digital outputs can be used as a start or stop command toward the microcontroller. The emulated voltage and current signals can be acquired using the microcontroller's ADC unit.

7.6 Field Oriented Control of Permanent Magnet Synchronous Motor

Permanent magnet synchronous motors play a very important role in advanced regulated electric drives, mostly because of high power density, relatively small rotor inertia, and high efficiency (Morimoto et al. 1994). In high-performance applications, the PMSM drives are ready to meet sophisticated requirements such as fast dynamic response, high power factor, and wide operating speed range. The most frequently used technique for PMSM control is FOC. In this subchapter, the test results of an electrical drive with PMSM are presented.

7.6.1 System Description

Figure 7.21 shows a block diagram of the setup which can be used for developing a sensorless/sensored FOC of a PMSM. The emulated phase current and DC-link voltage signals are sensed using the microcontroller's ADC unit. There is an option that one of the ADC channels is used to sense the emulated rotor position. The plant model is created using the HIL Schematic Editor by taking:

- The voltage source from the Sources library,
- The three-phase inverter (switches and diodes are modelled as ideal switches) component from the Converter library,
- The PMSM from the Machines library (parameters of used Nanotec motor can be found in Nanotec motor datasheet n.d.).

One of the HIL digital outputs is used as a start or stop command toward the microcontroller.

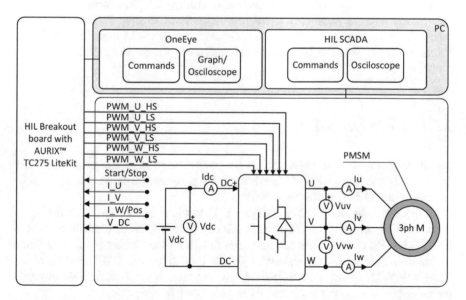

Fig. 7.21 Sensorless FOC of PMSM

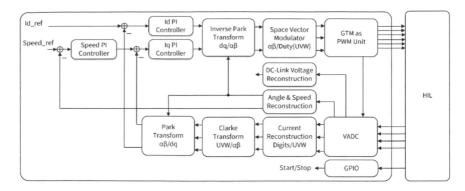

Fig. 7.22 FOC of PMSM: speed control scheme

The models of the PWM inverter, DC power supply, and electric motor are emulated with a time-step of 1 μs, while the control algorithm with a PWM carrier frequency of 20 [kHz] is implemented using a control platform based on the Infineon AURIX™ TC275 LiteKit. OneEye is used to send commands to microcontrollers and plot system variables using graph and oscilloscope functionality.

The block diagram of the PMSM drive control system is shown in Fig. 7.22. Speed is regulated by a proportional-integral regulator which outputs the q-axis current reference. The d-axis current reference can be set manually using OneEye.

The d- and q-axis currents are regulated by proportional-integral regulators and their outputs are d- and q-axis voltages. Those voltages are transformed to αβ reference frame using inverse Park transform. Outputs of Inverse Park transform are inputs for space vector block, which converts them to PWM signals duty cycles. Phase current signals are converted to dq reference frame using Clarke and Park transforms.

7.6.2 PWM Generation

The three-phase two-level inverter has six power switches and in order to control them, six synchronized PWM signals shall be generated. The dead time shall be inserted to avoid damage to power switches.

In this example, the GTM is used as a PWM unit. Based on the functionality of available pins, on AURIX™ TC275 LiteKit connectors X1 and X2, different sets of the GTM TOM channels can be used. The function of each port pin can be found in Infineon-TC27xDC-DataSheet-v01_00-EN n.d. Since the GTM module of the AURIX™ TC275 does not have a dead time module, seven TOM/ATOM channels are used, one as a master timer and six to generate PWM signals.

GTM TOM submodule with its channels and port pins used for PWM generation are listed in the table below (Table 7.3).

Table 7.3 AURIX™ TC275 peripherals used for PWM generation

Peripheral	Usage	Resource allocation
GTM	Base timer	GTM TOM1 Channel 0
	PWM generation for phase U—low side	GTM TOM1 Channel 1, P00.10
	PWM generation for phase U—high side	GTM TOM1 Channel 2, P00.11
	PWM generation for phase V—low side	GTM TOM1 Channel 3, P00.12
	PWM generation for phase V—high side	GTM TOM1 Channel 4, P33.0
	PWM generation for phase W—low side	GTM TOM1 Channel 5, P23.0
	PWM generation for phase W—high side	GTM TOM1 Channel 6, P33.2

Figure 7.23 shows the plant model in Schematic Editor.

Based on selected GTM TOM1 channels and ports and where they are connected, proper HIL digital inputs are used. Figure 7.24 shows the assignment of HIL digital inputs used for fetching the gate drive signals from the microcontroller.

Simple open-loop control is implemented to test the functionality of the space vector modulator and to validate that the motor rotates when PWM signals are applied to selected HIL inputs. Figure 7.25 shows the waveforms of signals like electrical angle, outputs of an open-loop algorithm (modulation index), and outputs of a space vector modulator. The outputs of the space vector modulator are PWM duty cycles. Open-loop amplitude and frequency are set to 0.5 [p.u]. and 20 [Hz], respectively.

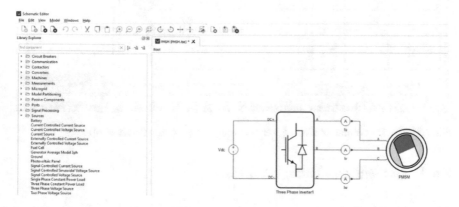

Fig. 7.23 Typhoon HIL schematic editor: plant model

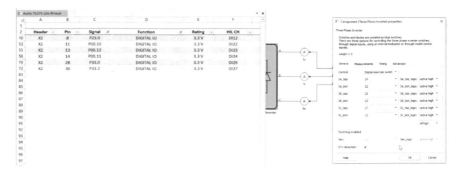

Fig. 7.24 Schematic editor: selection of digital inputs used in inverter model

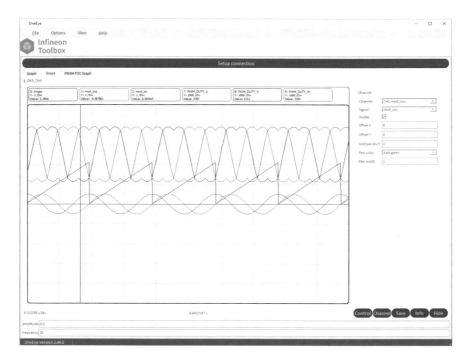

Fig. 7.25 OneEye: testing of PWM generation using scalar control (V/f control)

7.6.3 Current and Voltage Sensing

The traditional solution for vector-regulated medium and high-performance drive with a three-phase AC motor involves sensing of two or three-phase currents. The phase current sensing shall be synchronized with the PWM switching signals. The GTM TOM1 channel 7 is used as a source to trigger an ADC conversion at the specified point in time. Typically, three-phase current signals are sampled at the

Table 7.4 AURIX™ TC275 peripherals used for current and voltage sensing

Peripheral	Usage	Resource allocation
GTM	VADC Trigger	GTM TOM1 Channel 7
VADC	Phase U current sensing	VADC Group 7 Channel 4 Queue 0, P00.2
	Phase V current sensing	VADC Group 6 Channel 4 Queue 0, P00.8
	Phase W current sensing	VADC Group 5 Channel 7 Queue 0, AN47
	DC-link voltage sensing	VADC Group 0 Channel 1 Queue 0, AN1

same time. But, if we consider that system is symmetrical, then it is enough to sense two-phase currents synchronously.

AURIX™ TC275 peripherals used for the phase current and voltage sensing are shown in Table 7.4.

While some GPIO pins are used as analog inputs, the VADC Pull Down Diagnostics/Multiplexer Diagnostics are disabled for those pins.

The waveforms of three-phase currents, DC-link voltage, and open-loop electrical angular motor position are shown in OneEye (Fig. 7.26, left). The waveforms of three-phase currents, DC-link voltage, electrical and mechanical angular motor position, and mechanical speed are shown in HIL SCADA (Fig. 7.26, right).

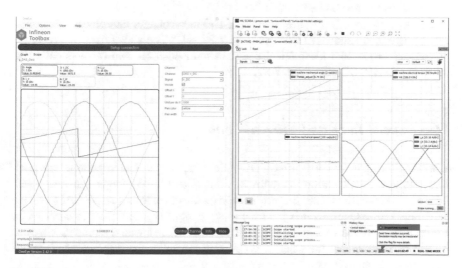

Fig. 7.26 Current and voltage sensing: OneEye(left) and HIL SCADA (right)

Table 7.5 AURIX™ TC275 Peripherals used for scalar control of PMSM or IM

Peripheral	Usage	Resource allocation
GTM	VADC Trigger	GTM TOM1 Channel 7
VADC	Phase U current sensing	VADC Group 7 Channel 4 Queue 0, P00.2
	Phase V current sensing	VADC Group 6 Channel 4 Queue 0, P00.8
	Electrical rotor position sensing	VADC Group 5 Channel 7 Queue 0, AN47
	DC-link voltage sensing	VADC Group 0 Channel 1 Queue 0, AN1

7.6.4 Position and Speed Sensing

In the FOC algorithm, it is necessary to have an accurate position of the rotor in order to achieve independent control of flux and torque. The emulated electrical position is scaled to range from 0 to 2π [rad] and sensed with the microcontroller's ADC unit. If one assumes that the system is symmetrical, then two-phase currents instead of three can be sensed, and instead of the third one, the ADC channel can be used to sense emulated electrical position. Based on the electrical angular motor position, speed can be recalculated.

AURIX™ TC275 peripherals used for phase current, voltage, and electrical position sensing are shown in Table 7.5.

The waveforms of two-phase currents, DC-link voltage, and electrical angular motor position are shown in OneEye (Fig. 7.27, left). The waveforms of two-phase currents, electrical and mechanical angular motor position, and mechanical speed are shown in HIL SCADA (Fig. 7.27, right).

7.6.5 Speed Loop and Current Loop

After confirming that phase current, DC-link voltage, electrical angular motor position, and speed calculation are correct, current and speed loops can be closed. Figure 7.28 shows the OneEye configuration on the left side and HIL SCADA on the right side. Using OneEye, it is possible to set controller references. In a particular example, the d-axis current reference is set to zero and the speed reference is changed in steps of 50 [rad/s]. In the OneEye graph, one can notice the response of the speed and current PI controllers.

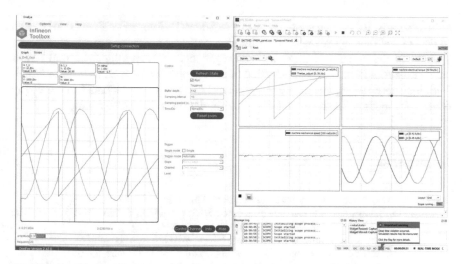

Fig. 7.27 Position and current sensing: OneEye (left) and HIL SCADA (right)

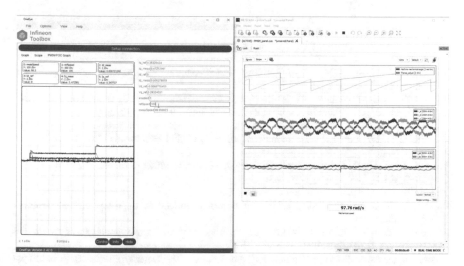

Fig. 7.28 OneEye (left): Response of the speed and current controllers and HIL SCADA (right)

7.7 Conclusion

The transportation sector transformation has accelerated in recent years. This has been and will be facilitated through the usage of specific development, testing, and verification tools. One of the most important tools in this regard is hardware-in-the-loop setups, and in particular, controller hardware-in-the-loop setups. This chapter has depicted one C-HIL setup that consists of a modern real-time simulator, a micro-controller commonly used in the automotive industry, and pertinent hardware and software tools. It has been shown how this safe and flexible environment can be used to design and develop software functionalities for various power electronics-based systems in the domain of xEV, without any concerns stemming from real power stage utilization. The material provided in the chapter can be used as a set of guidelines for interested engineers and researchers, but also by the students and the young developers that are interested in taking part in the transportation evolution.

References

https://softwaretools.infineon.com/tools/com.ifx.tb.tool.oneeye3

Anselma PG, Belingardi G (2019) Next generation HEV powertrain design tools: roadmap and challenges, pp 2019–01–2602. https://doi.org/10.4271/2019-01-2602

AURIX™ TC2xx microcontroller expert-training: https://www.infineon.com/aurix-expert-training

Electric Cars: Calculating the Total Cost of Ownership for Consumers (2021) The European Consumer Organisation, 2021. [Online]. Available: https://www.beuc.eu/publications/beuc-x-2021-039_electric_cars_calculating_the_total_cost_of_ownership_for_consumers.pdf. Accessed 6 Jan 2022

Infineon, "Infineon PIL." https://www.mathworks.com/products/connections/product_detail/infineon-tricore-microcontrollers.html

Infineon, "Virtualizer Development Kit." https://www.infineon.com/cms/en/product/promopages/virtualizer-development-kit/

Infineon-TC27x_D-step-UM-v02_02-EN: https://hitex.co.uk/fileadmin/uk-files/downloads/Shield Buddy/tc27xD_um_v2.2.pdf

Infineon-TC27xDC-DataSheet-v01_00-EN: https://www.infineon.com/dgdl/Infineon-TC27xDC-DataSheet-v01_00-EN.pdf?fileId=5546d462694c98b4016953972c57046a

Infineon-AURIX_TC275_Lite_Kit-UserManual: https://www.infineon.com/dgdl/Infineon-AURIX_TC275_Lite_Kit-UserManual-v01_02-EN.pdf?fileId=5546d46272e49d2a01730 5871f9464ab

Infineon Low Level Driver documentation: https://www.infineon.com/aurix-expert-training/TC27D_iLLD_UM_1_0_1_12_0.chm

Infineon Toolbox: https://www.infineon.com/toolbox

MathWorks, "C compilers for Matlab." https://www.mathworks.com/support/requirements/supported-compilers.html

Morimoto S, Tong Y, Hirasa T (1994) Loss minimization control of permanent magnet synchronous motor drives. IEEE Trans Ind Electron 41(5)

Nanotec motor datasheet: https://en.nanotec.com/fileadmin/files/Datenblaetter/BLDC/DB42/DB42S02.pdf

Report on EV charging pricing, regulatory framework and DSO role in the e-mobility development (2020) Interreg Europe, 2020. [Online]. Available https://www.interregeurope.eu/sites/default/files/inline/file_1628687446.pdf. Accessed 6 Jan, 2022

STMicroelectronics Microcontrollers & Microprocessors. https://www.st.com/en/microcontrollers-microprocessors/stm32-32-bit-arm-cortex-mcus.html

Texas Instruments C2000 real-time microcontrollers. https://www.ti.com/microcontrollers-mcus-processors/microcontrollers/c2000-real-time-control-mcus/overview.html

Typhoon HIL General Specification. https://www.typhoon-hil.com/documentation/typhoon-hil-hardware-manual/hil4-6_series_user_guide/References/hil4-6_general_specifications.html

World Energy Outlook (2020) International energy agency. [Online]. Available: https://iea.blob.core.windows.net/assets/a72d8abf-de08-4385-8711-b8a062d6124a/WEO2020.pdf. Accessed 6 Jan 2022

Chapter 8
Electric Vehicles Digital Twinning Using x-HIL Platforms

Ivan Todorović and Ivana Isakov

Abstract Contrary to how technology oftentimes evolves, the abrupt proliferation of electric vehicles in recent years has not been driven by disruptive technological developments, but rather by ecological, social, and cultural factors. To continue the accelerated development of electric vehicles and to make mass adoption of these vehicles feasible and viable, new devices, concepts, and tools are necessary. Among others, in recent years, the digital twin concept has emerged as a powerful and versatile environment that can facilitate said development. In this chapter, it will be shown how electric vehicles' digital twins can be developed and used as a platform for research and development of electric vehicle pertinent technologies. The presented digital twin consists of electrical (inverter and DC-DC converters), electromechanical (traction motor), and mechanical subsystems. In addition to the hardware aspect of the electric vehicle, the relevant control structures are analyzed, i.e., the chapter provides some details on how the hardware and software elements can be efficiently implemented and seamlessly integrated. The derived digital twin is modular which facilitates further improvements, development, and testing. The users will be enabled to change the subsystem that they want to improve, without needing to implement other EV systems that are not of primary interest to their research. Moreover, digital twins can be used as black boxes to develop and integrate other, emerging, systems and features. For example, they can be used for the development of energy management tools, examinations of advanced power electronics devices operation, tests regarding the interaction of communication systems with the rest of the vehicle's infrastructure, and so on. Besides details on how to synthesize relevant electric vehicle digital twins, the chapter provides the readers with the measurement data obtained during the tests and with all the data necessary to quickly run the digital twin and jump-start their research.

I. Todorović (✉) · I. Isakov
Faculty of Technical Sciences, University of Novi Sad, Trg Dositeja Obradovića 6, 21000 Novi Sad, Serbia
e-mail: ivan.todorovic@uns.ac.rs

I. Isakov
e-mail: ivana.isakov@uns.ac.rs

Keywords Electric vehicle · Digital twin · Hardware-in-the-loop ·
Virtual-harware-in-the-loop · Electric vehicle optimization

8.1 Introduction

The means for people and goods transportation are being increasingly transformed
since the end of the twentieth century and although new vehicles for public transportation are developed, more evident and rapid changes are introduced in the design
of vehicles used for personal transportation (Global EV Outlook 2021). The changes
are coming in many forms, but the following three aspects of vehicles are most
substantially changed—propulsion (i.e., what energy is used for propulsion), vehicle
utilization optimization, and security. The vehicles that will be used in the following
decades will be decreasingly based on oil derivatives and internal combustion engines
and increasingly on electric batteries (and other electric energy storage elements) and
electric drives. Also, mechanisms for energy utilization optimization are integrated
into vehicle control structures. Similarly, many novel systems for driving comfort and
entertainment during vehicle usage are developed. Hence, the vehicles are becoming
highly customizable, personalizable, and generally optimizable. This certainly was
not the case for traditional vehicles. Finally, it is expected that new vehicles will be
much safer to drive and consequently significant efforts are invested in developing
driving assistance and autonomous driving tools.

All these changes, and generally proliferation of hybrid and later pure electric
vehicles, have been initially instigated more by the shifts in ecological, social, and
cultural norms than by technological improvements. Arguably, the only advancement of fundamental importance for electric vehicles (EVs) recently was made in
the domain of batteries—the nickel-metal hydride and lithium-ion batteries have
matured sufficiently so that a reasonable driving range can be achieved, while driving
performance can be comparable with a traditional car (Khaligh and Li 2010). The rest
of the relevant technologies (power electronics devices, motors, mechanical devices)
were at least in principle well known and already in use in a wide range of similar
applications.

Still, to continue the trend of accelerated electric vehicles development and usage,
the ecological reasons and the market pull will not suffice. The EVs are still facing
many problems stemming from the EVs' performances, how they are built and how
they are handled once they reach the end-of-life product stage. These problems must
be addressed, i.e., a range of technological advancements must be achieved, before
the mass adoption of EVs can take place.

To facilitate and accelerate some of the necessary technological advancements,
one platform that has emerged in recent years as a powerful research and development
tool, in many different engineering domains, can be adopted—the digital twin.

Remark: It should be noted that although some researchers and engineers consider
the digital twin to be the same thing as what is traditionally called a (simulation)
model, here these two will be differentiated. Indeed, they are conceptually quite

similar, but it could be said that there is a subtle, yet important difference. The (simulation) model is a construct traditionally implemented on personal computers or some similar platforms. The digital twin is a construct that is implemented on *any* digital computer platform. In other words, all (simulation) models are digital twins, but not vice versa. Although this may seem like irrelevant semantic hairsplitting, in praxis it can result in significant differences. Namely, some modern digital computer platforms allow real pieces of hardware to be included in the digital twin. This can increase significantly the credibility and precision of research and development procedures (in comparison to traditional personal computer simulations). Moreover, many novel digital platforms are dedicated to the development of a specific kind of digital twins, i.e., are not general-purpose platforms such as personal computers. Consequently, the digital twin's execution can be much faster (real-time or faster than real-time execution is possible).

Digital twins, or replicas, are representations of addressed systems implemented on different digital computer platforms. They can be realized with varying precision and accuracy. Among others, the important advantage of using digital twins, in comparison to full power or scaled-down prototypes, is that digital twins are much safer and more flexible to operate, leading to accelerated testing and development procedures for novel concepts and devices.

The main goal of this chapter is to provide the reader with the digital twin of an EV that can be used to facilitate some of the crucial technological advancements, particularly those that are in broad terms related to EV traction. Correspondingly, the digital twin consists of all subsystems, relevant for the EV traction examination, which can be found in an EV:

- Electric energy storage elements.
- Power electronics devices.
- Traction motor(s).
- Vehicle's kinematic model.

The chapter provides theoretical and conceptual details necessary for digital twin synthesis. Relevant data and technical information are also given. Finally, together with the text, the digital twins that can be readily run, altered, and improved are provided.

In the following section of the chapter, the platform for digital twinning is described. The third section offers the theoretical background for digital twin synthesis. The hardware and software aspects of the EV are explicated with reasonable details. The fourth section contains details regarding the EV digital twin implementation. The fifth section is dedicated to the examination of the developed digital twin. The last section offers an overview of the chapter and concluding remarks.

8.2   The Platform for Digital Twining

The digital twins in general can be developed using a wide range of platforms, depending on the specificities of the research and development goals. Model-in-the-loop (MIL), processor-in-the-loop (PIL), controller-hardware-in-the-loop (C-HIL), power-hardware-in-the-loop (P-HIL), and other approaches are extensively in use nowadays. The details on these and other paradigms used for digital twinning can be found in Todorović and Isakov (2022).

Still, for EV digital twinning, the most meaningful approaches are MIL, C-HIL, and P-HIL.

MIL corresponds to the traditional personal computer (PC)-based modeling. Both power stage and control structure elements are implemented within the same environment (PC), oftentimes using similar elements native to the simulation tools used for digital twin development. Advantages of the MIL approach are safety (no real, especially high power, devices are used), flexibility (pertinent tools can be used to generate digital twins for various applications, with various levels of precision and complexity), and tools availability (only a PC with specific software is necessary). The disadvantages are that digital twins are run on PCs that are not optimized for any type of digital twins in particular. Hence, even for the twin of moderate complexity and level of precision, execution is significantly slower than real-time execution. This impedes development and testing processes significantly. Also, the MIL usually offers only crude information on how the real system is going to behave.

C-HIL is based on a dedicated digital computer platform, i.e., assumes the usage of specific devices designated as real-time simulators (RTSs) or emulators. These devices are designed to be capable of running the digital twin in real-time. The RTSs are used for power stage implementation, while the control structures are implemented on a real controller card. C-HIL got its name because a real piece of hardware (controller with the control code) is "in the loop", i.e., its behavior is inspected in more detail. The advantages of C-HIL are that fast development and testing procedures are possible, that a real controller with real control code is used, and that this environment is also quite safe since no high-power devices are used. There are no significant disadvantages of C-HIL usage. C-HIL does assume the usage of emulators, specific kind of devices, but the price of these devices is nowadays comparable to the price of some simulation software packages (such as MATLAB).

P-HIL is a platform that is conceptually quite similar to C-HIL. It is also based on real-time simulators or emulators. Hence, the digital twins are also run in real-time. The difference is that in P-HIL, a (high) power device is tested, i.e., "in the loop". The emulators again implement a certain power stage, but the emulators are interfaced with a high-power device-under-test and not a low-power device such as a controller board. This is a great way to examine the behavior of a power electronics device in different working conditions, for example. Still, as this approach assumes usage of a specific, usually custom-made, device that is tested the P-HIL approach will be disregarded in the remainder of the text (although P-HIL generally is relevant for

EV digital twinning). The provided material then would not be of interest to a wider audience and replicability of the provided digital model would be modest at best.

Here, the Typhoon HIL toolchain will be used to derive both MIL and C-HIL digital twins. The reasons why this toolchain is chosen are as follows:

- The same software can be used to derive MIL and C-HIL digital twins and switching from MIL to C-HIL approaches is quite easy.
- The software for digital twins' design is completely free (Typhoon n.d.). Hence, users can develop MIL digital twins free of charge, only a PC is needed (for C-HIL emulators are necessary and are not free).
- There is a rich library of readily available elements that can be used to build comprehensive digital twins. Also, many examples can facilitate twin development.

The two most important software tools are Schematic Editor and HIL SCADA. Schematic Editor is used to graphically design the digital twin, using the mentioned library of elements. In the case of MIL, the power stage and control structures are designed in the Schematic Editor. Afterward, the file created with the Schematic Editor is compiled and run on a PC. The MIL approach can be designated VHIL (virtual hardware-in-the-loop) here since the virtual hardware-in-the-loop paradigm is utilized. If the C-HIL setup is used, the power stage of the digital twin is again designed using Schematic Editor, but the control code is generated using other software tools (in accordance with the used controller card). The file created using the Schematic Editor, containing the digital twin's power stage, is compiled and downloaded to an emulator or emulators (depending on digital twin's complexity). HIL SCADA, as the name suggests, is used to collect and visualize the data from the digital twin. Also, the HIL SCADA can be used to generate control signals and generally manage the digital twin. The HIL SCADA is in both cases similarly used.

In Fig. 8.1, simplified schemes of the MIL and C-HIL setups can be found. Red arrows represent a model loading pathway. Blue arrows represent data exchanged between the HIL SCADA and the digital twin. Black arrows represent the analog and digital signals exchanged between the emulator and the interface board, i.e., the controller. "CC" stands for control code and "PS" stands for the power stage.

It should be emphasized that the power stages, built with the Schematic Editor, are rather similar for both types of digital twins. The only significant difference is how the signals exchanged between the power stage and control scheme are generated and routed. Consequently, once the digital twin for MIL (VHIL) is built, it is a trivial task to create one for the C-HIL setup. Regarding the Schematic Editor, only settings under *Control* for power electronics devices should be changed (*Internal modulator* or *Model* for MIL and *Digital input per switch* or *Digital input per leg* for C-HIL), as indicated in Fig. 8.2. In the first case, the modulating signals or gate signals are generated by the control structures executed on the PC, together with the power stage. In the second case, real gate signals are coming from the controller card, via the interface board, to the emulator.

In the remainder of the text, the focus will be more on MIL/VHIL than on the C-HIL approach since MIL digital twin is portable and replicable. Also, as indicated

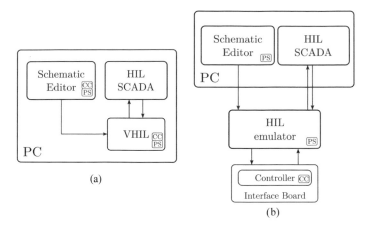

Fig. 8.1 Outlines of MIL (**a**) and C-HIL (**b**) setups

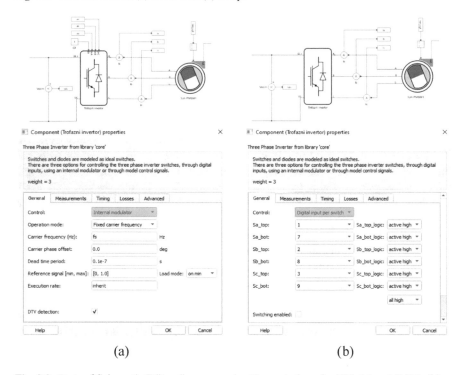

Fig. 8.2 Parts of Schematic Editor diagrams and settings windows for MIL (**a**) and C-HIL (**b**)

previously, once the MIL digital twin functions properly, the C-HIL digital twin can be derived in the manner of minutes. Finally, some details on how the C-HIL environment can be established, using commercially available controllers popular in the automotive industry (AURIX™ TriCore™ microcontrollers), can be found in Chap. 7 and will not be repeated here.

8.3 Electric Vehicle Modeling

The simplified scheme of the vehicle that will be modeled is given in Fig. 8.3. The electric vehicle with two electric energy storage systems (battery and supercapacitors) is considered. Those storage systems are connected to a common DC link via two bidirectional buck-boost converters. These two converters secure the stability of the DC-link voltage. The three-phase traction inverter is also connected to the DC link. The inverter is used to drive permanent magnet synchronous machines. The machine is "loaded" with the vehicle's kinematic model. The sensors and variables feedback lines are not depicted in the figure. Also, from the mentioned DC link other, smaller power, devices could be powered, but these devices are not regarded in the EV digital twin (they generally are of secondary importance for the vehicle behavior).

8.3.1 Electric Energy Storage Systems

The EV digital twin with two energy storage systems is considered since this kind of storage hybridization brings several operational and reliability-related advantages. In the context of EVs, the battery storage system is a high specific energy system (securing long driving range), while the supercapacitor system is a high specific power system (enabling high torque, i.e., acceleration, to be realized). Consequently,

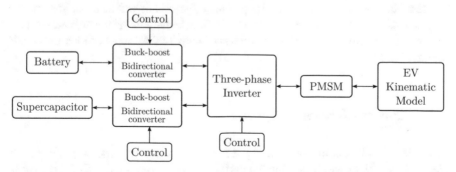

Fig. 8.3 Outline of the considered EV power stage

the vehicles with these two storage systems can deliver significantly better driving performances and energy recuperation. Similarly, the current profile that is necessary to be realized for a specific driving profile can be separated into two components. One, slow-changing component, is to be realized by the battery system and the other, fast-changing component, is to be realized by the supercapacitor system. The absence of abrupt changes in battery current extended the battery life significantly. Also, the battery could have smaller current and even capacity ratings, resulting in a smaller, cheaper, and lighter battery system, etc. (Vazquez et al. 2010).

The batteries and supercapacitors are complex electrochemical elements and their precise modeling can be quite complex. Within the Typhoon HIL toolchain, or more precisely within Schematic editor, three types of batteries and supercapacitors models can be utilized:

- Models from Schematic Editor Library. These elements are parametrizable, but the model's structure cannot be significantly changed. This is the fastest and the easiest way to include a battery or supercapacitor in a digital twin (by simply dragging and dropping the element in the digital twin), but can be less precise and accurate.
- Models based on look-up tables. These models are simple since there is no analytical battery of supercapacitor model derivation. Also, these models can be precise, but for specific operating conditions. If some variable of operational importance is changed (for example, ambient temperature), the look-up table could become inadequate.
- Custom-made models. These models are implemented using C-function blocks that are available in the Schematic Editor Library. C-functions can be used to implement almost arbitrarily complex analytical expressions that shall secure similar storage model behavior to a behavior of a real storage system. Also, look-up tables can be used here also, if necessary, to augment models' precision. This way of elements modeling is the most time-consuming, but the models are the most accurate and correspond precisely to real systems. Still, these models can be quite complex and special care should be given when choosing which model should be used, especially if a system-wide behavior is examined.

Here, the first type of battery and supercapacitor models will be used. The models that are available in the Library (out of the box) can be parametrizable to a satisfactory extent and they do capture the most relevant behavioral features of interest for EV digital twinning.

8.3.2 Power Electronics Devices

Following the EV outline given in Fig. 8.3, the two most important types of power electronics devices will be used in the digital twin—bidirectional DC-DC converter and three-phase inverter.

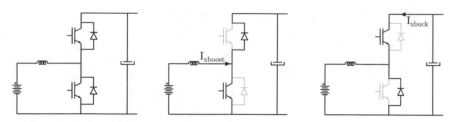

Fig. 8.4 Bidirectional DC-DC converter and necessary passive filters

The bidirectional DC-DC converters are used since two-way energy flow must be secured both for the battery and supercapacitor system. There are many novel bidirectional DC-DC converters, some of which are multiport DC-DC converters. Indeed, these converters seem suitable for usage in EVs with more than one energy storage system. Still, they tend to be complex, while not bringing particularly better performances. Consequently, they are not proven commercial solutions and still have not found a way to mass-produced EVs. Instead of these novel solutions, a traditional bidirectional DC-DC converter will be adopted—a transistor leg with passive elements (capacitors and inductors). This converter is shown in Fig. 8.4.

Regarding the control of the DC-DC converter, the current loop is realized using peak current mode control since it is a simple, fairly robust way of controlling the current flow in DC-DC converters (Erickson and Maksimović 2001). The control scheme for one DC-DC converter is represented in Fig. 8.5. In the figure, the I_x^{ref} is a reference value of the current that should be absorbed or generated by the energy storage system. The trigger signal is used to set the RS flip-flop periodically (at the frequency at which the transistors should be switched—pulse width modulation frequency). The flip-flops are reset at the moment when the storage system's instantaneous current becomes larger than the reference current value. Please note the orientations of I_{xboost} and I_{xbuck} currents in Fig. 8.4. Letter x stands either for battery or supercapacitor. For the peak current mode to function properly, a sufficiently large inductor must be used, so that the peak current does not differ too much from the average current level (Erickson and Maksimović 2001). The rest of the scheme activates or deactivates the appropriate transistor so that the appropriate current flow direction is secured. In the overall control structure, the DC-link voltage control is also assigned to the DC-DC converter. Hence, the DC-link voltage control loop, encompassing current control loop(s), generates the cumulative current reference (I^{ref}) to be produced by both storage systems. The currents produced by the two storage systems must be in a sum equal to the I^{ref} in a steady state. The I^{ref} is separated into the reference currents for battery and supercapacitor systems by employing a simple low-pass filter, as shown in Fig. 8.6. I_B^{ref} is then forwarded to the current control loop of the DC-DC converter connected to the battery system and I_{SC}^{ref} is forwarded to the current control loop of the DC-DC converter connected to the supercapacitor system.

For the traction converter, a standard three-phase two-level voltage source inverter is used. This is a reliable, simple, effective, and well-understood topology. Other

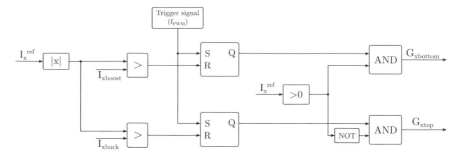

Fig. 8.5 Bidirectional DC-DC converter control scheme

Fig. 8.6 Derivation of
battery and supercapacitor
system current references

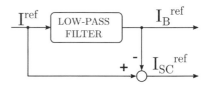

topologies have been proposed, but this converter is used in commercial vehicles, albeit with improvements made to increase the reliability, efficiency, and power density of this traditional converter.

The inverter is driven by the sensored field-oriented control, outlined in Fig. 8.7 (Vukosavic 2011). Speed and current regulators are implemented using proportional-integral regulators. The modulator based on the space-vector modulation technique generates gate signals. The d-axis current reference is held at zero, i.e., flux-weakening is not implemented, although it is an important feature in EVs (the drives must achieve high-speed values). This can be implemented using a look-up table or some analytical scheme for the derivation of i_d^{ref} for a specific speed and specific loading of the traction machine (Marčetić 2014).

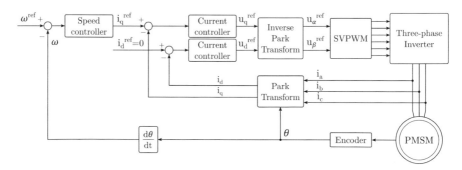

Fig. 8.7 Sensored field-oriented control for permanent magnet synchronous machine

It should be emphasized that all necessary power electronic devices for EV digital twinning can be readily found in the Schematic Editor Library, i.e., there is no need for additional modeling. Also, the elements present in the Library are made out of ideal switches. Still, it is possible to include the data for real (non-ideal) semiconductor devices, if losses analysis or some similar analysis should be conducted.

8.3.3   Traction Motor

The traction motor is chosen to be a permanent magnet synchronous machine (PMSM). Besides induction machines, the PMSMs are the most widely used for high-power EV drives. PMSMs are more expensive and slightly less reliable, but offer higher efficiency and less complicated cooling systems, in comparison with induction drives (Chau 2015).

The Schematic Editor Library offers a variety of drives that can be easily included in the digital twin—different PMSMs, induction drives, multi-phase, etc. The elements available in the Library are modeled with the same level of complexity and details. The transient, nonlinear, and slots-related phenomena are not considered.

As for other elements in the Schematic Editor Library, more complex phenomena can be implemented by the user. For example, a detailed model of PMSM, developed using finite element methods, can be imported if the machine's state-of-health tools are developed or examined. Still, such complex models cannot be simply integrated with a comprehensive digital twin of the electric vehicle. It is more meaningful to use such machine models when diagnostic tools are developed and the focus is only on the drive and its operation.

8.3.4   Vehicle's Kinematic Model

Depending on the vehicle's traction system (and instantaneous torque and power produced by the traction system) and mechanical characteristics of the vehicle (mass of the vehicle, chassis type, center of mass, etc.) different operating points (speed and acceleration) and performances can be realized. Consequently, to create a comprehensive electric vehicle digital twin, a vehicle's kinematic model must be derived and implemented.

Although the pertaining analysis can become quite involved, here the five basic forces exerted onto the moving vehicle will be addressed in simplified form.

The cumulative force exerted onto the vehicle can be expressed in the following form (Ehsani 2018):

$$F_c = F_a + F_r + F_g + F_{la} + F_{\omega a} \tag{8.1}$$

- F_c—cumulative force developed by the traction system,
- F_a—aerodynamic drag,
- F_r—rolling resistance,
- F_g—gravitational pull (hill climbing force),
- F_{la}—force necessary for linear acceleration (of the whole vehicle),
- $F_{\omega a}$—force necessary for angular acceleration (of the vehicle's rotating masses).

Aerodynamic drag is estimated using the expression (8.2):

$$F_a = \frac{1}{2}\rho A C_D (v - v_{Va})^2 \qquad (8.2)$$

- ρ —air density (1.1–1.5 kg/m3),
- A—vehicle's frontal area (expressed in m^2),
- C_D—drag coefficient (usually between 0.15 and 0.3),
- v—vehicle's speed (expressed in m/s),
- v_{Va}—relative airspeed (expressed in m/s).

Rolling resistance is expressed using expression (8.3):

$$F_r = \mu_k \cdot m \cdot g \qquad (8.3)$$

- μ_k—coefficient of rolling resistance,
- m—vehicle mass (expressed in kg),
- g—gravitational acceleration (9.8 m/s^2).

Gravitational pull can be simply expressed using (8.4):

$$F_g = m \cdot g \cdot \sin\alpha \qquad (8.4)$$

- α—slope angle,
- m—vehicle mass (expressed in kg),
- g—gravitational acceleration (9.8 m/s^2).

The force necessary for linear acceleration is given in expression (8.5):

$$F_{la} = m \cdot a \qquad (8.5)$$

- a—Linear acceleration of the vehicle (expressed in m/s^2),
- m—vehicle mass (expressed in kg).

The force necessary for angular acceleration of the vehicle's rotating masses can be calculated using (8.6):

$$F_{\omega a} = \frac{J \cdot G^2 \cdot a}{r^2 \cdot \eta_g} \qquad (8.6)$$

- J—moment of inertia of rotating masses (motor shaft, transmission system, etc.— expressed in kgm^2),
- G—gear ratio (in simplified terms),
- a—linear acceleration of the vehicle (expressed in m/s^2),
- r—wheel diameter (expressed in m),
- η_g—transmission system efficiency.

Hence, the cumulative force developed by the traction system can be written in the following form:

$$F_c = \frac{1}{2}\rho A C_D (v - v_{Va})^2 + \mu_k mg + mg\sin\alpha + ma + \frac{JG^2 a}{r^2 \eta_g} \qquad (8.7)$$

After substituting F_c with $\frac{G}{r}T_c$ (T_c is cumulative torque produced by the traction system) and substituting a with $\frac{dv}{dt}$, (8.7) becomes

$$\eta_g \frac{G}{r} T_c = \frac{1}{2}\rho A C_D (v - v_{Va})^2 + \mu_k mg + mg\sin\alpha + m\frac{dv}{dt} + \frac{JG^2}{r^2 \eta_g}\frac{dv}{dt} \qquad (8.8)$$

Finally, after rearranging (8.8), the expression that can be simply implemented is obtained as follows:

$$\eta_g \frac{G}{r} T_c = \mu_k mg + mg\sin\alpha + \frac{1}{2}\rho A C_D (v - v_{Va})^2 + \left(m + \frac{JG^2}{r^2 \eta_g}\right)\frac{dv}{dt} \qquad (8.9)$$

The digital twin is organized so that the output of the vehicle's kinematic model is the speed that is then forwarded and imposed on the traction motor. The motor generated a torque (T_c) which is in turn forwarded to the vehicle's kinematic model. Besides the T_c, regarding (8.9), the inputs to the vehicle's kinematic model are air speed and slope angle, while the rest of the variables are considered constant.

8.4   Electric Vehicle Digital Twin Implementation

In this chapter, the implementation of the EV digital twin introduced in the previous chapter and outlined in Fig. 8.3 will be presented. The digital twin is realized using Typhoon HIL VHIL software (MIL paradigm). Still, as indicated previously, once VHIL is operating correctly, it is a trivial task to create a digital twin that can run on real-time simulators (taking into account that the control code, in that case, has to be implemented on a dedicated microcontroller).

8.4.1 Electric Energy Storage Systems

The settings for battery and supercapacitor systems are shown in Fig. 8.8. The settings were chosen in accordance with standard values used in commercial vehicles. The current limits are set within the pertaining control loops. In the *Battery type* menu, other battery types can be chosen, including a *User-defined* battery. In the *Signal Processing* tab, the user can enable the battery's state-of-charge to become available for further calculations. This information is useful for energy management strategies. Similarly, the supercapacitor voltage is measured in order to dispatch the supercapacitor only when its state-of-charge is within defined limits.

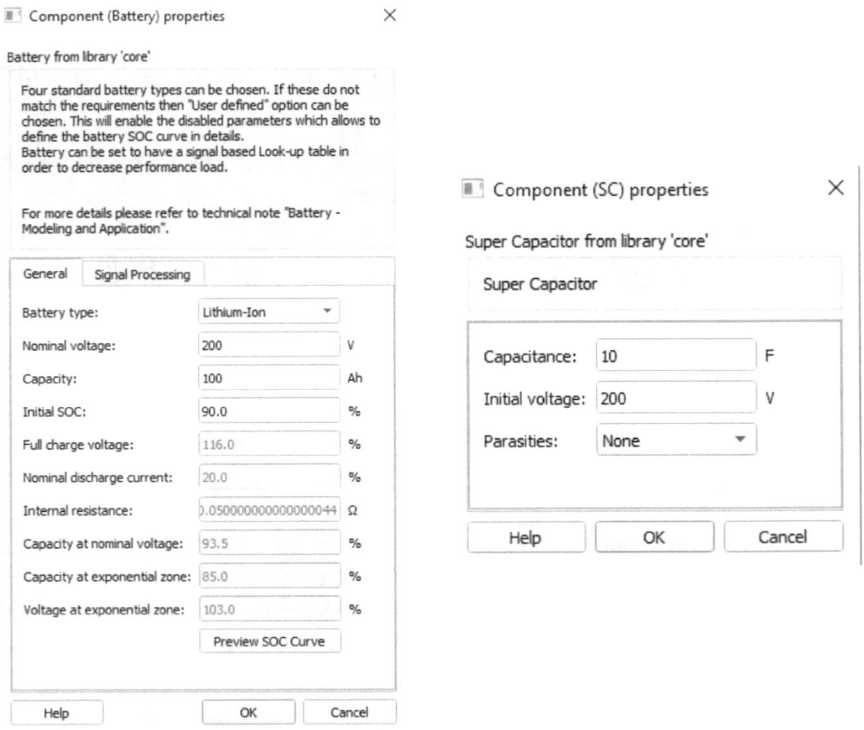

Fig. 8.8 Battery (a) and supercapacitor (b) settings

8.4.2 Power Electronics Devices

The part of the schematic that contains DC-DC converters and other relevant elements is shown in Fig. 8.9. How signals (gate signals and measured currents and voltages) are routed can be also seen in the figure.

Figure 8.10 depicts the DC-link voltage control loop. The proportional gain of the regulator is set to 0.2 and the integral gain is set to 20. The low-pass filter cutoff frequency is set to 1 Hz. The C-function block seen in the figure implements an energy management scheme. In the provided model, rudimental functions are implemented and it is up to readers to expand and improve the functionalities of this block and implement a more comprehensive energy management strategy. The outputs of the "energy management" block are reference values for battery and supercapacitor currents. The DC-link voltage control loop is executed at 10 kHz (this frequency is marked T_s in the *Model initialization script*).

The peak current mode control for the DC-DC converters, which corresponds to Fig. 8.5, is shown in Fig. 8.11. The control is the same for the battery and supercapacitor system. Only the current references are different. The flip-flops are set at 100 kHz frequency. The outputs of this loop are the gate signals for transistors.

Figure 8.12 depicts the three-phase inverter and how the relevant signals are routed. Since the control structures are implemented in the same environment as

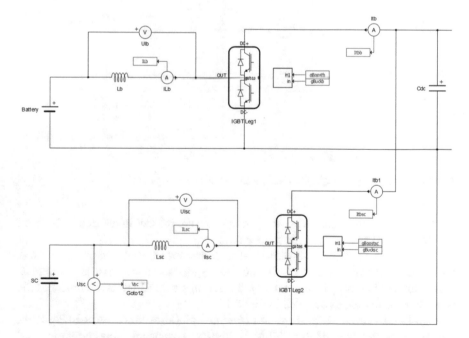

Fig. 8.9 DC-DC converters for battery and supercapacitor systems

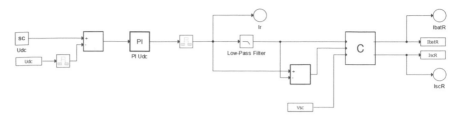

Fig. 8.10 DC-link voltage control loop with current reference management

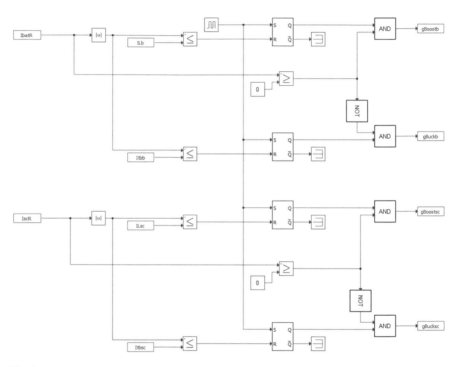

Fig. 8.11 Control for DC-DC converters driving SC and battery system

the power stage, the signals that drive transistors are present in the figures both for DC-DC converters and three-phase inverters.

There is a slight difference—for DC-DC converters, the gate signals are routed, while for the three-phase inverter, the modulating signals are routed (the gate signals are calculated internally). Accordingly, the settings for DC-DC converters and the inverter are different, as depicted in Fig. 8.13.

Figures 8.14 through 8.16 are relevant for three-phase inverter control. In Fig. 8.14, the speed control loop of the PMSM is shown. A standard proportional-integral regulator with anti-windup protection is implemented. The reference speed is set by the look-up table that is defined by the Common Artemis Driving Cycle (Urban).

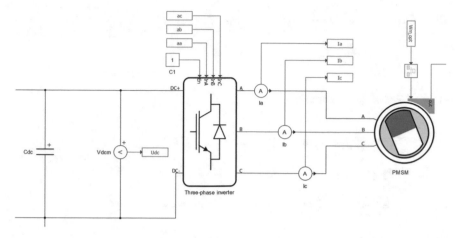

Fig. 8.12 Three-phase inverter with accompanying elements

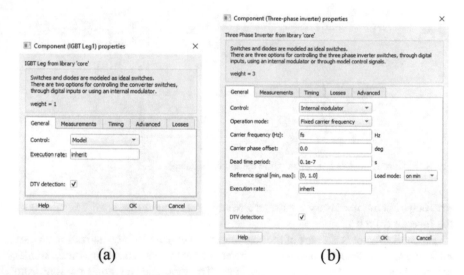

(a)   (b)

Fig. 8.13 Settings for the IGBT legs (**a**) and the three-phase inverter (**b**)

Since the control scheme for the inverter is normalized, the data coming from the look-up table had to be also "normalized". Hence, the gain is placed after the look-up table.

Figure 8.15 depicts the current loop, regulating the PMSM direct current component. The reference value for this loop can be set from the HIL SCADA if desired (default value is zero). The output is the necessary direct voltage component that shall be realized by the inverter (in p.u.).

Figure 8.16 depicts the current loop, regulating the PMSM quadrature current component. The reference value for this loop is defined by the speed control loop.

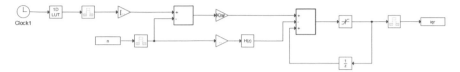

Fig. 8.14 PMSM speed control loop

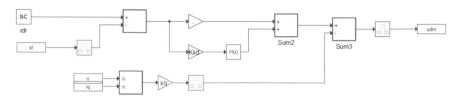

Fig. 8.15 Current control loop (direct current component)

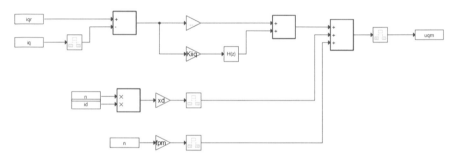

Fig. 8.16 Current control loop (quadrature current component)

The output is the necessary quadrature voltage component that shall be realized by the inverter (in p.u.).

Figure 8.17 shows the part of the schematic that consists of normalization blocks, Park and Clarke transformations, and inverse transformations. Also, block implementing space-vector modulator is depicted. The modulating signals coming from the space-vector modulator are fed to the three-phase inverter. Please note that the power-invariant Clarke transformation is used. The data for the PMSM and other data relevant for the converter's power stages and control structures are provided in the schematic accompanying this chapter and are not given here for esthetic purposes. Specifically, the relevant data can be found in the *Model initialization script*.

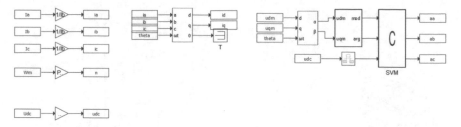

Fig. 8.17 Normalization, transformation, and space-vector modulator elements

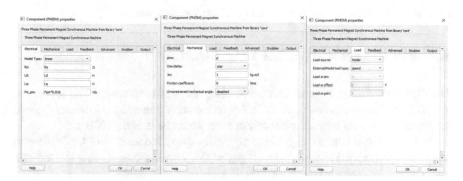

Fig. 8.18 Settings for the PMSM

8.4.3 Traction Motor

The settings for the PMSM traction motor are given in Fig. 8.18. It is important to note that in tab *Load*, the *Load source* is set to *Model,* and the *Model load type* is set to *speed*. This enables the vehicle kinematic model to define the speed of the machine shaft. Additionally, in tab *Output*, the observation of electrical torque, mechanical speed, and mechanical angle is enabled (these variables are available for processing).

8.4.4 Vehicle's Kinematic Model

Figure 8.19 depicts the part of the schematic related to the vehicle's kinematic model. The C-function block contains only expressions derived from (8.9). Also, in Fig. 8.19, it can be seen that the torque produced by the PMSM is the input to the vehicle's kinematic model. The output of the vehicle's kinematic model is the mechanical speed of the PMSM shaft (actually the output is the vehicle speed in m/s, but this can be converted into the motor shaft speed using a simple gain block). This speed is input in the PMSM block.

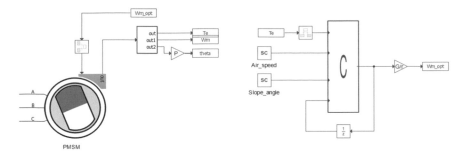

Fig. 8.19 Part of the schematic implementing vehicle's kinematic model

8.5 Electric Vehicle Digital Twin Behavior

The digital twin's behavior was recorded using the VHIL approach. The test lasted for 1000 s (simulation time). It took 7 h to execute. Naturally, if the model was deployed on the emulators, the execution time would be exactly 1000 s.

Still, considering that the digital twin contains a detailed model of PMSM, several converters, controlled using closed loops, with switching frequencies in medium, high ranges, etc., even VHIL execution time is quite reasonable.

Figures 8.20 and 8.21 depict the responses recorded when both battery and supercapacitor systems were activated and when only the battery system was activated, respectively. The sole (simple) goal was to investigate what will be the differences in battery current when the supercapacitors are deployed and when they are not. The differences in the two waveforms should provide clear evidence that supercapacitors could extend battery lifetime.

Figures depict reference speed (defined by the Common Artemis Driving Cycle (Urban) (Common Artemis Driving Cycles (CADC) n.d.)) and vehicle speed, the power delivered to the vehicle by the traction system, inverter phase currents, torque produced by the traction system, battery current, supercapacitors current (if there is one), and DC-link voltage.

The first waveforms testify that the vehicle speed follows the speed reference precisely. The waveform depicting traction motor power output indicates that the PMSM is adequate since the power spikes reach 50 kW since the vehicle speed was not at maximum and the maximum power output is 70 kW. Similar remarks can be given for the currents and torque waveforms. Considering battery and supercapacitor currents, although the supercapacitor current consists of substantial high-frequency components, the battery spikes are still significant. Also, amplitudes of the supercapacitor system current are much smaller than the battery system current counterparts. This signifies that the low-pass filter (Fig. 8.6) cutoff frequency could have been set to an even lower value. In Fig. 8.22, the battery system waveforms are superimposed. It is evident that the amplitudes of the battery currents are generally higher and high-frequency changes are present when the supercapacitor system is deactivated, as expected. Finally, from the last waveforms in Figs. 8.21 and 8.22, it can

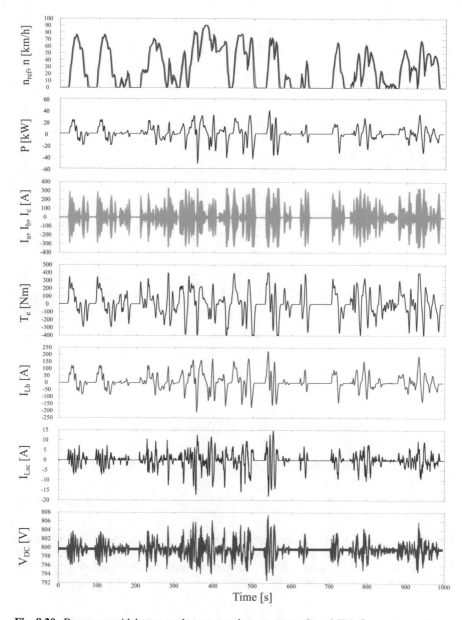

Fig. 8.20 Responses with battery and supercapacitor systems activated. This figure shows, respectively, reference speed (n_{ref}) and vehicle speed (n), the power delivered to the vehicle by the traction system (P), inverter phase currents (I_a, I_b, and I_c), the torque produced by the traction system (T_c), battery current (I_{Lb}), supercapacitors current (I_{Lsc}), and DC-link voltage (V_{DC})

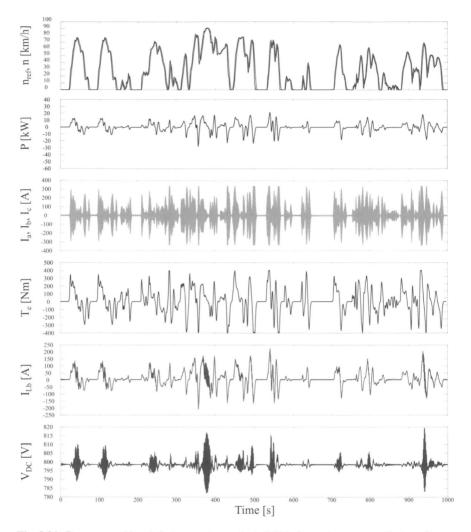

Fig. 8.21 Responses with only battery system activated. This figure shows, respectively, reference speed (n_{ref}) and vehicle speed (n), the power delivered to the vehicle by the traction system (P), inverter phase currents (I_a, I_b and I_c), the torque produced by the traction system (T_c), battery current (I_{Lb}), and DC-link voltage (V_{DC})

be concluded that the DC-link voltage regulation loop operates properly (taking into account that a significantly larger DC-link capacitor was used in the second case).

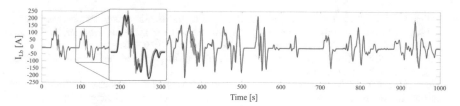

Fig. 8.22 The difference in the battery currents with (red) and without (green) supercapacitors

8.6 Conclusions

This chapter provides the theoretical and conceptual introduction to what constitutes an electric vehicle digital twin. The focus of the discussion were the power stage and the vehicle's kinematics. Also, it is explicated how particular electric vehicle parts can be implemented on a VHIL platform and the details of the implementation are provided. The last part of the chapter offers a set of waveforms capturing the most important electric vehicle behavior features.

The presented digital twin is not by any means the most comprehensive digital twin possible. Improvements can be made. For example, a more detailed transmission system could be regarded, tire slippage can be included, a vehicle with more than one traction motor can be developed, an energy management scheme can be significantly enhanced, etc.

Still, it should be a good starting point for a more thorough analysis, branching into any domain that regards electric vehicle power stage or vehicle movement dynamics. For example, the efficiencies of different electric vehicle subsystems can be addressed. Advanced power electronics control structures can be examined as well. Moreover, the digital twin can be used as a foundation for the development of different topologies of hybrid-electric vehicles' digital twins.

The authors hope that the readers will find the presented chapter useful and are willing to join efforts in continued research on this topic, with all interested parties.

References

Chau KT (2015) Electric vehicle machines and drives: design, analysis and application. Singapore: IEEE, Wiley

"Common Artemis Driving Cycles (CADC)" https://dieselnet.com/standards/cycles/artemis.php

Ehsani M (2018) Modern electric, hybrid electric, and fuel cell vehicles, 3rd edn. Taylor & Francis, CRC Press, Boca Raton

Erickson RW, Maksimović D (2001) Fundamentals of power electronics, 2nd edn. Kluwer Academic, Norwell, Mass

Global EV Outlook 2021. IEA, Paris, 2021. [Online]. Available: https://www.iea.org/reports/global-ev-outlook-2021

Khaligh A, Li Z (2010) Battery, ultracapacitor, fuel cell, and hybrid energy storage systems for electric, hybrid electric, fuel cell, and plug-in hybrid electric vehicles: state of the art. IEEE Trans Veh Technol 59(6):2806–2814, https://doi.org/10.1109/TVT.2010.2047877

Marčetić DP (2014) Mikroprocesorsko upravljanje energetskim pretvaračima, II. Novi Sad, Serbia: Fakultet tehničkih nauka u Novom Sadu

Todorović I, Isakov I (2022) Advances and prospects in distributed generation sources digital twins design. In: Smart grids technology and applications [Working Title], IntechOpen, https://doi.org/10.5772/intechopen.102703, https://www.intechopen.com/online-first/80842

Typhoon HIL, "VHIL." https://www.typhoon-hil.com/products/virtual-hil-device/

Vazquez S, Lukic SM, Galvan E, Franquelo LG, Carrasco JM (2010) Energy storage systems for transport and grid applications. IEEE Trans Ind Electron 57(12):3881–3895, https://doi.org/10.1109/TIE.2010.2076414

Vukosavic SN (2011) Digital control of electrical drives. Springer, New York; London

Chapter 9
Microgrid Primary Controller Performance Characterization

Alexandre T. Pereira, Humberto Pinheiro, Márcio Stefanello,
Jorge R. Massing, Henrique Magnago, and Fernanda Carnielutti

Abstract This chapter reviews the main types of primary controllers for grid-forming converters found in microgrids with multiple distributed converter-based energy resource units. The main type of primary controllers are droop, virtual synchronous generator and dispatchable virtual oscillator, and some variations are described aiming to reveal their dynamic behavior and select their control parameters. This establishes the foundations for a fair comparison among the primary control alternatives considered. The large and small signal models for the primary controllers are derived, and it is demonstrated how the primary controller parameters impact the steady-state and transient behaviors; in addition, time domain simulation on Hardware-in-the-Loop (HIL) illustrates their performance. Since the microgrid controller presents different scenarios of operation, an automated Test-Driven Design (TDD) reveals from extensive simulations in the time domain the strong and weak points of each primary controller. Initially, metrics for both steady-state and transient performances are defined. Then, the key scenarios based on each operating mode such as grid-connected, islanded and unplanned islanding are selected to carry out the tests. Finally, an automated report is given, revealing the strengths and weaknesses of each considered a primary controller.

Keywords Primary controller · Hardware-in-the-loop · Microgrid · Automated test driven design · Droop control · Typhoon HIL

A. T. Pereira (✉)
UFSM, Av. Roraima, 1000 Santa Maria, Brazil
e-mail: alexandretpereira@gmail.com

H. Pinheiro · J. R. Massing · H. Magnago · F. Carnielutti
UFSM, Santa Maria, Brazil
e-mail: jorgemassing@gepoc.ufsm.br

M. Stefanello
UNIPAMPA, Av. Tiaraju, 810 Alegrete, Brazil

© The Author(s), under exclusive license to Springer Nature Singapore Pte Ltd. 2023   211
S. M. Tripathi and F. M. Gonzalez-Longatt (eds.), *Real-Time Simulation and Hardware-in-the-Loop Testing Using Typhoon HIL*, Transactions on Computer Systems and Networks, https://doi.org/10.1007/978-981-99-0224-8_9

9.1   Introduction

Renewable energy is providing a larger share of electrical energy production each day. Distributed Energy Resources (DERs) such as wind and solar power and battery energy stored systems are connected to the grid using static converters, whose behaviors depend on their controllers (Olivares et al. 2014). Renewable energy is providing a larger share of electrical energy production each day. Distributed Energy Resources (DERs) such as wind and solar power are connected to the grid using converters, whose behaviors are mainly dictated by their primary control as compared with the classical synchronous machines (Olivares et al. 2014). There are two ways to control inverter-based distributed energy resources: grid-following and grid-forming modes (Matevosyan et al. 2019). The vast majority of inverters found in photovoltaic (PV) and wind turbines operating nowadays are controlled in grid-following mode. It has an inner current loop to make the grid side currents follow their references and an synchronization algorithm that estimates the angular position and frequency of the voltage at the point of connection (PoC). The current references are usually generated to ensure the desired active and reactive power exchange with the grid. For instance, the active power reference can come from the DC bus voltage outer loop, but it may also incorporate a frequency support action. The reactive power reference can be obtained from grid supporting functionalities, such as voltage-reactive power characteristic, or from an external signal sent by the microgrid control system (MGCS). Grid-forming inverters, on the other hand, usually have an inner control loop to synthesize the voltage at the DER PoC. The voltage reference comes from the primary controller and can also incorporate a virtual impedance. Therefore, the inverter can be seen as an ideal voltage source behind an impedance, whose voltage amplitude and frequency are obtained from the active or reactive powers, depending on the type of line impedance. The primary controller is responsible for providing active and reactive power-sharing, which is a key characteristic when paralleling multiple inverter-based distributed energy resources.

Many studies have been performed concerning the impact of renewable generation on the electrical system, some of them focusing on technical issues due to the lack of inertia (Long et al. 2021), as almost all inverters connected to the transmission grid are presently controlled in a grid-following mode. For a power system based on inverters, grid-following controls cannot be used for all devices. If there are no synchronous machines connected to the network, there will be no voltage to synchronize to. At least one of the inverters will need to be operated in grid-forming mode (Denis et al. 2018). It has been demonstrated that it is possible for grid-forming inverters to emulate inertia on power systems in the presence of a storage element. There are different alternatives for grid-forming inverters' implementation. The main ones are Droop Control, Virtual Synchronous Machine Control (VSM), also known as Virtual Synchronous Generator Control (VSG), and Virtual Oscillator Control (VOC), which demands the need to perform a fair comparison of these techniques from a standard perspective.

Droop Control is the most popular primary control (Kawabata and Higashino 1988; Tuladhar et al. 1997; Chandorkar et al. 1994). It has been reported for power-sharing in UPS inverters since the late 80s. Some authors have proposed different control schemes to overcome several limitations of the droop method (Guerrero et al. 2006, 2004). Important contributions were also made by pointing out that the output impedance of the inverters plays a critical role in power- sharing (Guerrero et al. 2005, 2004; Beheshtaein et al. 2019) and proposing the use of hierarchical control structures (Guerrero et al. 2011; Coelhoet al. 2016).

The second primary controller method presented is the Virtual Synchronous Machine (VSM), where an inverter and its controller emulate the dynamic behavior of a Synchronous Generator (Bevrani and Ise 2014; Sakimoto and Miura 2011; Zhong and Weiss 2011). Synchronous generators are widely used in power systems, and they have inherently active and reactive power- sharing characteristics. As a result, the VSM integration increased the flexibility of the distribution system, once it was established using the well-known synchronous generator model (Alipoor et al. 2015; Zhong et al. 2014; Ashabani and Mohamed 2014; Shintai et al. 2014). In particular, a comparison between the dynamic performance of VSMs and traditional droop controllers is presented in Liu et al. (2016). These two approaches have been developed in two different contexts but show strong similarities. They are equivalent under certain conditions as demonstrated in D'Arco and Suul (2014). Although different types of VSMs have been proposed in the literature, the idea of operating inverters to simulate the complete dynamic behavior of synchronous generators was also developed with the concept of synchronverters (Zhong and Hornik 2013; Zhong and Zeng 2011; Zhong and Weiss 2009), where some of the system parameters, such as inertia, friction coefficient and field and mutual inductances, can be appropriately chosen to improve dynamic performance (Zhong et al. 2018). Moreover, it can emulate the virtual machine with interesting properties, like the limitation of the voltage and frequency, vital in a real-life implementation. In Stallmann and Mertens (2020), sequence impedance modeling is carried out for a grid-forming control based on matching inverter and synchronous generator dynamics showing that both control concepts, although being different, present similar small-signal dynamics.

An alternative way to implement the primary controller is the Virtual Oscillator Control (VOC). Its roots lie on the Liénard-type oscillators (Dhople et al. 2013; Johnson et al. 2014; Sinha et al. 2015). When used to control parallel-connected inverters, they have power-sharing characteristics. It has been demonstrated that the average voltage and frequency dynamics have similar characteristics to a classic droop law (Johnson et al. 2016). VOC is generally programmed as nonlinear differential equations that exhibit cycle-limit behavior, and it is usually implemented in microcontrollers. A comparison between VOC and traditional droop controllers is presented in Johnson et al. (2017). It is noteworthy that VOC is an approach derived in the time domain different from the droop control that considers phasor quantities and assumes the existence of a quasi-sinusoidal steady state (Sinha et al. 2017).

This chapter presents in Sects. 9.2, 9.3 and 9.4 a brief description of the main primary control techniques for grid-forming inverters, presenting methodologies for defining the parameters of the controllers. Sections 9.5 and 9.6 present the concept of the Test-Driven Design and the case study system. Sections 9.7, 9.8 and 9.9 show an extensive analysis of the performance of the main primary control techniques using automatic testing, and finally, Sect. 9.10 summarizes the main points of this chapter.

9.2 Controllers Based on Droop Control

The Droop Control has its principle based on the operation of synchronous generators where the active power is linked with the frequency and the reactive power with the voltage since they have an inductive output impedance and are usually connected to inductive networks; see Fig. 9.1. In an inverter-based distributed energy resource, the droop control performs this link. A local or remote secondary controller sends the setpoints of active and reactive power, as well as the frequency and voltage references to the primary controller, which, in this case, is implemented by the Droop Controller. The droop controller measures the active and reactive power at the PoC and compares them with their setpoints. The droop coefficients m link the inverter output active power to the frequency, whereas droop coefficients n link the reactive power to the inverter output voltage. It is important to notice that in low voltage networks, the equivalent impedance seen by the inverter is not always predominantly inductive, and voltage source inverters do not have significant inductive output impedance as synchronous generators. For these reasons, a virtual inductor can be included in the controller. It provides an inductive output impedance for the inverter in the frequency range of the droop control operation. In Fig. 9.1, a block diagram of Droop Control is presented with voltage and current loops required to damp the LC filter resonance, compensate for the inverter nonlinearities and limit the inverter output current.

In order to understand the behavior of the Droop Control and tune its parameters, it is possible to assume that the converter is operating in a sinusoidal steady state where the dynamics of the inner voltage and current loops are neglected. As a result, the DER can be represented by the equivalent circuit shown in Fig. 9.1. It is easy to demonstrate that active and reactive power exchanged with the grid at the PoC are given, respectively, by (9.1) and (9.2):

$$p = \frac{V}{X_L} E \sin(\delta) \tag{9.1}$$

$$q = \frac{V}{X_L} E \cos(\delta) - \frac{V^2}{X_L} \tag{9.2}$$

where X_L is the reactance of the virtual inductor and δ is the difference between the angular position of the inverter and the voltage at the PoC.

Considering that the voltage amplitude at the PoC, V, is constant and the power angle, δ, is small, the active power p is proportional to the power angle, δ. On the other hand, the reactive power depends on the voltage V. These relations, in turn, motivate the droop control implementation shown on the block diagram of Fig. 9.1.

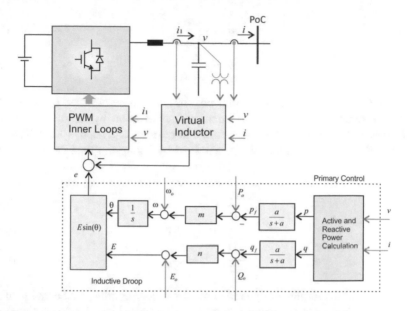

Fig. 9.1 Block diagram representing droop control

The active and reactive powers measured at the PoC pass through low-pass filters that attenuate the high-frequency components above the grid nominal frequency. A state-space representation of these filters can be expressed in (9.3) and (9.4):

$$\frac{dp_f}{dt} = -ap_f + ap \tag{9.3}$$

$$\frac{dq_f}{dt} = -aq_f + aq \tag{9.4}$$

9.2.1 Conventional Droop Control

The filtered active and reactive powers are used in the droop laws to generate the inverter frequency (9.5) and voltage amplitude (9.6). It is important to notice that the actual inverter output voltage must also include the virtual inductor drop:

$$\omega = \omega_o - m(p_f - P_o) \tag{9.5}$$

$$E = E_o - n(q_f - Q_o) \tag{9.6}$$

The droop coefficient, m, is usually defined to make the inverter deliver its rated active power for a frequency deviation $\Delta\omega$ around the defined setpoint, P_o (9.7). Similarly, the droop coefficient n is designed to make the converter exchange the rated reactive power for a voltage deviation ΔE (9.8):

$$m = \frac{\Delta\omega}{P_{rated}} \tag{9.7}$$

$$n = \frac{\Delta E}{Q_{rated}} \tag{9.8}$$

The inverter output voltage angular position, θ, can be obtained by integrating the droop control variable, ω, as shown in (9.9). In addition, by assuming a constant grid frequency ω_g, the inverter voltage angular position is given by (9.10). The state equation for the power angle can be obtained by taking the time derivative of θ, (9.11):

$$\frac{d\theta}{dt} = \omega \tag{9.9}$$

$$\theta = \omega_g t + \delta \tag{9.10}$$

$$\frac{d\delta}{dt} = \omega - \omega_g \tag{9.11}$$

As a result, the nonlinear dynamic equations that describe the behavior of the power angle, frequency and amplitude of an inverter with inductive droop connected to a stiff grid are presented in (9.12), (9.13) and (9.14), respectively. In order to simplify the notation, constants $a_{21}, a_{22}, a_{31}, b_{11}, b_{21}$ and b_{31} are defined. It is noticed that the dynamic equation is nonlinear and that the nonlinearities are associated with the product of the state variable e, which is the inverter background voltage amplitude, with the sine and cosine of the power angle δ:

$$\frac{d\delta}{dt} = \omega + b_{11} \tag{9.12}$$

$$\frac{d\omega}{dt} = a_{21}\omega + a_{22}e \sin(\delta) + b_{21} \tag{9.13}$$

$$\frac{de}{dt} = a_{21}e + a_{31}e \cos(\delta) + b_{31} \tag{9.14}$$

where

$$a_{21} = -a$$
$$a_{22} = -aVm/X_L$$
$$a_{31} = -nV/X_L$$
$$b_{11} = -\omega_g$$
$$b_{21} = a(\omega_o + mP_o)$$
$$b_{31} = a(E_o + nQ_o) + \frac{nV^2a}{X_L}$$

In order to have a feeling of the qualitative behavior of the inverter with inductive droop connected to a stiff grid, the equilibrium points of the derived nonlinear dynamic equation can be found. This is done by setting the derivative of the state variables to zero and then solving the resulting nonlinear algebraic equation for the state variables, which results in the possible operating points. The inverter local qualitative behavior is defined by the eigenvalues of the Jacobian matrix (9.15). The eigenvalues can be easily found from the roots of the determinant of $(\lambda \mathbf{I} - \mathbf{J})$:

$$\mathbf{J} = \begin{bmatrix} 0 & 1 & 0 \\ a_{22}e^* \cos(\delta^*) a_{21} & a_{22} \sin(\delta^*) \\ -a_{32} \sin(\delta^*) & 0 & a_{21} + a_{32} \cos(\delta^*) \end{bmatrix} \tag{9.15}$$

As an example, a 33 kVA single-phase inverter connected to a 220 V 60 Hz grid is considered. The droop coefficients can be set as 5% for the frequency and 10% for the voltage. In addition, the low-pass filter has been selected as 12 Hz and the virtual inductor as 2 mH, which, in this case, corresponds to 0.51 pu. The equilibrium points are $(\delta^*, \omega^*, e^*) = (27.207°, 377 \,\mathrm{rad/s}, 247.37 \,\mathrm{V})$ and $(\delta^*, \omega^*, e^*) = (157.44°, 377 \,\mathrm{rad/s}, 294.76 \,\mathrm{V})$. The later is unstable while the former is stable. The stable eigenvalues are given in (9.16):

$$\begin{bmatrix} \lambda_1 \\ \lambda_2 \\ \lambda_3 \end{bmatrix} = \begin{bmatrix} -37.01 + j37.125 \\ -37.01 - j37.125 \\ -83.234 \end{bmatrix} \tag{9.16}$$

It is possible to obtain the steady-state performance of the grid-forming inverter connected to the grid with the considered droop control using the stable equilibrium point previously derived.

In Fig. 9.2a is presented the reactive and active powers exchanged with the grid as a function of the DER PoC voltage when the setpoints are $E_o = 1$; $\omega_o = 1$; $P_o = 0$; $Q_o = 0$; $n = 10\%$; $m = 5\%$; and $X_L = 0.25$ pu. It is possible to see that there is a considerable error between the expected voltage-var droop characteristic and the one obtained. The droop coefficient has been set to 10% in this case; however, the voltage-var curve obtained has a slope of less than 3%. In Fig. 9.2b, it is seen that frequency-watt droop characteristics are as expected. When the grid frequency deviates 5% from the nominal value, the inverter delivers plus or minus 1 pu of active power. This difference between the volt-var and frequency-watt characteristics is attributed to the fact that the Reactive power-Voltage loop does not have an integrator like the active power-frequency loop. In addition, reactive power-voltage droop characteristic also depends on the virtual reactance, as demonstrated by the red ($X_L = 0.25$ pu) and green ($X_L = 0.13$ pu) curves in Fig. 9.2a.

The relation between the states and the modes could be accessed from the right and left eigenvectors of the Jacobian matrix individually. However, the entries of the eigenvectors depend on the scaling and units of the variables. To overcome this limitation, it is possible to use the participation factors (9.17), which combines both

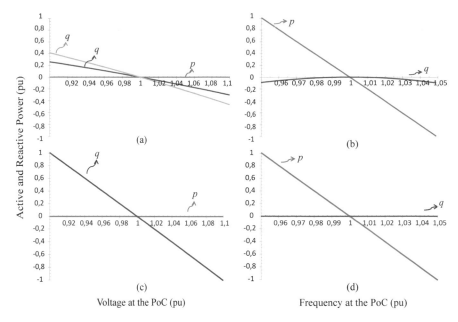

Fig. 9.2 Steady-state performance of the grid-forming inverter connected to the grid with the traditional droop control (**a**) and (**b**) and Improved droop control (**c**) and (**d**)

left and right eigenvectors associated with each eigenvalue of Jacobian matrix (9.16). It is seen that the pair of complex eigenvalues are associated with the frequency command and the power angle, while the real eigenvalue is related to the amplitude of the inverter background voltage:

$$
\begin{array}{cccc}
\lambda_1 & \lambda_2 & \lambda_3 & \\
0.706 & 0.706 & 0.03 & \delta \\
0.690 & 0.690 & 0.03 & \omega \\
0.018 & 0.018 & 0.964 & e
\end{array}
\tag{9.17}
$$

9.2.2 Improved Droop Control

The steady-state performance of the reactive power-voltage droop can be improved by introducing an integrator in the voltage-var loop. In this case, the main difference from the conventional droop controller is the additional state associated with the integrator included in the voltage-var loop. Therefore, there are two parameters to adjust: k_q and k_v; see Fig. 9.3.

The ratio between them defines voltage-reactive power droop characteristics, and specific values of k_q and k_v have a direct impact on the dynamics of the amplitude

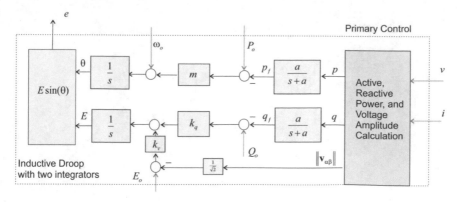

Fig. 9.3 Improved drop control block diagram

of the voltage synthesized by the inverter. By increasing k_v, the transient response of the voltage amplitude loop speeds up. The dynamic equation in this case has four states: (9.12), (9.13), (9.18) and (9.19):

$$\frac{de_d}{dt} = a_{21}e_d + a_{32}e\cos(\delta) + b_{32} \qquad (9.18)$$

$$\frac{de}{dt} = e_d \qquad (9.19)$$

where

$$a_{32} = -k_q a V / X_L$$
$$b_{32} = k_q a \left[Q_o + V^2/X_L + (E_o - V)/n \right]$$
$$n = k_q/k_v$$

It is possible to obtain the steady-state performance of the grid-forming inverter connected to the grid with the considered droop controller with the additional integrator in the voltage-reactive power loop using the stable equilibrium point of the derived nonlinear model. It is seen that both droop characteristics are decoupled and they have the slope as designed, 10% for the volt-var and 5% for the frequency-watt, as shown in Fig. 9.2c and d. In this case, the voltage-var characteristics do not depend on the virtual reactance. The impact of the states on the mode system can be revealed from the participation factors (9.21). The pair of complex eigenvalues are associated with the frequency command and the power angle, while the two real eigenvalues are related to the amplitude of the inverter voltage e and its derivative, the larger one being associated with the derivative of the amplitude (9.20). It is important to notice that this is a local result which depends on the setpoints and grid voltage and frequency:

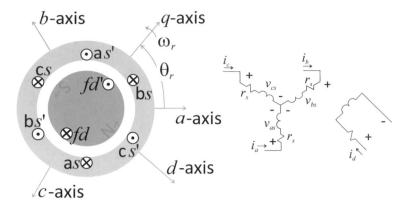

Fig. 9.4 Non-salient pole three-phase uniformly gaped synchronous machine

$$
\begin{bmatrix} \lambda_1 \\ \lambda_2 \\ \lambda_3 \\ \lambda_4 \end{bmatrix} = \begin{bmatrix} -94.25 + j68.55 \\ -94.25 - j68.55 \\ -8.39 \\ -180.10 \end{bmatrix} \tag{9.20}
$$

$$
\begin{array}{cccc}
\lambda_1 & \lambda_2 & \lambda_3 & \lambda_4 \\
0.86 & 0.86 & 0 & 0 & \delta \\
0.86 & 0.86 & 0 & 0 & \omega \\
0 & 0 & 0.05 & 1.058 & e_d \\
0 & 0 & 1.058 & 0.05 & e
\end{array} \tag{9.21}
$$

9.3 Controllers Based on Virtual Synchronous Machine

Virtual Synchronous Machine is a broad term and covers several implementations that mimic the dynamics of synchronous machines using a static converter and its controller. The primary motivation behind the adoption of virtual synchronous generators is to make use of the conventional power system legacy that has dominated power generation for over one century. Then, the transition from a synchronous generator-dominated system to a multiconverter-dominated system can be accomplished using well-established concepts. One of the main properties of the Virtual Synchronous Generator is the reaction to frequency transients as if it had a rotating mass. It is worth mentioning that inverter-based DERs can exhibit inertia and damping properties similar to conventional generation units; however, it depends on the storage energy capacity available from the DC bus. In practice, to mimic inertia, it is necessary to provide energy to or absorb energy from the grid.

In order to implement a VSM, the dynamic model of a synchronous machine has to be derived. Figure 9.4 presents a two non-salient pole three-phase uniformly gaped synchronous machine. The stator windings are identically sinusoidally distributed, displaced by 120 degrees. The rotor is round and equipped with field winding. For simplicity, magnetic saturation and rotor damping windings, usually present in the conventional synchronous machine, are not included in this elementary representation. Since the stator does not have a neutral connection, it is common to represent the machine in an orthogonal stationary reference frame, here named alpha-beta. The voltage equations in stationary alpha-beta reference frame can be expressed in matrix form as (9.22), whereas the stator flux linkage vector is expressed in (9.23). The inductance matrixes are presented in (9.24) and (9.25):

$$\mathbf{v}_{\alpha\beta} = r_s \mathbf{i}_{\alpha\beta} + \frac{d\boldsymbol{\lambda}_{\alpha\beta}}{dt} \tag{9.22}$$

$$\boldsymbol{\lambda}_{\alpha\beta} = \begin{bmatrix} \mathbf{L}_s & \mathbf{l}_{sr} \end{bmatrix} \begin{bmatrix} \mathbf{i}_{\alpha\beta} \\ i_d \end{bmatrix} \tag{9.23}$$

$$\mathbf{L}_s = \begin{bmatrix} L_s & 0 \\ 0 & L_s \end{bmatrix} \tag{9.24}$$

$$\mathbf{l}_{sr} = \begin{bmatrix} L_{sfd} \sin(\theta_r) \\ -L_{sfd} \cos(\theta_r) \end{bmatrix} \tag{9.25}$$

Combining Eqs. (9.22)–(9.25) results in (9.26), where the electromotive force (EMF) voltage vector as (9.27) and the rotor flux as $\psi_d = L_{sfd} i_d$ have been defined. From the derivative of the energy stored in the coupling field with respect to the rotor angular position, it is possible to obtain the electromagnetic torque (9.28):

$$\mathbf{v}_{\alpha\beta} = r_s \mathbf{i}_{\alpha\beta} + \mathbf{L}_s \frac{d\mathbf{i}_{\alpha\beta}}{dt} + \mathbf{e}_{\alpha\beta} \tag{9.26}$$

$$\mathbf{e}_{\alpha\beta} = \psi_d \omega_g \begin{bmatrix} \cos(\theta_r) \\ \sin(\theta_r) \end{bmatrix} \tag{9.27}$$

$$T_e = \psi_d \mathbf{i}_{\alpha\beta}{}^T \begin{bmatrix} \cos(\theta_r) \\ \sin(\theta_r) \end{bmatrix} \tag{9.28}$$

Once in power system analysis, it is usual to assume positive stator currents flowing out of the synchronous machine as it often operates as a generator; the voltage equation can be rewritten as (9.29). In (9.28), the electromagnetic torque, T_e, is positive for generation action. From Newton's second law, it is possible to obtain the torque and rotor speed relationship (9.30), where $d\theta_r dt = \omega_r$, T_i is the input torque, J is the moment of inertia and D_p is a damping coefficient. Finally, for the virtual implementation of a synchronous generator, the reactive power associated with the EMF can be obtained from the magnitude of the vector product of the stator current and EMF vectors (9.32):

$$\mathbf{v}_{\alpha\beta} = -r_s \mathbf{i}_{\alpha\beta} - \mathbf{L}_s \frac{d\mathbf{i}_{\alpha\beta}}{dt} + \mathbf{e}_{\alpha\beta} \tag{9.29}$$

$$\frac{d\omega_r}{dt} = \frac{1}{J}(T_i - T_e - D_p\omega_r) \tag{9.30}$$

$$Q_{gap} = \psi_d \omega_r \mathbf{i}_{\alpha\beta}^T \begin{bmatrix} -\sin(\theta_r) \\ \cos(\theta_r) \end{bmatrix} \tag{9.31}$$

$$Q_{PoC} = \left(\mathbf{v}_{\alpha\beta} \begin{bmatrix} 0 & 1 \\ -1 & 0 \end{bmatrix} \right)^T \mathbf{i}_{\alpha\beta} \tag{9.32}$$

9.3.1 Virtual Synchronous Machine

From the previously derived dynamic model, it is possible to implement a controller for a grid-forming inverter that emulates the synchronous generator. A fundamental characteristic in any synchronous machine is the relation between the accelerating power and angular acceleration, called the swing equation. This is highlighted at the top of the VSM control block diagram in Fig. 9.5.

To compensate for the resulting steady-state error, introduced by the damping coefficient D_p, a feedforward action is seen from the setpoint of frequency to the torque. The frequency droop characteristic of the VSM is defined by the damping coefficient D_p. Moreover, there is an active power setpoint that defines input torque T_i that represents the torque supplied by the prime mover in a real synchronous generator. At the bottom of the VSM block diagram is presented the flux control loop. The inputs are the setpoint for the reactive power and the PoC voltage magnitude. The volt-var droop characteristic is defined by D_q. It is worth mentioning that there is a virtual impedance—highly inductive in this case—that makes the reactive power at the VSM gap different from the reactive power exchanged with the grid at the PoC.

From the derived synchronous generator equations and assuming operation in sinusoidal quasi-steady-state operation, it is possible to obtain the nonlinear dynamic equation that describes the VSM connected to a stiff grid (9.33)–(9.35). The selected state variables are the angular frequency command ω, the power angle δ and the VSM flux, ψ. As a single-phase system is being considered, it should be noticed that the V, ψ and E_o are the rms values of their fundamental component:

$$\frac{d\omega}{dt} = \frac{1}{J} \left(-D_p\omega - \frac{V\psi\sin(\delta)}{X_L} + \frac{P_o}{\omega_o} + D_p\omega_o \right) \tag{9.33}$$

$$\frac{d\delta}{dt} = \omega - \omega_r \tag{9.34}$$

$$\frac{d\psi}{dt} = \frac{1}{K} \left(-D_q V + Q_o - \frac{\omega^2\psi^2}{X_L} + \frac{\omega\psi V}{X_L}\cos(\delta) + D_q E_o \right) \tag{9.35}$$

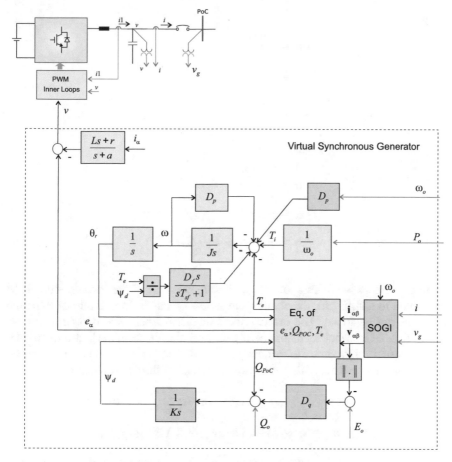

Fig. 9.5 VSM connected to a strong grid—reactive power computed at the PoC

In order to simplify the notation, some constants are defined (9.36)–(9.38). The VSM's nonlinearities are associated with the products of state variables and the sine and cosine of the power angle δ:

$$\frac{d\omega}{dt} = a_{11}\omega + a_{12}\psi\sin(\delta) + b_{11} \tag{9.36}$$

$$\frac{d\delta}{dt} = \omega + b_{12} \tag{9.37}$$

$$\frac{d\psi}{dt} = a_{32}\omega^2\psi^2 + a_{33}\omega\psi\cos(\delta) + b_{31} \tag{9.38}$$

where

$$a_{11} = \frac{-D_p}{J}$$

$$a_{12} = \frac{-V}{JX_L}$$

$$a_{32} = \frac{-1}{KX_L}$$

$$a_{33} = \frac{V}{KX_L}$$

$$b_{11} = \frac{1}{J}\left(\frac{P_o}{\omega_o} + D_p\omega_o\right)$$

$$b_{21} = -\omega_r$$

$$b_{31} = \frac{1}{K}\left(Q_o + D_q\left(E_o - V\right)\right)$$

In order to have a feeling of the qualitative behavior of the inverter with VSM connected to a stiff grid, the equilibrium points of the derived nonlinear dynamic equation can be found. The inverter local qualitative behavior is defined by the eigenvalues of the Jacobian matrix (9.39), which is the matrix of the first-order partial derivative of the right side of the ordinary differential equation at an equilibrium point. The eigenvalues can be easily found from the roots of the determinant of $(\lambda \mathbf{I} - \mathbf{J})$:

$$\mathbf{J} = \begin{bmatrix} a_{11} & a_{12}\psi_o\cos(\delta_o) & a_{12}\sin(\delta_o) \\ 1 & 0 & 0 \\ 2a_{32}\omega_o\psi_o^2 + a_{32}\psi_o\cos(\delta_o) & -a_{33}\psi_o\omega_o\sin(\delta_o) & 2a_{32}\omega_o^2\psi_o + a_{33}\omega_o\cos(\delta_o) \end{bmatrix} \tag{9.39}$$

The damping action (9.40) is used to emulate the damping component found in the classical synchronous generator model. As a result, it provides an additional tunable parameter that allows adjusting the damping ratio associated with the active power loop without affecting the steady-state frequency droop characteristic (Dong and Chen 2018):

$$D_f \frac{d}{dt}\left(\frac{T_e}{\psi_d}\right) \propto (\omega - \omega_r) \tag{9.40}$$

As an example, the same 33 kVA single-phase inverter connected to a 220 V, 60 Hz grid that was presented in the previous section is considered. The coefficients are 5% for the frequency droop ($D_p = \Delta T/\Delta\omega = 4.64$) and 10% for the voltage droop ($D_q = \Delta Q/\Delta E = 1500$). One desired feature of the VSM is that the moment of inertia appears explicitly as a control parameter. The moment of inertia $J = \tau_f D_p$ and the flux loop constant $K = \tau_v \omega_r D_q$ can be obtained from the frequency (τ_f) and voltage (τ_v) time constants. For instance, by selecting τ_f as 100 ms and τ_v as 500 ms, J becomes 0.46 kilogram per square meter and K more than a hundred thousand ($K = 113, 110$). Since the VSM is nonlinear, the qualitative behavior depends not only on the circuit parameters but also on the operating point. The participation matrix (9.42) reveals that the real eigenvalue (9.41) is associated with the magnitude of the flux of the VSM. The complex pair of eigenvectors are associated with the VSM frequency and power angle:

$$
\begin{bmatrix} \lambda_1 \\ \lambda_2 \\ \lambda_3 \end{bmatrix} = \begin{bmatrix} -5.01 + j15.47 \\ -5.01 - j15.47 \\ -0.41 \end{bmatrix}
\tag{9.41}
$$

$$
\begin{array}{ccc}
\lambda_1 & \lambda_2 & \lambda_3 \\
0.525 & 0.525 & 0 \quad \omega \\
0.525 & 0.525 & 0 \quad \delta \\
0 & 0 & 0.999 \quad \psi
\end{array}
\tag{9.42}
$$

9.3.2 Improved Virtual Synchronous Machine

In order to improve the steady-state performance in terms of setpoint tracking, it is possible to feedback the reactive power at the DER PoC. Figure 9.5 shows how the VSM implementation is in this case. The main difference is found in the voltage-var loop. The dynamic equation that describes the VSM connected to a stiff grid is (9.37), (9.43) and (9.44). When compared with the VSM with the feedback of the reactive power at the gap, it is noticed that the flux equation does not have the square product of the flux and frequency:

$$
\frac{d\omega}{dt} = \frac{1}{J} \left(-D_p \left(\omega - \omega_o \right) - \frac{V\psi \sin(\delta)}{X_L} + \frac{P_o}{\omega_o} \right)
\tag{9.43}
$$

$$
\frac{d\psi}{dt} = \frac{1}{K} \left(D_q \left(E_o - V \right) + Q_o - \frac{V\omega\psi \cos(\delta)}{X_L} + \frac{V^2}{X_L} \right)
\tag{9.44}
$$

Once more, the impact of the states on the modes can be revealed from the participation factors. The pair of complex eigenvalues (9.45) are associated with the VSM frequency and the power angle, while the real eigenvalue is associated with the VSM flux magnitude (9.46):

$$
\begin{bmatrix} \lambda_1 \\ \lambda_2 \\ \lambda_3 \end{bmatrix} = \begin{bmatrix} -9.92 + j38.24 \\ -9.92 - j38.24 \\ -3.62 \end{bmatrix}
\tag{9.45}
$$

$$
\begin{array}{ccc}
\lambda_1 & \lambda_2 & \lambda_3 \\
0.517 & 0.517 & 0 \quad \omega \\
0.516 & 0.516 & 0 \quad \delta \\
0 & 0 & 0.999 \quad \psi
\end{array}
\tag{9.46}
$$

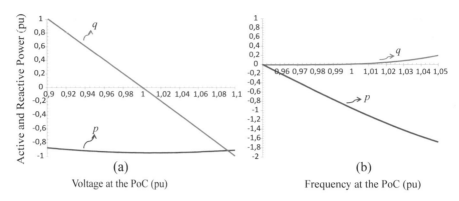

Fig. 9.6 Steady-state performance of the grid-forming inverter connected to a strong grid with the VSM control with reactive power computed at the PoC

From the stable operating point, it is possible to obtain the droop characteristics. In this example, the active power setpoint is −1 pu and the reactive power setpoint is 0 pu. The frequency and voltage droop coefficients have been set to 5 and 10%, respectively. Figure 9.6a shows the voltage-var characteristic as expected. Figure 9.6b shows the frequency-active power characteristic.

9.4 Dispatchable Virtual Oscillator Control

The Droop and the VSM have their roots in well-known principles of operation of conventional electric power systems with synchronous generators. On the other hand, there are controllers whose foundations lay on oscillators. In this section, two of these controllers are described, namely the Virtual Oscillator Control (VOC) and the Dispatchable Virtual Oscillator Control (dVOC).

The Virtual Oscillator Control (VOC) follows an analogous path but aiming to emulate a nonlinear limit-cycle dead-zone oscillator (Johnson et al. 2014). The main attributes of the VOC are global asymptotic synchronization among all parallel-connected VOC controlled inverters, power-sharing capacity in the islanded mode without measurements of powers directly and programmable droop behavior in steady state.

Although VOC has been conceived for single-phase systems—for which the application is straightforward—it is possible to implement the VOC for three-phase systems as well. The main issue is that the original VOC is not dispatchable since there are no inputs to assign active and reactive power setpoints. In addition, it presents odd harmonics on the voltage used to modulate the power converter (Awal et al. 2020). It is worth mentioning that recent modifications have been proposed to mitigate these shortcomings, such as extra control loops that can be added for power dispatching allowing the use of the VOC in the control of voltage source converters (VSC) in grid-connected mode. Considering the distortions, it is possible to mitigate the impact of the nonlinear oscillator on the harmonic content of the output voltages via filtering.

The dispatchable Virtual Oscillator Control (dVOC) solves the main issues of the VOC. As mentioned, the VOC is not a dispatchable controller because there are no setpoints for active and reactive powers. Furthermore, the nonlinear voltage-controlled element h (v_{voc}) introduces distortions on the inverter output voltage. On the other hand, the dVOC has input setpoints for active and reactive powers that are P_0 and Q_0 and does not present distortions like the VOC. The roles of the three terms on the right-hand side of the dVOC law are to define the nominal frequency of the dVOC; to regulate the frequency; and to regulate the voltage:

$$
\frac{d}{dt}\begin{bmatrix} v_\alpha \\ v_\beta \end{bmatrix} = \begin{bmatrix} 0 & -\omega_0 \\ \omega_0 & 0 \end{bmatrix}\begin{bmatrix} v_\alpha \\ v_\beta \end{bmatrix}
$$

$$
+ \eta \underbrace{\begin{bmatrix} 0 & -1 \\ 1 & 0 \end{bmatrix}\left(\frac{1}{V_0^2}\begin{bmatrix} P_0 & Q_0 \\ -Q_0 & P_0 \end{bmatrix}\begin{bmatrix} v_\alpha \\ v_\beta \end{bmatrix} - \begin{bmatrix} i_{1\alpha} \\ i_{1\beta} \end{bmatrix}\right)}_{\begin{bmatrix} e_{\theta\alpha} \\ e_{\theta\beta} \end{bmatrix}}
$$

$$
+ \eta\alpha \underbrace{\frac{V_0^2 - \left\| \mathbf{v}_{\alpha\beta} \right\|^2}{V_0^2}\begin{bmatrix} v_\alpha \\ v_\beta \end{bmatrix}}_{\begin{bmatrix} e_{v\alpha} \\ e_{v\beta} \end{bmatrix}} \quad (9.47)
$$

The dVOC was introduced in Colombino et al. (2017), and its implementation for inductive grids is given by (9.47). There are three main objectives of a grid-forming inverter control for microgrid applications that are accounted, namely frequency setpoint, which is implemented by the first term on the right-hand side of the dVOC law; voltage magnitude control that comes directly from the third term on the right-hand side by computing the error between the actual amplitude and its setpoint V_0; and power dispatch. It is important to notice that the power regulation objective is not readily seen in the second term. After some algebraic manipulations, it is possible to uncover the active and reactive powers exchanged at the PoC. In addition, no inner-loops are being modeled. Since the dVOC has setpoints for active and reactive powers, it is possible to express the second term of the dVOC law according to (9.48):

$$
\frac{\eta}{\left\| \mathbf{v}_{\alpha\beta} \right\|^2}\left(\begin{bmatrix} 0 & -\Delta p \\ \Delta p & 0 \end{bmatrix}\begin{bmatrix} v_\alpha \\ v_\beta \end{bmatrix} + \begin{bmatrix} \Delta q & 0 \\ 0 & \Delta q \end{bmatrix}\begin{bmatrix} v_\alpha \\ v_\beta \end{bmatrix}\right) \quad (9.48)
$$

where

$$\Delta p = \frac{\|\mathbf{v}_{\alpha\beta}\|^2}{V_0^2} P_0 - p$$

$$\Delta q = \frac{\|\mathbf{v}_{\alpha\beta}\|^2}{V_0^2} Q_0 - q$$

$$\frac{d}{dt}\begin{bmatrix} v_\alpha \\ v_\beta \end{bmatrix} = \begin{bmatrix} 0 & -\omega_0 - \frac{\eta}{V_0^2}\left(P_0 - \frac{V_0^2}{\|\mathbf{v}_{\alpha\beta}\|^2}p\right) \\ \omega_0 + \frac{\eta}{V_0^2}\left(P_0 - \frac{V_0^2}{\|\mathbf{v}_{\alpha\beta}\|^2}p\right) & 0 \end{bmatrix}\begin{bmatrix} v_\alpha \\ v_\beta \end{bmatrix}$$
$$+ \frac{\eta\alpha}{V_0}\left(\frac{1}{\alpha V_0}\left(Q_0 - \frac{V_0^2}{\|\mathbf{v}_{\alpha\beta}\|^2}q\right) + \frac{V_0^2 - \|\mathbf{v}_{\alpha\beta}\|^2}{V_0}\right)\begin{bmatrix} v_\alpha \\ v_\beta \end{bmatrix}$$

$$(9.49)$$

The previous development demonstrated that, although the powers are not directly measured, they are captured by the dVOC law. Replacing (9.48) in (9.47), it is possible to recognize that the characteristics of an inductive droop are present: the active power has a direct impact on the frequency of the system, whereas the reactive power impacts the voltage (9.49). The droop coefficients recovered from these droop laws are $m = \eta/V_0^2$ and $n = 1/\alpha V_0$. The relations $f \times P$ and $V \times Q$ are not entirely decoupled due to the actual voltages present in all terms. However, as the dVOC voltage amplitude approaches the setpoint V_0, the droop equations become more and more decoupled.

The dVOC in its original formulation achieves its objectives using the output current of the grid-forming inverter and the voltage signals generated internally by the dVOC. However, following the path of making the dVOC compatible with inductive lines, a virtual inductor is added. Furthermore, inner-loops are implemented. Therefore, the equivalent circuit seen by the dVOC is a voltage-controlled voltage source followed by the virtual inductor behind the PoC. The inner-loops are assumed to perform well enough so that the voltage at the capacitor is equal to the output voltage from the dVOC. For this reason, an improved dVOC for single-phase systems with inner-loops is presented in Fig. 9.7. The voltage at the PoC may be used in place of the magnitude voltage of the dVOC. This choice is motivated by the fact that the dVOC behaves better after the clearing of a short circuit. As in the case of the VOC, the output current here is no longer the output current of the inverter but the current delivered at the PoC.

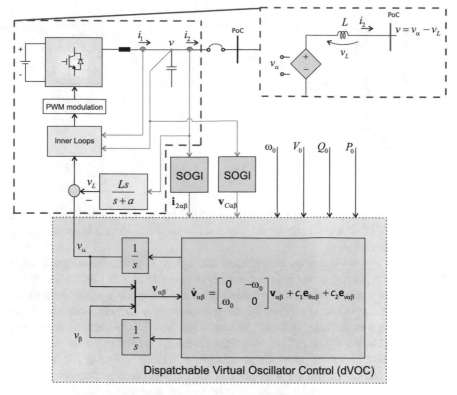

Fig. 9.7 dVOC for a single-phase DER with inner-loops block diagram

9.5 Test-Driven Design

For grid-tied applications, the inverter control system is usually implemented as a firmware in a microprocessor or Digital Signal Processor (DSP). Prior to connecting the inverter to the grid, the firmware should be exhaustively tested and validated—which is usually a time-consuming and complex process. The use of an automated virtual test environment could bring flexibility and safety to cover a wide range of operating conditions and contribute to reduce design/redesign costs and time-to-market (Janzen and Saiedian 2005; Fucci et al. 2017).

In power electronics and applied control areas, the TDD has a great potential to be used to systematically characterize the performance of the aforementioned grid-tied inverter control blocks, and verify whether they satisfy predefined metrics. In addition, the TDD test results also provide the means to benchmark different implementations of a given control block, for instance, different current controller synchronization algorithms. The Test-Driven Design (TDD), also called Test-Driven Development, is an approach for software development where instead of testing after development, like traditional validation methods (Kumar and Bansal 2013), the

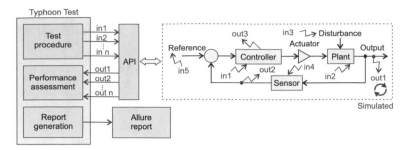

Fig. 9.8 Block diagram of the Typhoon test IDE

developer writes some tests for a given application before the application or device under test (DUT) is actually designed. Then, software development is carried out in a recursive manner.

Among the software development methodologies, TDD can be adopted when designing and testing controllers for grid-tied converters. The basic idea of the TDD is to perform a series of automated tests, each one for a specific scenario. The scenarios are defined a priori by the designer via a software that drives the simulation as a supervisory system. The results presented in this chapter have been implemented in the Typhoon Test Integrated Development Environment (IDE), the framework that integrates pytest for running tests written in Python, as shown in Fig. 9.8. In addition, there are a number of application protocol interfaces, APIs, that allow quicker and easier testing. Typhoon Test IDE also integrates with the Allure Framework reporting tool, providing holistic interactive reports. Each scenario is simulated systematically and the results can be collected and compared with each other. Examples of scenarios include, for instance, structural and/or parametric changes in the system to be simulated. All the schematics and test scripts mentioned in this chapter can be found under the examples provided with Typhoon HIL Control Center installation (examples\courses\digital control of grid-tied converters\TDD_TestDrivenDesign\TDD5_Droop) or within the Control of Microgrids course at https://hil.academy/.

9.6 System Description

In order to evaluate and compare the performance of the different primary controllers derived, a case study has to be defined: a microgrid comprised of two identical DERs sharing a load that combines linear resistive and nonlinear inductive, Fig. 9.9.

The case study considers three different operating modes:

- Grid-connected Mode;
- Transition Mode;
- Islanding Mode.

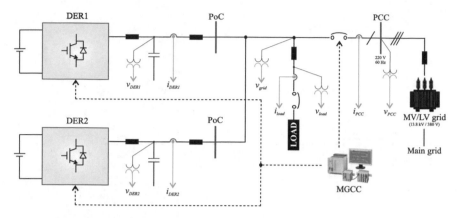

Fig. 9.9 Full simulation block diagram

The primary control algorithms to be evaluated are

- Traditional Droop control;
- Improved Droop control—with the additional state associated with the integrator included in the voltage-var loop;
- Dispatchable Virtual Oscillator control;
- Virtual Synchronous Machine control—with the reactive power computed at the gap;
- Improved VSM—with the reactive power computed at the DER PoC.

Once there are multiple scenarios to evaluate and compare, an automated tool for testing and report generation is desirable. As mentioned, the Typhoon test Integrated Development Environment is an alternative to create test scenarios, capture signals and apply transformations in a Python environment. The main system schematic developed in Typhoon Schematic Editor is presented in Fig. 9.10.

The DERs are connected to the main system through two identical subsystems, presented in Fig. 9.11. Each subsystem comprises a DC source, a single-phase inverter and a LC filter. The current and voltage measurement signals are sent to C-Blocks where the primary controller is implemented. Signals for active and reactive power references are sent from the Microgrid Central Controller (MGCC), presented in Fig. 9.14.

The main system presents three identical meter subsystems, Fig. 9.12. These subsystems are responsible for the current and voltage measurements on the DERs and on the load, as well as for the active and reactive power calculation.

The load subsystem is presented in Fig. 9.13 and comprises multiple contactors that connect and disconnect the linear and nonlinear load, and the short circuit. The Microgrid Central Controller (MGCC) subsystem receives the power reference signals from the Scada or Test IDE and sends them to DERs subsystems. The MGCC also receives the contactor signals to connect or disconnect the system from the grid. The MGCC subsystem schematic is presented in Fig. 9.14.

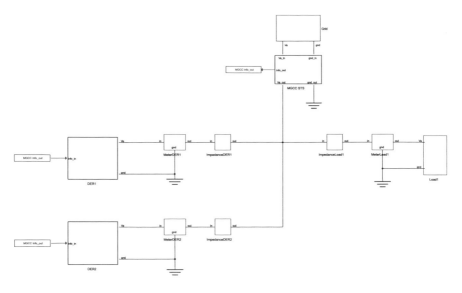

Fig. 9.10 Main system schematic

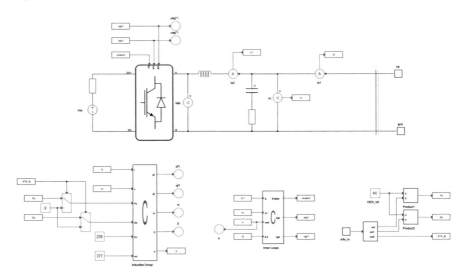

Fig. 9.11 DERs subsystem schematic

Fig. 9.12 Meters subsystem
schematic

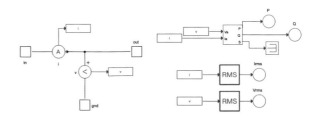

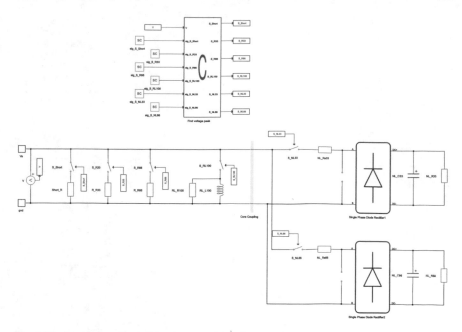

Fig. 9.13 Load subsystem schematic comprising linear and nonlinear load, and short circuit

Finally, the grid subsystem schematic is presented in Fig. 9.15 and comprises the grid whose voltage amplitude, frequency and phase can be controlled via Scada input signals that can be sent from Scada or Test IDE.

Initially, the test is parameterized by defining, for instance, the microgrid grid-rated voltage and frequency, DERs rated power, as well as characterizing the events to be captured in the different operating modes, such as voltage, frequency and power thresholds; simulation time intervals; and so on. Once the test is finished, the results can be inspected using the test report tool Allure, Fig. 9.16. This report contains all the test parameters, DUT and OCs, as well as the test results and performance indexes.

9.7 Grid-Connected Mode

This test evaluates whether or not the active and reactive power exchanged by the DERs are within a predefined envelope of 15% around the ideal droop curves. These tests were evaluated for all five controllers, and the results are summarized in Fig. 9.16a. All the events (connection, disconnection, load change, etc.) occur at the same instant on both DERs.

The grid frequency changed from 57 Hz to 63 Hz, as shown in Fig. 9.17. For each case, the active power was compared with the expected frequency droop curve

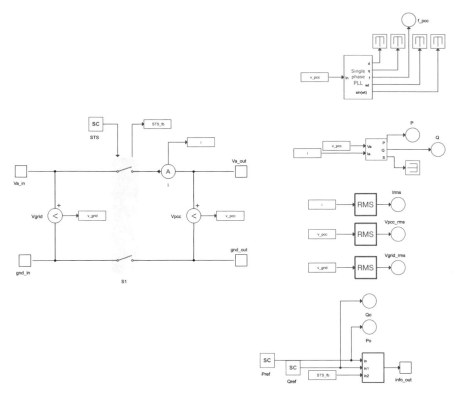

Fig. 9.14 Microgrid Central Controller (MGCC) subsystem schematic

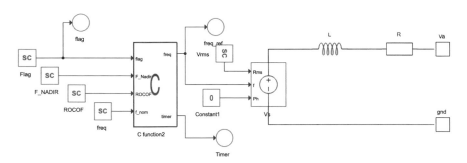

Fig. 9.15 Connection to the grid subsystem schematic

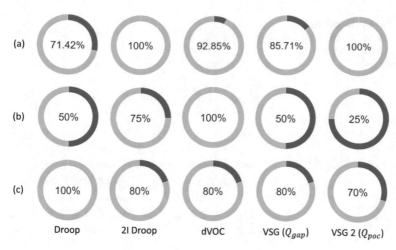

Fig. 9.16 Summary of automated testing results for **a** grid-connected mode; **b** unplanned islanding mode; and **c** islanding mode

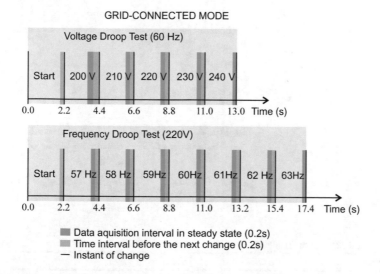

Fig. 9.17 Voltage and frequency change during grid-connected mode test

described by a continuous line. The parameters of the frequency test are presented in Table 9.1.

The grid voltage was changed from 200 to 240 Volts with a step voltage of 10V, Fig. 9.17. For each case, the reactive power and the voltage were compared with the expected voltage droop curve described by a continuous line. The parameters of the voltage test are presented in Table 9.2.

Fig. 9.18 Grid-connected
mode test results for droop
control: **a** frequency test; and
b voltage test

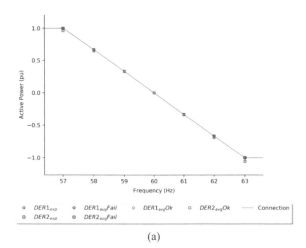

(a)

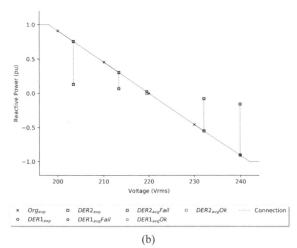

(b)

Table 9.1 Parameters of frequency test

Parameter	Value
Frequency range (Hz)	57–63
Active power reference (pu)	0
Time interval before the next change (s)	0.2
Time interval to reach steady state (s)	2
Desired droop (pu/Hz)	−0.334
Acceptable active power tracking error (pu)	0.15
Acceptable active power ripple (kW)	1

Fig. 9.19 Grid-connected mode test results for improved droop control: **a** frequency test; and **b** voltage test

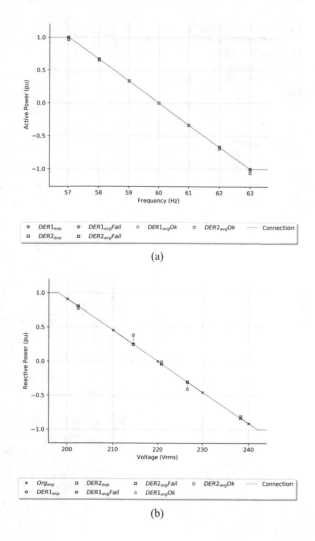

(a)

(b)

Table 9.2 Parameters of voltage test

Parameter	Value
Voltage range (V)	200–240
Reactive power reference VAr (pu)	0
Time interval before the next change (s)	0.2
Time interval to reach steady state (s)	2
Desired droop (pu/V)	−0.0455
Acceptable reactive power tracking error (pu)	0.15
Acceptable reactive power ripple (kVAr)	1

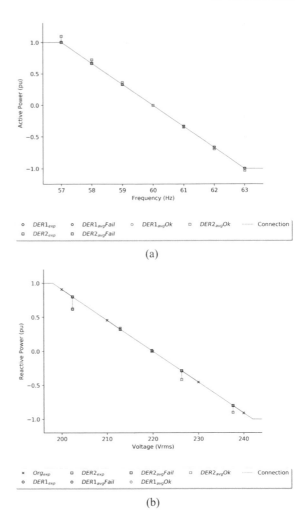

Fig. 9.20 Grid-connected mode test results for dispatchable virtual oscillator control: **a** frequency test; and **b** voltage test

     The points marked in green squares in Figs. 9.18, 9.19, 9.20, 9.21 and 9.22 mean the active power delivered by the DERs is within the predefined threshold. In order to compute the voltage-var curves, the voltages are measured at the DER PoC, to avoid the voltage drop in the feeder. The frequency droop characteristic of Droop control is as expected, Fig. 9.18a. For the Droop control nominal grid-voltage, the reactive power is within the thresholds, which is, in this case, 0.15 pu, Fig. 9.18b. However, for all other voltage values, the reactive power exceeds the maximum allowable limits. For each case, the reactive power and the voltage are plotted and compared with the expected voltage droop curve. The Droop control with the integrator in the voltage loop, Fig. 9.19, passes in all cases: for all frequencies and voltages, the active and reactive powers delivered by the DERs are within the defined envelope. Similarly as for the Droop control, the dVOC passes in association with the frequency

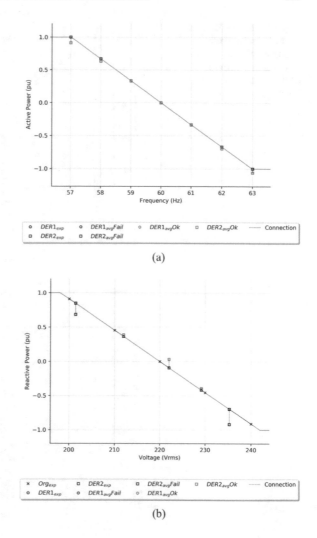

Fig. 9.21 Grid-connected mode test results for virtual synchronous machine control: **a** frequency test; and **b** voltage test

support in steady state, Fig. 9.20a. However, regarding the voltage support, the dVOC fails close to the lower voltage limit, Fig. 9.20b. The VSM passes the steady-state frequency support curve test, Fig. 9.21a. As expected, the VSM with the reactive power computed at the gap fails voltages close to the lower and upper voltage limit considered, Fig. 9.21b. This is attributed to the voltage drop on the virtual inductor. Finally, the VSM with the reactive power computed at the DER PoC passes in all tests, Fig. 9.22.

All primary controllers considered pass the frequency support steady-state tests. The Droop presents a significant error on the voltage-var curve among all other primary controllers considered. The VSM with reactive power on the gap fails in

Fig. 9.22 Grid-connected mode test results for improved virtual synchronous machine control: **a** frequency test; and **b** voltage test

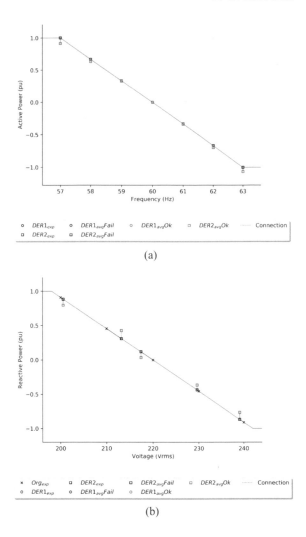

(a)

(b)

the voltage-var curve, which is attributed to the voltage on the virtual inductor. Even though the dVOC fails at a point of the voltage-var curve, its performance is satisfactory.

9.8 Transition Mode

The transition mode tests evaluate the steady-state and the transient behavior of the microgrid voltage, frequency and power-sharing when unintentionally disconnecting it from the main grid. The microgrid is unintentionally disconnected from the main

Fig. 9.23 Load and power setpoints change during transition mode test

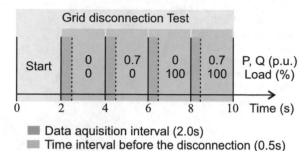

TRANSITION MODE

Grid disconnection Test

Table 9.3 Parameters of the unplanned islanding test

Parameter	Value
Time interval before the next change (s)	0.5
Time interval after change (s)	1.5
Voltage sag before disconnecting	false
Maximum rms voltage after disconnecting (V)	30
Maximum rms voltage settling time (s)	1
Maximum frequency settling time (s)	1
Maximum rms voltage steady-state error[a] (V)	20
Maximum frequency steady-state error[a] (Hz)	3

[a] Compared with nominal value, after transition

grid at $t = 0.5\,\mathrm{s}$. The disconnections are performed under four operating conditions: with power setpoints set to zero and 0.7 pu combined with operation under no load and full load, Fig. 9.23. The unplanned islanding test evaluates the power-sharing among the DERs, the voltage and frequency settling time after the disconnection, as well as the steady-state errors for both frequency and voltage. The parameters of the unplanned islanding test are presented in Table 9.3. Several waveforms have been obtained, but just the ones that fail are presented. Since the DERs have similar behavior, just the rms voltage of one of them is shown in this paper.

Droop control fails as rms voltage after the disconnection with full load and power setpoints set to zero exceeds the threshold of 20V below the nominal value, Fig. 9.24a. Similar behavior is observed when the active and reactive power setpoints are 0.7 pu, Fig. 9.24b. In the case of the Improved Droop control, it is not able to share active power soon after the transition to the island mode, Fig. 9.25a. The yellow envelope defines the active power tolerance region for the DER2. The power-sharing error between the DERs is slightly out of the defined envelope. The VSM with reactive power computed at the gap fails in the unplanned islanding test with active and reactive power setpoints set to zero pu and 0.7 pu, Fig. 9.25b and (c), as the steady-state frequency after the disconnection is below 5% of the nominal microgrid

Fig. 9.24 Transition mode test results for droop control with unintentional disconnection from the main grid at $t = 0.5s$ with full load and power setpoints set to **a** $P_o = 0$ pu and $Q_o = 0$ pu; and **b** $P_o = 0.7$ pu and $Q_o = 0.7$ pu

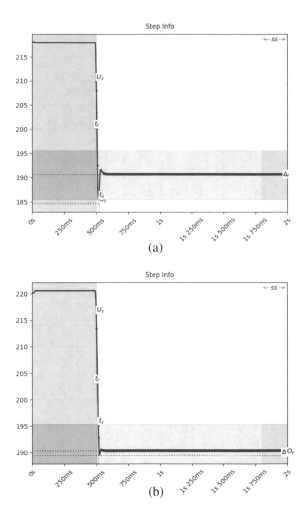

frequency. Finally, the VSM with the reactive power computed at the DER PoC fails with the power setpoints set to zero and disconnection with full load. Both settling time and steady-state error do not meet the defined limits, Fig. 9.26a. With power setpoints set to 0.7 pu and disconnection under no load, the active power-sharing requirement is almost achieved. There is only a small time interval that the defined threshold is violated, Fig. 9.26b. In unplanned islanding with setpoints set to 0.7 pu and full load, the frequency steady-state error exceeds the 3 Hz tolerance defined, Fig. 9.26c.

The automated test points out that the dVOC presents the best results, Fig. 9.16b. A detailed analysis reveals that the Improved Droop and the VSMs could also lead to good results in meeting the defined requirements by adjusting their control parameters. The communication delay between the MGCC and the DERs could change the

Fig. 9.25 Transition Mode test results for **a** Improved Droop control with no load and $P_o = 0.7$ pu and $Q_o = 0.7$ pu; Virtual Synchronous Machine with full load and **b** $P_o = 0$ pu and $Q_o = 0$ pu; and **b** $P_o = 0.7$ pu and $Q_o = 0.7$ pu

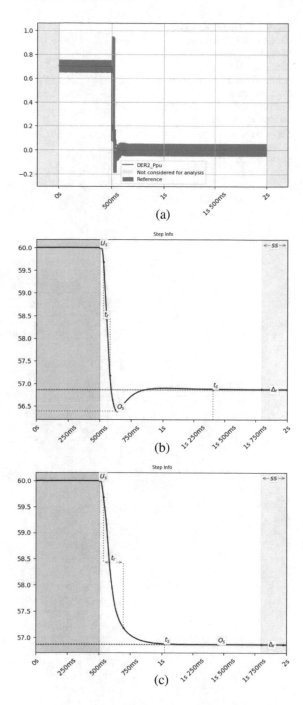

Fig. 9.26 Transition Mode test results for Improved Virtual Synchronous Machine control with unintentional disconnection from the main grid at $t = 0.5s$ and power setpoints set to **a** $P_o = 0$ pu and $Q_o = 0$ pu; **b** and **c** $P_o = 0.7$ pu and $Q_o = 0.7$ pu

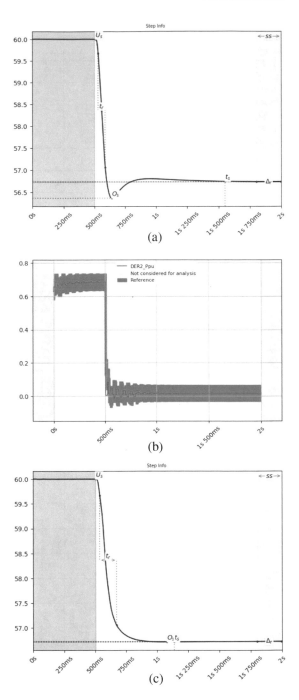

results of this test as the MGCC can inform the DERs that the operation microgrid is in islanding mode.

9.9 Islanding Mode

This test evaluates the performance of the considered primary controllers when operating in islanding mode. Three load conditions are considered: linear (33%, 66% and 100% of full load); nonlinear (33%, 66% and 100% of full load); and short-circuit (at $t = 0.05$ s and $t = 0.25$ s), Fig. 9.27. Steady-state and transient metrics for the voltage and frequency are analyzed, and the power-sharing between the DERs is also considered.

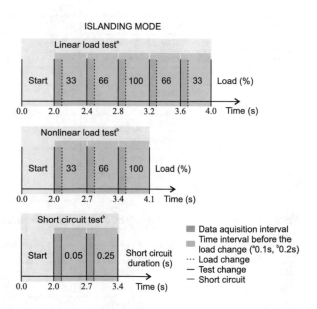

Fig. 9.27 Load change during linear load test, nonlinear load test and short-circuit test during islanding mode

Table 9.4 Parameters of the islanding mode test with linear load

Parameter	Value
Time interval before the next change (s)	0.1
Time interval after the change (s)	0.3
Maximum rms voltage settling time[a] (s)	0.5
Maximum rms voltage steady-state error[b] (V)	35
Maximum frequency steady-state error[b] (Hz)	3

[a] Settling time region is ±5Vrms of last value in capture
[b] Compared with nominal value, after load transition

The test script allows the operator to set the thresholds for overshoot, settling time and steady-state error for the voltage and frequency during events resulting from load changing, Tables 9.4 and 9.5. The microgrid frequency steady-state error is also evaluated. In addition, power-sharing among the DERs is also considered. The nonlinear load is defined as in IEC 62040-3 standard that characterizes the performance of UPSs. In addition, it is possible to parameterize the short-circuit durations, Table 9.6. In the following results, 50 ms and 250 ms short-circuit durations are considered. As previously stated, just the results that fail are analyzed.

The Improved Droop control fails in both short-circuit tests, Fig. 9.28. Despite the power-sharing mismatch during the 50 ms short circuit, after it is cleared the voltage and frequency are restored to their nominal values. However, for the 250 ms short-circuit duration, after the clearing, the voltage and frequency are not restored to their nominal values. On the other hand, the voltage at the DER PoC during a 50

Table 9.5 Parameters of the islanding mode test with nonlinear load

Parameter	Value
Time interval before the next change (s)	0.2
Time interval after the change (s)	0.5
Maximum voltage after removing the load (V)	357
Maximum peak current during load (A)	400
Maximum rms voltage overshoot during load (V)	100
Maximum rms voltage settling time[a] (s)	0.5
Maximum rms voltage steady-state error[b] (V)	20
Maximum frequency steady-state error[b] (Hz)	1.5

[a] Settling time region is ±5 Vrms of last value in capture
[b] Compared with nominal value, until load is removed

Table 9.6 Parameters of the islanding mode test with short circuit

Parameter	Value
Short-circuit duration test points (s)	0.05, 0.25
Time interval before the next change (s)	0.2
Time interval after the change (s)	0.5
Maximum voltage after returning from short circuit (V)	800
Maximum peak current during short circuit (A)	400
Maximum rms voltage overshoot during short circuit (V)	300
Maximum rms voltage settling time[a] (s)	0.5
Maximum rms voltage steady-state error[b] (V)	100
Maximum frequency steady-state error[b] (Hz)	5

[a] Settling time region is ±5 Vrms of last value in capture
[b] Compared with nominal value, after the short-circuit event

Fig. 9.28 Islanding mode test results for improved droop control: active power in pu during **a** 50 ms short circuit; and **b** 250 ms short circuit

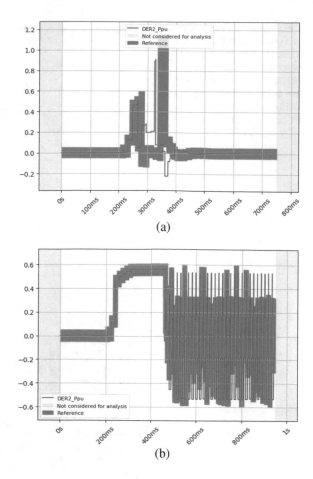

ms short-circuit when the dVOC is used is shown in Fig. 9.29a. It fails on the test because of the overvoltage soon after the short-circuit clearing. Similar behavior is observed for the 250ms short circuit, Fig. 9.29b. Figure 9.29c shows the DERs output currents during the short circuit. They are limited by the DER current inner loop.

The VSM with the reactive power computed at the gap fails in the linear load test since the steady-state frequency error is slightly below the considered 3 Hz threshold when operating at full load, Fig. 9.30a. In addition, it also fails in 250 ms short-circuit test due to the overvoltage soon after the short-circuit clearing, Fig. 9.30b, as well as the active and reactive sharing after the 250 ms short-circuit clearing, Fig. 9.31.

Fig. 9.29 Islanding mode
test results for dispatchable
virtual oscillator control:
voltage at the DER PoC
during **a** 50 ms short circuit;
and **b** 250 ms short circuit; **c**
DERs output currents during
the short circuit

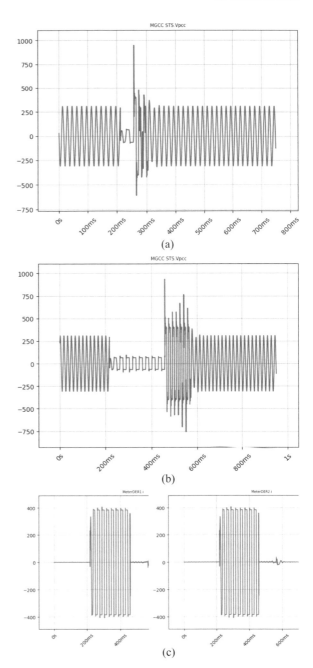

Fig. 9.30 Islanding mode test results for virtual synchronous machine with the reactive power computed at the gap: results for **a** frequency at full linear load test; and **b** voltage at 250 ms short-circuit test

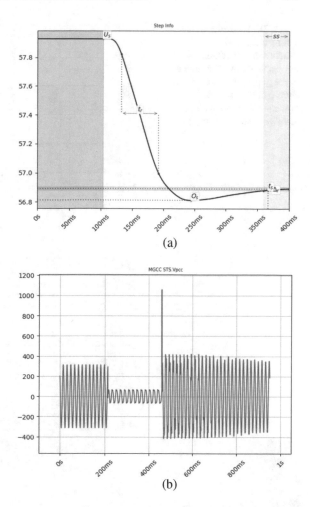

(a)

(b)

Finally, the VSG with the reactive power computed at the DER PoC fails at the full load test, Fig. 9.32a. The frequency is slightly below the considered threshold. In addition, overvoltage, Fig. 9.32b, and power-sharing issues, Fig. 9.33, are also observed for this VSG.

Fig. 9.31 Islanding mode test results for virtual synchronous machine with the reactive power computed at the gap: results for 250 ms short circuit **a** active power-sharing (pu); and **b** reactive power-sharing (pu)

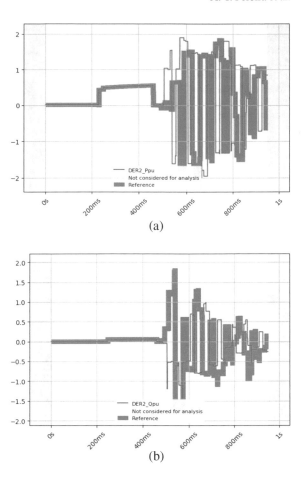

(a)

(b)

9.10 Conclusion

Methodologies for the main primary control techniques for grid-forming inverter parameterization were presented. The test-driven design provided a holistic way to evaluate the performance of primary controllers for microgrid and pointed out the strong and weak points of each alternative. The Droop Control is not only simply implemented and presents excellent behavior during short circuit but it also presents large reactive power dispatch errors when connected to the grid. The 2I Droop Control reduces the Droop Control steady-state error at the price of penalizing the behavior after the short circuit. The dVOC demonstrated to be a strong candidate for the primary controller, just requiring attention for short-circuit cases in islanding mode. The VSM not only makes it possible to select the droop characteristic and moment of inertia as in conventional synchronous generators but also presents a critical point to be addressed in the power-sharing after the short-circuit clearing.

Fig. 9.32 Islanding mode test results for virtual synchronous machine with the reactive power computed at the PoC: **a** frequency at full linear load test; voltage at **b** 50 ms short circuit; and **c** 250 ms short circuit

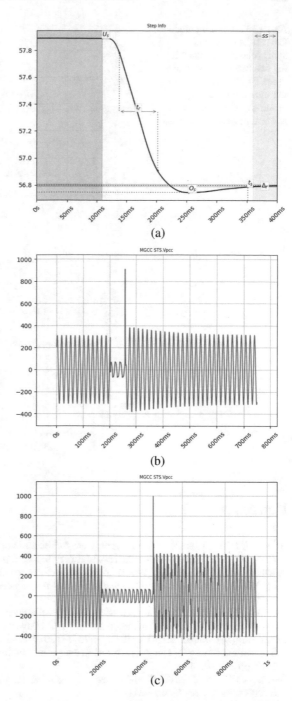

Fig. 9.33 Islanding mode test results for virtual synchronous machine with the reactive power computed at the PoC: results for 250 ms short circuit **a** active power-sharing (pu); and **b** reactive power-sharing (pu)

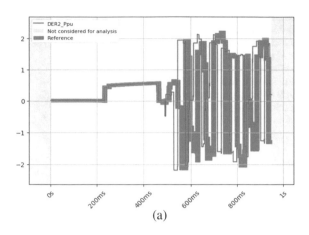

(a)

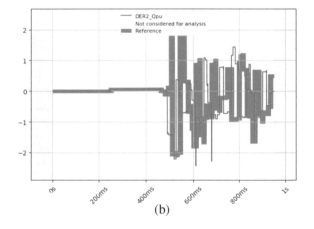

(b)

References

Alipoor J, Miura Y, Ise T (2015) Power system stabilization using virtual synchronous generator with alternating moment of inertia. IEEE J Emerg Select Top Power Electron 3(2):451–458

Ashabani M, Mohamed YAI (2014) Novel comprehensive control framework for incorporating VSCS to smart power grids using bidirectional synchronous-vsc. IEEE Trans Power Syst 29(2):943–957

Awal MA, Yu H, Tu H, Lukic SM, Husain I (2020) Hierarchical control for virtual oscillator based grid-connected and islanded microgrids. IEEE Trans Power Electron 35(1):988–1001

Beheshtaein S, Golestan S, Cuzner R, Guerrero JM (2019) A new adaptive virtual impedance based fault current limiter for converters. In: IEEE energy conversion congress and exposition (ECCE) 2019, pp 2439–2444

Bevrani YMH, Ise T (2014) Virtual synchronous generators: a survey and new perspectives. Int J Electr Power Energy Syst 54:244–254

Chandorkar MC, Divan DM, Hu Y, Banerjee B (1994) Novel architectures and control for distributed ups systems. In: Proceedings of 1994 IEEE applied power electronics conference and exposition - ASPEC'94, 1994, vol 2, pp 683–689

Coelho EA, Wu D, Guerrero JM, Vasquez JC, Dragicevic T, Stefanovic C, Popovski P (2016) Small-signal analysis of the microgrid secondary control considering a communication time delay. IEEE Trans Indust Electron 63(10):6257–6269. Bevrani and Ise

Colombino M, Groß D, Dörfler F (2017) Global phase and voltage synchronization for power inverters: a decentralized consensus-inspired approach. In: 2017 IEEE 56th annual conference on decision and control (CDC), 2017, pp 5690–5695

D'Arco S, Suul JA (2014) Equivalence of virtual synchronous machines and frequency-droops for converter-based microgrids. IEEE Trans Smart Grid 5(1):394–395

Denis G, Prevost T, Debry M, Xavier F, Guillaud X, Menze A (2018) The migrate project: the challenges of operating a transmission grid with only inverter-based generation. a grid-forming control improvement with transient current-limiting control. IET Renew Power Gener 12(5):523–529

Dhople SV, Johnson BB, Hamadeh AO (2013) Virtual oscillator control for voltage source inverters. In: 2013 51st annual allerton conference on communication, control, and computing (Allerton), 2013, pp 1359–1363

Dong S, Chen C (2018) Adjusting synchronverter dynamic response speed via damping correction loop. In: IEEE power energy society general meeting (PESGM) 2018, p 1

Fucci D, Erdogmus H, Turhan B, Oivo M, Juristo N (2017) A dissection of the test-driven development process: does it really matter to test-first or to test-last? IEEE Trans Softw Eng 43(7):597–614

Guerrero JM, de Vicuna LG, Matas J, Castilla M, Miret J (2005) Output impedance design of parallel-connected ups inverters with wireless load-sharing control. IEEE Trans Indust Electron 52(4):1126–1135

Guerrero JM, de Vicuna LG, Matas J, Miret J, Castilla M (2004) Output impedance design of parallel-connected ups inverters. In: 2004 IEEE international symposium on industrial electronics, vol 2, pp 1123–1128

Guerrero JM, Matas J, De Vicunagarcia Garcia, De Vicuna L, Castilla M, Miret J (2006) Wireless-control strategy for parallel operation of distributed-generation inverters. IEEE Trans Indust Electron 53(5):1461–1470

Guerrero JM, de Vicuna LG, Matas J, Castilla M, Miret J (2004) A wireless controller to enhance dynamic performance of parallel inverters in distributed generation systems. IEEE Trans Power Electron 19(5):1205–1213

Guerrero JM, Vasquez JC, Matas J, de Vicuna LG, Castilla M (2011) Hierarchical control of droop-controlled ac and dc microgrids-a general approach toward standardization. IEEE Trans Industr Electron 58(1):158–172

Janzen D, Saiedian H (2008) Does test-driven development really improve software design quality? IEEE Softw 25(2):77–84

Janzen D, Saiedian H (2005) Test-driven development concepts, taxonomy, and future direction. Computer 38(9):43–50

Jeffries R, Melnik G (2007) Guest editors' introduction: Tdd-the art of fearless programming. IEEE Softw 24(3):24–30

Johnson B, Rodriguez M, Sinha M, Dhople S (2017) Comparison of virtual oscillator and droop control. In: 2017 IEEE 18th workshop on control and modeling for power electronics (COMPEL), 2017, pp 1–6

Johnson BB, Dhople SV, Hamadeh AO, Krein PT (2014) Synchronization of parallel single-phase inverters with virtual oscillator control. IEEE Trans Power Electron 29(11):6124–6138

Johnson BB, Dhople SV, Hamadeh AO, Krein PT (2014) Synchronization of nonlinear oscillators in an LTI electrical power network. IEEE Trans Circuits Syst I Regul Pap 61(3):834–844

Johnson BB, Sinha M, Ainsworth NG, Dörfler F, Dhople SV (2016) Synthesizing virtual oscillators to control islanded inverters. IEEE Trans Power Electron 31(8):6002–6015

Kawabata T, Higashino S (1988) Parallel operation of voltage source inverters. IEEE Trans Ind Appl 24(2):281–287

Kumar S, Bansal S (2013) Comparative study of test driven development with traditional techniques. Int J Soft Comput Eng 3(1):352–360

Liu J, Miura Y, Ise T (2016) Comparison of dynamic characteristics between virtual synchronous generator and droop control in inverter-based distributed generators. IEEE Trans Power Electron 31(5):3600–3611

Long B, Liao Y, Chong KT, Rodríguez J, Guerrero JM (2021) Mpc-controlled virtual synchronous generator to enhance frequency and voltage dynamic performance in islanded microgrids. IEEE Trans Smart Grid 12(2):953–964

Matevosyan J, Badrzadeh B, Prevost T, Quitmann E, Ramasubramanian D, Urdal H, Achilles S, MacDowell J, Huang SH, Vital V, O'Sullivan J, Quint R (2019) Grid-forming inverters: are they the key for high renewable penetration? IEEE Power Energy Mag 17(6):89–98

Olivares DE, Mehrizi-Sani A, Etemadi AH, Cañizares CA, Iravani R, Kazerani M, Hajimiragha AH, Gomis-Bellmunt O, Saeedifard M, Palma-Behnke R, Jiménez-Estévez GA, Hatziargyriou ND (2014) Trends in microgrid control. IEEE Trans Smart Grid 5(4):1905–1919

Sakimoto TIK, Miura Y (2011) Stabilization of a power system with a distributed generator by a virtual synchronous generator function. IEEE 8th international conference power electronics and ECCE Asia, pp 1498–1505

Shintai T, Miura Y, Ise T (2014) Oscillation damping of a distributed generator using a virtual synchronous generator. IEEE Trans Power Delivery 29(2):668–676

Sinha M, Dörfler F, Johnson BB, Dhople SV (2017) Uncovering droop control laws embedded within the nonlinear dynamics of van der pol oscillators. IEEE Trans Control Netw Syst 4(2):347–358

Sinha M, Dörfler F, Johnson BB, Dhople SV (2015) Virtual oscillator control subsumes droop control. In: American control conference (ACC) 2015, pp 2353–2358

Stallmann F, Mertens A (2020) Sequence impedance modeling of the matching control and comparison with virtual synchronous generator. In: 2020 IEEE 11th international symposium on power electronics for distributed generation systems (PEDG), 2020, pp 421–428

Tuladhar A, Jin H, Unger T, Mauch K (1997) Parallel operation of single phase inverter modules with no control interconnections. In: Proceedings of APEC 97 - applied power electronics conference, vol 1, pp 94–100

Williams L, Maximilien E, Vouk M (2003) Test-driven development as a defect-reduction practice. In: 14th international symposium on software reliability engineering, 2003. ISSRE 2003, pp 34–45

Zhong Q-C, Hornik T (2013) Control of power inverters in renewable energy and smart grid integration. Wiley

Zhong Q, Weiss G (2009) Static synchronous generators for distributed generation and renewable energy. In: 2009 IEEE/PES power systems conference and exposition, March 2009, pp 1–6

Zhong Q, Zeng Y (2011) Can the output impedance of an inverter be designed capacitive? In: IECON 2011 - 37th annual conference of the IEEE industrial electronics society, Nov 2011, pp 1220–1225

Zhong Q, Weiss G (2011) Synchronverters: Inverters that mimic synchronous generators. IEEE Trans Industr Electron 58(4):1259–1267

Zhong Q, Nguyen P, Ma Z, Sheng W (2014) Self-synchronized synchronverters: inverters without a dedicated synchronization unit. IEEE Trans Power Electron 29(2):617–630

Zhong Q, Konstantopoulos GC, Ren B, Krstic M (2018) Improved synchronverters with bounded frequency and voltage for smart grid integration. IEEE Trans Smart Grid 9(2):786–796

Chapter 10
Effect of Diverse Harmonic Order Frequencies on Dynamic PV Hosting Capacity Assessment in Active Distribution Network: A Typhoon HIL Based Approach

Swati Kumari, Sourav Kumar Sahu, and Debomita Ghosh

Abstract Increase in the per capita energy consumption across the globe is show-casing the use of a wide range of electrical devices, machineries and appliances in domestic as well as in industries. Major portions of these loads are either dynamic or non-linear. These dynamic and non-linear loads often cause distortion in the sinu-soidal signals leading to the generation of unwanted harmonics. These harmonics cause various operational as well as long-term permanent damage to the equipment connected to the distribution system such as variation in load profile, increased line losses, overheating of machines and devices, false tripping of the protection system and generation loss. To decrease the gap between generation and demand, integra-tion of new renewable to the existing grid is ineluctable. To use these renewable in the existing grid, power system parameters need to be verified if they are within the specified utility standards. Although the maximum value of the renewable inte-grations is estimated during integration, the presence of unexpected non-sinusoidal signals diminishes such estimated values. In this work, the estimation of the effect of individual harmonics on line parameters and subsequently its effect on the hosting capacity (HC) of the network is analyzed extensively. In this context authors used Typhoon HIL real-time simulation platform for modeling the reconfigured IEEE-33 bus distribution system and accurate and cost effective analysis for this study. Additionally, ETAP is used for calculating THD and MATLAB is used for scripting of load flow techniques and measurement of current THD using MATLAB's FFT analysis tool on Simulink to support the claim of the real-time results.

Keywords Active distribution network · Distribution load flow · Harmonics · Harmonic load flow · Harmonic voltage · Hosting capacity · Real-time simulation · Total harmonic distortion

S. Kumari · S. K. Sahu · D. Ghosh (✉)
Birla Institute of Technology, Mesra, Jharkhand, India
e-mail: debomita.ghosh@bitmesra.ac.in

© The Author(s), under exclusive license to Springer Nature Singapore Pte Ltd. 2023   255
S. M. Tripathi and F. M. Gonzalez-Longatt (eds.), *Real-Time Simulation and Hardware-in-the-Loop Testing Using Typhoon HIL*, Transactions on Computer Systems and Networks, https://doi.org/10.1007/978-981-99-0224-8_10

10.1 Introduction

With the growing power demand and depletion of fossil fuels, the conventional power system needs up-gradation to meet the demand. One of the ways to meet the demand is to use renewable sources in the existing grid (Bollen and Hassan 2022). Among various distributed energy resources (DERs), photovoltaic distributed generation (PVDG) is widely used due to various technical, economical as well as environmental benefits (Tsikalakis and Hatziargyriou 2007). To balance the demand, DER integration is considered one of the solutions but the amount of DER that can be added needs to be calculated; hence the concept of hosting capacity (HC) comes into picture. The amount of DER penetration into the power grid that may be permitted without compromising the system's reliability or power quality is determined using the network's HC (Chathurangi et al. 2018). PV HC is governed by various operational limiting factors which include bus voltage limits, ampacity, harmonic limits, variation in load parameter, feeder length, R/X ratio, Thevenin impedance, and distribution network topology (Chathurangi et al. 2018; Sahu and Ghosh 2020; Jothibasu et al. 2019; Alturki et al. 2018; Santos et al. 2015). The voltage limitations are imposed to reduce the losses in the system and prevent faulty operation of the protection devices whereas the ampacity reduces the risk of overheating of conductors, and insulation failures (Sahu et al. 2020). Numerous researches have been done to maximize the PV HC which focuses on the several approaches for the estimation and maximization of HC including active distribution network (ADN) management, a stochastic approach using Monte Carlo simulation and optimization-based calculations. The optimization-based simulation focuses on the deployment of PV with optimal size at optimal locations (Alturki et al. 2018). To maximize the PV HC, random allocation of PV is done at various locations keeping fundamental voltage as a constraint at each node in a stochastic approach through Monte Carlo simulation (Jothibasu et al. 2019). However, the ADN can be managed to maximize the PV HC by optimally using control devices and smart inverters (Ding and Mather 2017). By installing new distribution lines with tie-switches, adding switching capacitors, altering voltage regulator taps, managing controllable branch switches, and monitoring smart PV inverters, the PV HC can be optimized (Chathurangi et al. 2018; Ding and Mather 2017; Alturki and Khodaei 2018). The optimal reconfiguration will also lead to a better voltage profile and reduced losses. Therefore, for this work, a reconfigured IEEE-33 bus system is considered (Rajaram et al. 2015; Wazir and Arbab 2016). To assess the effect of diverse harmonic order frequency on network parameters and HC, various literatures were studied which are reviewed in the next section.

10.2 Literature Survey

To analyze the effect of diverse harmonic order frequency on HC assessment in a distribution network, various literatures reviewed indicate that the loads in the distribution system are mainly non-linear loads which include power electronic equipment like rectifiers that draws current in pulses which introduce harmonics in the distribution system. With such harmonics present in the network, various losses, premature aging, insulation failure, communication interference, and equipment damage are some of the obvious impacts (Abdelrahman and Milanović 2019; Abbas and Saqib n.d.). The harmonics injected into the system will lead to voltage and current distortion. These distortions should abide by the IEEE standard i.e., IEEE 519–2014. This standard provides criteria for the current and voltage distortions i.e., THD (Blooming et al. 2007; IEEE Recommended Practice 2014). With the presence of multiple non-fundamental frequencies in the network, various frequency-dependent power system parameters may deteriorate. The presence of non-fundamental frequency will also lead to a change in the line impedances thereby resulting in the change of bus voltages. Bus voltages being an important aspect of analyzing the HC, the effect of frequency on bus voltages needs to be analyzed in a descriptive manner. Harmonic load flow (HLF) can be used as one of the ways to observe the change in line impedances and node voltages due to non-fundamental frequency orders. HLF is used by researchers to analyze the bus voltage and current injections at different nodes due to different harmonic orders. This method is effective because it considers the net effect of various types of loads present in a distribution network (Teng and Chang 2007). While HLF is used for the harmonic analysis of the network, distribution load flow (DLF) is used to effectively observe the steady state bus voltages of the distribution network. DLF is a much faster and more reliable method compared to the conventional load flow analysis (Teng 2003).

To assess the HC, various available literatures consider a wide range of power system indicators like THD, voltage level (Sahu and Ghosh 2020) and distribution network topology (Jothibasu et al. 2019) using simulation platforms like DIgSILENT PowerFactory (Chathurangi et al. 2018), Typhoon HIL, etc. Typhoon HIL is a real-time simulation platform that allows a user to perform real-time analysis on the plant. It has found its application in various fields including power systems and micro-grid (Sahu and Ghosh 2020, 2021a, b; Ahmad 2021; Bagudai et al. 2019). It is emerging as a powerful tool as it allows testing and validation of results in the real-time environment at a much faster rate and with better accuracy (Bagudai et al. 2019). HIL testing helps in eliminating the need for hardware equipment, thus reducing the cost of performance analysis of the plant. Although, THD has been used previously by researchers as the governing parameter of HC assessment for fundamental frequency using Typhoon HIL (Sahu and Ghosh 2020), the effect of harmonic orders on HC is not analyzed in a descriptive manner. Further, HC assessment considering various harmonic orders is not performed using HLF analysis and Typhoon HIL as the real-time simulator in the existing literatures according to the best knowledge of the authors. HC assessment using Typhoon HIL real-time simulator will allow accurate estimation as it will mimic the real plant that will consider the system harmonics. The integration of various non-linear devices into the network injects higher order of harmonics which thereby results in a change of network parameters which will eventually affect HC. Therefore, it needs to be studied in an elaborate manner.

Below are the contributions of the work to illustrate the effect of diverse harmonic orders of frequency on HC.

i.   A 5-bus distribution network and a reconfigured IEEE-33 bus system are modeled in Typhoon HIL and ETAP software for the analysis of diverse harmonic components on HC.
ii. The currents and voltages at the PCC are extracted from Typhoon HIL, thereafter, the FFT analysis is done on these signals using FFT analysis tool to calculate the THD% and dominant harmonics using MATLAB to ensure current THD abides by the IEEE 519–2014 standard.
iii. HLF is performed on the distribution systems using MATLAB to calculate the changes in the network parameters due to the presence of non-fundamental currents in the system.
iv. DLF is performed considering the dynamic effect of the harmonic current on the varying network parameters to check the operational constraints such as node voltage.
v.   Lastly, HLF is performed to confirm the variation of HC with a change in harmonic contents in the distribution system.

The organization of this chapter is as follows. The effect of power system parameters on HC is highlighted in Sect. 10.3. Section 10.4 gives the methodology for the determination of harmonic impedance. Section 10.5 illustrates the concept of hosting capacity. Section 10.6 describes the real-time simulation platform used to analyze the effect of harmonics on HC. Section 10.7 details other tools used for HC assessment. The case study and outcomes are depicted in Sect. 10.8. Lastly, Sect. 10.9 concludes the findings of the study.

10.3 Brief Overview of the Effect of Power System Parameters on HC

With the increasing introduction of power electronic devices like rectifiers, inverters, static VARs, etc. into the existing power system, the sinusoidal nature of the voltages and currents in the system gets disturbed resulting in harmonic variation in the system. These harmonics hamper the voltage and current profile of the system which eventually leads to deterioration of the power quality of the system. The injection of harmonics also leads to changes in parameters like the node voltage, current injection, line impedances, and power injection into the system. Since the harmonic frequency keeps on changing, it results in the change in line impedance Z_{line} as the reactance of the line X_{line} is dependent on frequency f.

$$Z_{line} = R_{line} + jX_{line} \tag{10.1}$$

where R_{line} is the line resistance and the line reactance $X_{line} = 2\pi fL$ (L is the line inductance). Due to changes in line parameters, the node voltages start to change. Parameters such as node voltage and current injection into the distribution system play a pivotal role in the HC assessment. The distribution side may experience a number of problems as a result of violating the fundamental voltage restrictions, including an uneven voltage profile that may contribute to ohmic losses and erroneous tripping of protective devices. The next decisive factor should be the current since variations in the current can cause equipment failure, device malfunction, and equipment and line thermal breakdown. Hence, voltage and current are the deciding factors for the estimation of HC. Hence these factors need to be maintained according to the IEEE standards.

In this context, IEEE 519 standard ensures the harmonic component content in voltage as well as current in the power system network at the Point of Common Coupling (PCC). As per the IEEE-519 standard, THD is defined as the ratio of the root mean square (rms) of the harmonic content, considering harmonic components up to the 50th order and specifically excluding inter-harmonics, expressed as a percent of the fundamental (IEEE Recommended Practice 2014). This standard limits the total distortions in currents and voltages at the PCC. This work focused on the current distortion limits at the PCC across the network. Current THD (THD_I) is defined as the ratio of the root mean square value of the harmonic currents to the fundamental current.

$$THD_I = \frac{\sqrt{I_2^2 + I_3^2 + \cdots + I_N^2}}{I_1} \tag{10.2}$$

where I_1 is the fundamental rms value of current signal and I_N is the nth rms harmonic component of current signal.

THD plays a significant role in maintaining the distortions under specified limits as per IEEE-519. Therefore, to find the THD of the current signal, the Fast Fourier Transform (FFT) analysis is performed on those signals. The discrete Fourier transform (DFT) of a sequence, or its inverse (IDFT), is calculated by the FFT method (Brigham and Morrow 1967). Fourier analysis converts a signal from its primary domain (often time) to a representation in the frequency domain and vice versa.

The FFT of current signal $i(n)$ can be represented as:

$$I(k) = \sum_{n=0}^{N-1} i(n).w_N^{kn} \tag{10.3}$$

where $k = 0, 1, 2, 3, \ldots, N-1$, $w_N = e^{-\frac{j2\pi}{N}}$ is the twiddle factor.

In this work, FFT analysis is performed using MATLAB Simulink. FFT analysis of the signals will provide the THD% values and the harmonic order of frequency that is dominant. Upon finding the THD and dominant harmonic frequencies, load flow is performed on the distribution system to determine harmonic impedance which is detailed in the next section.

10.4 Methodology for Determination of Harmonic Impedance and Harmonic Node Voltage

The distribution system is rich in harmonics which is contributed by several loads in the system. DER integration can also be considered a source of harmonics in the system. These harmonics have a huge impact on the system resulting in the change of parameters like the line impedance which will eventually change the voltage profile and current injection into the system. Therefore, to study the effect of harmonic order of frequency, DLF and HLF methods are performed on the networks. These methods are further elaborated as follows.

10.4.1 Distribution Load Flow

To find the change in voltages at different buses, load flow analysis is done on the distribution networks. To analyze the fundamental voltage profile, Distribution Load Flow (DLF) analysis is performed. DLF method of analysis is preferred over various other methods because of its lower complexity, quick convergence, and higher computation speed. This method eliminates the tedious work of calculation of Y_{bus} matrix or Jacobian matrix. DLF method only requires the use of line and load parameters (Teng 2003). With the line parameters, two fixed matrices are formed named BIBC and BCBV matrix, and the combination of both the matrices is named as DLF matrix. The calculation of these matrices is detailed as follows:

For a radial distribution system, the relationship between bus injection currents and branch currents is evaluated which uses the concept of Kirchhoff's Current Law (KCL) and backward sweep methodology. The Eqs. (10.4)–(10.15) used for DLF analysis is referred from Teng (2003). For a distribution system with N number of buses and M branch sections, the relationship between bus injection currents I_{node} and branch currents I_{branch} is expressed in Eq. (10.4).

$$[I_{branch}] = [BIBC].[I_{node}] \tag{10.4}$$

where BIBC matrix is bus-injection to branch-current matrix. The dimension of the BIBC matrix is $(M \times (N - 1))$.

The relationship between branch currents I_{branch} and bus voltages are given by Eq. (10.5).

$$[\Delta V] = [BCBV][I_{branch}] \tag{10.5}$$

where BCBV is the branch-current to bus-voltage matrix. The dimensions of the BCBV matrix are $((N - 1) \times M)$.

Therefore from Eqs. (10.4) and (10.5), the voltage drop ΔV at each node can be expressed in Eq. (10.6).

$$[\Delta V] = [BCBV][BIBC][I_{node}] \tag{10.6}$$

$$[\Delta V] = [DLF][I_{node}] \tag{10.7}$$

where DLF is the distribution load flow matrix.

This DLF matrix remains constant for the entire load flow unless the configuration of the network changes, hence the computation time is low.

The algorithm can be summarized as follows:

Step 1: Initialization of bus voltage V_j

$$V_j^{(0)} = V_s \angle 0° \tag{10.8}$$

$$for \ \ j = 2, 3, ..., N$$

where V_s is the nominal grid voltage.

Step 2: Initialization of iteration count i.e., k = 1.

Step 3: Calculation of nodal load current at each node I_j

$$I_j^{(k)} = \left(\frac{P_j + jQ_j}{V_j^{(k-1)}} \right)^* \tag{10.9}$$

$$for \ \ j = 2, 3, ..., N$$

$$I_{node} = \begin{bmatrix} I_2^{(k)} \\ I_3^{(k)} \\ \vdots \\ I_N^{(k)} \end{bmatrix} \tag{10.10}$$

For nodes where PVDG is connected, the net apparent power S_{net} at the node in Eq. (10.9) can be considered as

$$S_{net} = P_j + jQ_j = (P_{load} - P_{PV}) + jQ_{load} \tag{10.11}$$

where apparent power of the load is $S_{load} = P_{load} + jQ_{load}$ and the real power of PVDG is P_{PV}.

Step 4: Compute $[\Delta V]$

$$\left[\Delta V^{(k)}\right] = [DLF][I_{node}^{(k)}]$$ (10.12)

where $[\Delta V] = \begin{bmatrix} \Delta V_2^{(k)} \\ \Delta V_3^{(k)} \\ \vdots \\ \Delta V_N^{(k)} \end{bmatrix} = \begin{bmatrix} V_s - V_2 \\ V_s - V_3 \\ \vdots \\ V_s - V_N \end{bmatrix}$

Step 5: Update node voltage

$$V_j^{(k)} = V_s - \Delta V_j^{(k)}$$ (10.13)

$$for \ j = 2, 3, ..., N$$

Step 6: Calculate the error

$$e_j^{(k)} = \left| V_j^{(k)} - V_j^{(k-1)} \right|$$ (10.14)

$$for \ j = 2, 3, ..., N$$

Step 7: Compute maximum error

$$e_{max}^{(k)} = max(e_2^{(k)}, e_3^{(k)}, ..., e_N^{(k)})$$ (10.15)

Step 8: If $e_{max}^{(k)} \le \varepsilon(tolerance)$, then compute node voltages else update the iteration as $k = k + 1$ and go to Step 3.

The above steps will calculate the node voltages with a distribution network with DG at steady state. The following section explains the harmonic load flow technique which helps in observing the harmonic node voltages at different order of frequencies.

10.4.2 Harmonic Load Flow

To analyze a wide range of harmonics, in this work Backward/Forward Sweep based Harmonic Analysis method is used. This is a powerful method to perform load flow analysis at different harmonics in the distribution system. It computes the harmonic load flow without the cumbersome work of calculating of bus admittance matrix at different harmonics. The generalized harmonic load flow considers the net effect of harmonic currents contributed by linear loads, non-linear loads, and shunt capacitors as shown in Teng and Chang (2007).

However, for the concerned networks of this book chapter, there is no inclusion of shunt capacitors hence the matrices concerning them in Teng's paper (Teng and Chang 2007) can be omitted. Therefore the simplified HLF based on the forward/backward sweep method used here can be represented as follows:

The system harmonic currents $[I^h]$ can be expressed as in Eq. (10.16):

$$[I^{h,k}] = [Ih^{h,k}]^T \tag{10.16}$$

where $[Ih^h]$ is the matrix associated with net harmonic currents contributed by different loads connected. The superscript k denotes the order of iteration.

Further, the backward sweep method is used to find the coefficient matrix. The coefficient matrix for the branch between bus i and j is defined as given in Eq. (10.17).

$$\left[A_{ij}^{h,k}\right] = [Ah_{ij}^{h,k}]^T \tag{10.17}$$

where $[Ah_{ij}^h]$ is the coefficient matrix associated with harmonic currents contributed by different loads connected.

Hence the branch currents due to harmonic currents can be represented by Eq. (10.18).

$$\left[B_{ij}^{h,k}\right] = [A_{ij}^{h,k}]^T [I^{h,k}] \tag{10.18}$$

The branch voltage drop due to system harmonics can be represented by the expression in Eq. (10.19).

$$\Delta V_{ij}^{h,k} = Z_{ij}^h [A_{ij}^{h,k}]^T [I^{h,k}] \tag{10.19}$$

where $Z_{ij}^h = R_{ij} + j(X_{ij} * h)$ is the branch impedance between buses i and j at hth harmonic order.

Now to calculate the bus voltages, the forward voltage sweep method is used. Therefore, the bus voltage at hth harmonic order can be expressed by Eq. (10.20).

$$[V^{h,k}] = [HA^{h,k}][I^{h,k}] \tag{10.20}$$

where $[HA^h]$ represents the relationship matrix between bus voltage matrix and current harmonic matrix.

Equation (10.20) will calculate the harmonic bus voltages at a different order of frequencies.

The voltage variations due to harmonic injection can then be observed in different harmonic orders. Since voltage variation plays a major role in the power injection into the system, variation in voltage can cause a change in the PV hosting capacity of the entire system as explained in Sect. 10.5.

10.5 Concept of Hosting Capacity

To meet the growing power demand, over the years, there has been an increase in PV usage in the distribution system. Although usages of distributed energy resources (DERs) like solar photovoltaic (SPV) is widely adopted, with harmonic rich power systems, power injection gets limited into the system. With the usage of DERs, there is a bi-directional power flow in the system which complicates the conventional power flow. Thus, the PV penetration into the radial distribution system may result in deteriorating the power system parameters beyond the specified utility standards. This deviation eventually raises the concern of the quality maintenance, hence the PV integration needs to be done according to the network capacity, i.e., according to hosting capacity (HC). Hosting capacity is defined as the maximum amount of DER that can be connected to the existing distribution network without violating the voltage and current limits (Sahu and Ghosh 2020).

Different types of DER integration have different effects on the hosting capacity (Sahu et al. 2020). PV hosting capacity is also dependent on several factors which include the system's Thevenin impedance, network characteristics, size and location of solar PV, and voltage control regulating devices. HC is also dependent on the length of the feeder and the conductor type used. To allow additional PV integration into the existing system optimally adding new distribution lines with tie-switches, adding switching capacitors, regulating voltage regulator taps, monitoring controllable branch switches, reactive power control, and controlling smart PV inverters could be some of the various ways in which PV HC can be maximized (Chathurangi et al. 2018; Sahu and Ghosh 2020; Ding and Mather 2017; Alturki and Khodaei 2018). SPV hosting is also determined by critical factors such as the maximum feeder voltage and excessive line loading (Jothibasu et al. 2019).

The real power injection at the PCC can be estimated using Eq. (10.21) (Sahu and Ghosh 2020) as follows:

$$P = \frac{V_{pcc}(V_{pcc} - V)}{(R_{th} + X_{th}tan\varphi)} + P_{load} \tag{10.21}$$

where V_{pcc} is the maximum voltage at PCC, V is the grid voltage, R_{th} is the Thevenin resistance of the line, X_{th} is the Thevenin reactance of the line, P_{load} is the active power of the load and φ denotes the power factor angle of the load.

Equation (10.21) suggests a relation between the power injected into the system, node voltage, and Thevenin impedance. Therefore, it can be inferred that voltage and current play an important role in the estimation of HC. In this work, the authors

Fig. 10.1 Proposed strategy for harmonic-based HC assessment

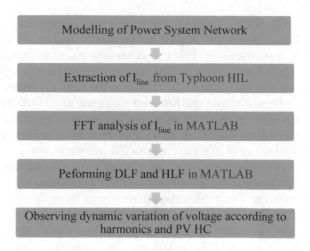

Fig. 10.1 Proposed strategy for harmonic-based HC assessment

focused on the prior mentioned parameters and the effect of harmonic order on those parameters. Also, due to the various types of load integration across nodes of the network, there is a certainty of a wide range of harmonics injection into the distribution system. Additionally, the variation in load can't be ignored which leads to changes in the HC at different nodes of the network. Hence to further analyze the effect of harmonics on the indicators of HC, multiple high fidelity real-time as well as offline simulation software are used namely Typhoon HIL, ETAP, and MATLAB. Typhoon HIL has been used by several researchers to analyze HC for real-time simulation (Sahu and Ghosh 2020, 2021a, b), but the application of Typhoon HIL for HC considering harmonic impedance as an indicator is not been reported in any literature so far according to the best knowledge of the authors. A detailed explanation of the simulation platforms is explained in the next section.

The flowchart shown in Fig. 10.1, summarizes the workflow to achieve the objectives of this chapter.

10.6 Real-Time Simulation Platform to Analyze the Effect of Harmonics on HC

Real-time simulation has gained its importance over the years. It allows users to mimic the actual complex model onto a platform on the computer and execute it at the same rate as an actual system. It enables users to efficiently study various controllers and networks in an actual environment at a reduced cost, reduced risk of damaging the hardware, and quick testing and simulation of the networks. Therefore, the effect of system's diverse harmonics on the estimation of HC can be analyzed using a real-time simulation platform, the Typhoon HIL real-time simulator which enables the authors to perform accurate estimation of the effect of harmonics.

Typhoon HIL is a software and hardware package used for real-time simulation which is not limited to micro-grids and distribution systems. The hardware and software both being co-developed enable the researchers to save time on the integration of the two. The real-time simulator whose primary processor is FPGA and general-purpose CPUs is used for slow dynamics. Typhoon is enriched with different libraries and firmware configurations for diverse applications. It is known for its high fidelity and speed and high switching frequency with the latest 4th generation devices. It can allow a high switching frequency and very low simulation time step as low as 500 ns. Typhoon HIL software allows users to work at different step times according to the type of simulation, alternatively, if the user doesn't have a fair idea about the time step the internal process selects an optimal time step. This software allows researchers/engineers to test their networks at a reduced cost. With the expanse of the HIL series and its libraries, the testing process is simplified and saved time for engineers working on micro-grids, distribution networks, and solar PV inverters. Typhoon HIL real-time emulators help in testing and validating the result with real-time simulation. The results from the emulator are very efficient.

The advantages of Typhoon HIL can be listed as follows:

(1) High fidelity
(2) Very small simulation time step
(3) Ultra-low latency Analog IO
(4) Ultra-fast Digital IO.

Typhoon HIL finds application in various fields which are not limited to:

(1) Power Systems
(2) Power electronics and drives
(3) Micro-grid and Smart grids
(4) Power Quality Analysis
(5) HVDC and FACTS
(6) Power system protection
(7) Phasor Measurement Unit(PMU) and Synchrophasor study, etc.

Typhoon HIL offers various device settings such as HIL 402, 404 (4th generation), 602+, 603, 604, 606 (4th generation), and VHIL+. Along with that, it has several configurations which vary in memory, machine support, etc.

Below steps are followed to create and simulate a model in Typhoon HIL real-time simulator:

Step 1: Choosing Hardware settings
The hardware device is chosen along with the configuration according to the number of SPCs and converter weight needed. Typhoon HIL can also detect the hardware if an emulator is connected to the plant. The simulation method and time step can also be set in this window.

Different simulators offer different hardware configurations. One such real-time simulator is HIL 604 whose configuration is shown in Fig. 10.2.

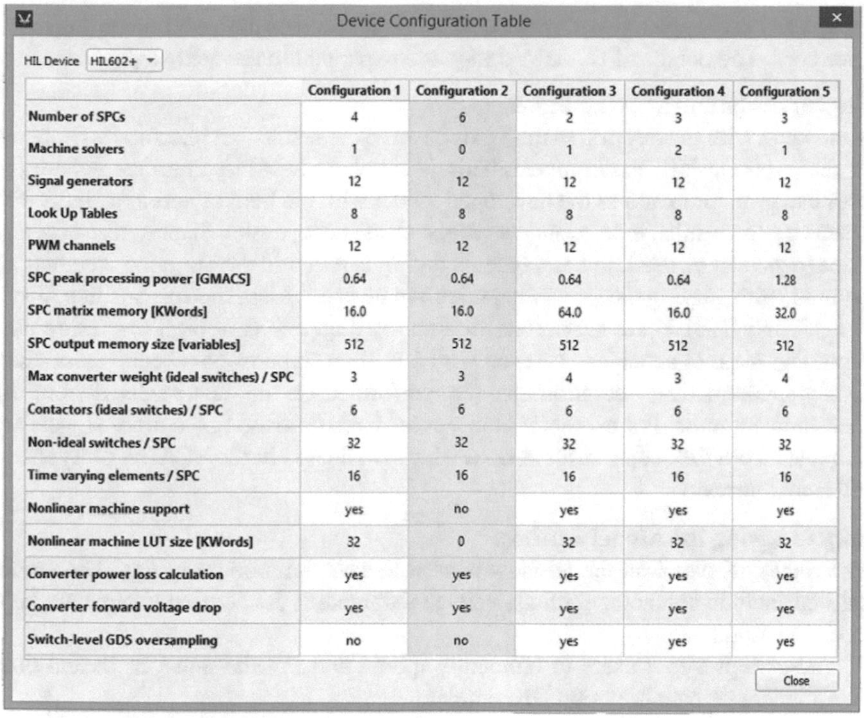

	Configuration 1	Configuration 2	Configuration 3	Configuration 4	Configuration 5
Number of SPCs	4	6	2	3	3
Machine solvers	1	0	1	2	1
Signal generators	12	12	12	12	12
Look Up Tables	8	8	8	8	8
PWM channels	12	12	12	12	12
SPC peak processing power [GMACS]	0.64	0.64	0.64	0.64	1.28
SPC matrix memory [KWords]	16.0	16.0	64.0	16.0	32.0
SPC output memory size [variables]	512	512	512	512	512
Max converter weight (ideal switches) / SPC	3	3	4	3	4
Contactors (ideal switches) / SPC	6	6	6	6	6
Non-ideal switches / SPC	32	32	32	32	32
Time varying elements / SPC	16	16	16	16	16
Nonlinear machine support	yes	no	yes	yes	yes
Nonlinear machine LUT size [KWords]	32	0	32	32	32
Converter power loss calculation	yes	yes	yes	yes	yes
Converter forward voltage drop	yes	yes	yes	yes	yes
Switch-level GDS oversampling	no	no	yes	yes	yes

Device Configuration Table

HIL Device HIL602+

Close

Fig. 10.2 Hardware configuration selection

Step 2: Creating Model
The leftmost icon is used to create a model. Different libraries can be used to create the model effectively. A large system can be separated using the core coupling from the library based on the weights of the component.

Step 3: Compiling Model
The rightmost icon which is shown in Fig. 10.3 is used to compile the model. Compilation ensures that the matrix and probes are utilized properly depending on the hardware configuration used.

Step 4: Loading of the Model
The compiled model is loaded onto virtual HIL(V-HIL) or the real Typhoon HIL device (if connected to the PC). In Typhoon HIL real-time platform to observe the

Schematic Editor

File   Edit   View   Model   Windows   Help

Fig. 10.3 Toolbar to create and compile the model

results and to change any parameter a SCADA needs to be designed according to the user need. The details of SCADA design are explained in the next step.

Step 5: Designing SCADA Panel

To interact with the developed model or observe the results, an interface is designed in the Typhoon HIL environment. This is called as SCADA interface, where the user can give command to the developed model with the help of action weights. For observing the results, monitoring weights such as scope, digital display, test-box, and trace graph can be used. In case, data recording is required data logging weights can be used, with varying data sampling rates can be used. Also, file formats like '.csv', '. mat', and 'HDF5' are supported for data logging. For very high-resolution data recording internal scope can be used with 1 million samples per second resolution. Lastly, mathematical operation can also be formed on the data received from the measurement units. For example: in this considered case the line current signals are extracted from the scope in.csv format which is situated in the SCADA of Typhoon HIL environment.

Step 6: Saving the Model Settings

The rightmost panel on the window is used to save the model settings. The model settings include contactor settings, voltage parameters, etc. The contactors are kept in closed mode.

These steps allow users to efficiently model and run the network in real-time environment using a Typhoon HIL emulator.

The next section briefly explains the use of other tools required for the analysis of distribution networks in presence of harmonics.

10.7   Other Tools Used for the Analysis of Distribution Networks

10.7.1   Electrical Transient Analyzer Program (ETAP)

This software is used for electrical network modeling and simulation software tools used by power systems engineers to analyze electrical power system dynamics, transients, and protection. This is industry-standard software that is very useful for performing complex analyzes on electric power systems. It can be efficiently used for load flow analysis and harmonic analysis on large systems.

In this chapter, this software will be used for the estimation of line current THD% in different distribution networks, using the harmonic analysis module in this software.

10.7.2 MATLAB

This software is a mathematical programming platform that allows matrix operations and helps in analyzing data by creating Simulink models. It offers various tools and libraries to perform simulations. In this chapter, this software is used for performing harmonic load flow on the distribution systems with the help of MATLAB code.

10.8 Modeling and Analysis of Diverse Harmonic Order Frequency on ADN

For analysis of attaining the desired objective, the case study is done on two models: 5 bus system and reconfigured IEEE 33 bus system. Both the distribution systems are modeled and simulated on Typhoon HIL and ETAP software. To test the effect of diverse harmonic order on hosting capacity, SPV loads are connected at different nodes in the distribution system.

10.8.1 5-Bus System

In the 5 bus ADN, PVDG are considered to be connected to bus 2. It is assumed that the substation is not responsible for any harmonic injection into the system. The input line and load data for the 5 bus systems used are given in Table 10.1 (Fig. 10.4).

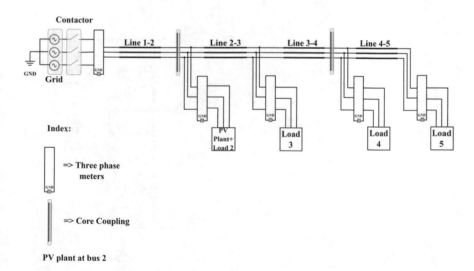

Fig. 10.4 Modeling of 5-bus distribution system on Typhoon HIL real-time platform

Table 10.1 Line and load data of 5-bus distribution system

Sl. No	From bus	To bus	Line data		Load data	
			R(Ω)	X(Ω)	P(kW)	Q(kVAR)
1	1	2	0.0922	0.0470	100	60
2	2	3	0.4930	0.2512	90	40
3	3	4	0.3661	0.1864	120	80
4	4	5	0.3811	0.1941	60	30

Further, the same distribution network is modeled in ETAP software as shown in Fig. 10.5.

The distribution system was modeled and simulated on Typhoon HIL. The Capture/Scope widget in HIL SCADA is used to extract the currents of different lines. The signals are extracted in '.csv' format. Thereafter, the FFT analysis is performed on the signals extracted from Typhoon HIL, in MATLAB which gives the THD% values of the current signals. Also, the same network is modeled in ETAP, and the current THD is calculated using the Harmonic Analysis module. The current THD% values evaluated on both the software is tabulated in Table 10.2 as follows.

These THD% values in Table 10.2 are well within the IEEE 519 standard as all the values are below 5% for the current distortion limit. The difference in THD

Fig. 10.5 Modeling of a 5-bus distribution system in ETAP

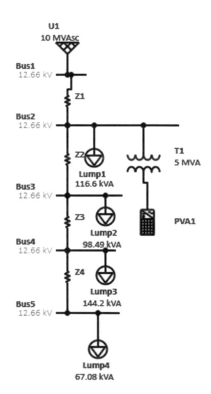

Table 10.2 Current THD% values on ETAP and Typhoon HIL of 5-bus distribution network	Line	THD%	
		ETAP	Typhoon HIL
	1–2	4.78	2
	2–3	2.82	2
	3–4	2.76	1.8
	4–5	2.88	1.6

calculation from both the software is due to the fact that ETAP is an industry-based software that tends to provide a safe limit of THD value. Hence, the values in ETAP are slightly higher as compared to that of Typhoon HIL. Typhoon provides a real-time simulation environment making it instantaneously available for further analysis. Rigorous simulation data in both the platforms i.e., Typhoon HIL and ETAP also confirms that for the same condition, the THD values in case of ETAP are slightly greater than the Typhoon HIL real-time values. These phenomenon can also be observed in Tables 10.2 and 10.5.

It is found from the FFT analysis that the odd order of harmonics upto 11th order is dominant. Further to analyze the effect of individual harmonic orders on the system, harmonic analysis using the forward/backward based method is performed using MATLAB scripting. To perform HLF, the node current signal harmonic components are fed into MATLAB scripting for accurate analysis. From the harmonics analysis, it is witnessed that the line reactance increases with the harmonic order, and is tabulated in Table 10.3. The individual harmonic voltages corresponding to various harmonic orders are depicted in Fig. 10.6.

It can be observed from Fig. 10.6 that if we consider individual harmonic order, there will be a change in voltage at all buses. It can also be observed that the dominant harmonic frequencies contribute to higher node voltages compared to the non-dominant harmonic order. Therefore, it can be concluded that the change in voltage will result in a change in the power flow of the system. This will result in a variation in the PV hosting capacity of the network.

10.8.2   Reconfigured IEEE-33 Bus System

A reconfigured IEEE-33 bus system is used for the analysis purpose in this chapter. Various literatures (Rajaram et al. 2015; Wazir and Arbab 2016), suggest this network over the non-reconfigured network due to the following advantages:

i.   Reduced losses
ii. Improved voltage profile
iii. Bus voltage magnitudes are under the acceptable limits.

Table 10.3 Harmonic line impedance for different harmonics of 5-bus system

Sl No	From bus	To bus	R(Ω)	Fundamental X(Ω)	3rd harmonic	5th harmonic	7th harmonic	9th harmonic	11th harmonic
1.	1	2	0.0922	0.0470	0.1410	0.2350	0.3290	0.4230	0.517
2.	2	3	0.4930	0.2512	0.7536	1.2560	1.7584	2.2608	2.7632
3.	3	4	0.3661	0.1864	0.5592	0.9320	1.3048	1.6776	2.0504
4.	4	5	0.3810	0.1941	0.5823	0.9705	1.3587	1.7469	2.1351

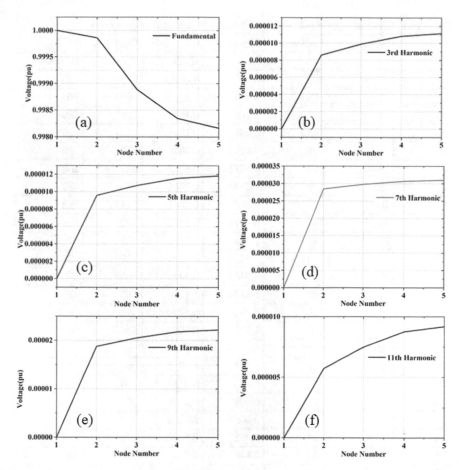

Fig. 10.6 Harmonic node voltages of the 5-bus distribution network **a** Fundamental frequency, **b** 3rd harmonic order, **c** 5th harmonic order, **d** 7th harmonic order, **e** 9th harmonic order, and **f** 11th harmonic order

Therefore, the reconfigured IEEE-33 bus ADN is modeled considering PVDG connected at buses 2, 7, 10, 14, 25, 32, and 33 (Sahu and Ghosh 2021). The modeling is done using the line and load data given in Table 10.4.

Considering the tabulated parameters as in Table 10.4, a distribution system model in Typhoon HIL real-time platform is built for further analysis as shown in Fig. 10.7. Multiple DER integration can be seen at various nodes. Additionally, the total distribution circuit is partitioned for effective core utilization to minimize the sampling time for the simulation.

With the identical values that are mentioned in Table 10.4, a distribution system is modeled in the ETAP environment, as in Fig. 10.8.

The reconfigured IEEE-33 bus distribution system was modeled and simulated on Typhoon HIL to extract the currents of different lines and nodes using the

Table 10.4 Line and load data of reconfigured IEEE-33 bus distribution system

Sl. No	From bus	To bus	Line data		Load data	
			R(Ω)	X(Ω)	P(kW)	Q(kVAR)
1.	1	2	0.0922	0.0470	100	60
2.	2	3	0.4930	0.2512	90	40
3.	3	4	0.3661	0.1864	120	80
4.	4	5	0.3811	0.1941	60	30
5.	5	6	0.8190	0.7070	60	20
6.	6	7	0.1872	0.6188	200	100
7.	8	9	1.0299	0.7400	60	20
8.	10	11	0.1967	0.0651	45	30
9.	11	12	0.3744	0.1298	60	35
10.	12	13	1.4680	1.1549	60	35
11.	13	14	0.5416	0.7129	120	80
12.	15	16	0.7462	0.5449	60	20
13.	16	17	1.2889	1.7210	60	20
14.	17	18	0.7320	0.5739	60	40
15.	18	33	0.5000	0.5000	60	40
16.	2	19	0.1640	0.1565	90	40
17.	19	20	1.5042	1.3555	90	40
18.	20	21	0.4095	0.4784	90	40
19.	21	22	0.7089	0.9373	90	40
20.	12	22	2.0000	2.0000	90	40
21.	3	23	0.4512	0.3084	90	50
22.	23	24	0.8980	0.7091	420	200
23.	24	25	0.8959	0.7071	420	200
24.	6	26	0.2031	0.1034	60	25
25.	26	27	0.2842	0.1447	60	25
26.	27	28	1.0589	0.9338	60	20
27.	28	29	0.8043	0.7006	120	70
28.	29	30	0.5074	0.2585	200	600
29.	30	31	0.9745	0.9629	150	70
30.	31	32	0.3105	0.3619	210	100
31.	8	21	2.0000	2.0000	90	40
32.	9	15	2.0000	2.0000	60	10

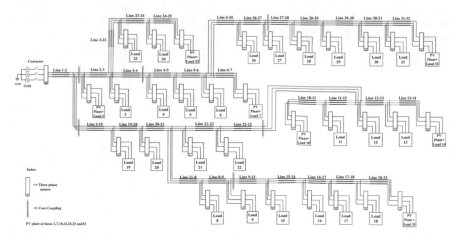

Fig. 10.7 Modeling of reconfigured IEEE-33 bus distribution system in Typhoon HIL real-time platform

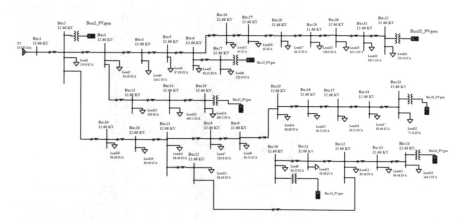

Fig. 10.8 Modeling of reconfigured IEEE-33 bus distribution network in ETAP

Capture/Scope widget in HIL SCADA in '.csv' format. Thereafter, the FFT analysis is performed in MATLAB's FFT Analysis tool on those extracted signals which give the THD% values. The identical network harmonic analysis is performed on ETAP using the Harmonic Analysis module to find the current THD%. The current THD% values evaluated on both the software is tabulated in Table 10.5.

It is seen from Table 10.5 that the current THD% values are well within the IEEE-519 standards of current distortions. It is found from the FFT analysis that the odd harmonic orders are dominant. FFT analysis on node currents are also performed for signals extracted from Typhoon HIL. It helps in providing the harmonic content at the nodes which is further fed into the MATLAB scripting for HLF analysis. Similar steps are considered as previously explained for this reconfigured IEEE-33

Table 10.5 Current THD%
values on Typhoon HIL and
ETAP for the reconfigured
IEEE-33 bus system

Line	THD%	
	ETAP	Typhoon HIL
1–2	0.703	1
2–3	0.64	0.29
2–19	0.73	0.2556
3–4	0.48	0.1848
3–23	1.08	0.1292
4–5	0.48	0.1814
5–6	0.48	0.1802
6–7	1.1	0.129
6–26	0.41	0.058
8–9	0.54	0.0579
8–21	0.51	0.067
9–15	0.55	0.1536
10–11	1.61	0.173
11–12	0.84	0.196
12–13	0.38	0.12
12–22	0.70	0.1696
13–14	0.27	0.104
15–16	0.56	0.1522
16–17	0.57	0.17
17–18	0.61	0.181
18–33	0.78	0.159
19–20	0.76	0.253
20–21	0.79	0.25
21–22	0.91	0.2014
23–24	1.14	0.1292
24–25	0.58	0.0472
26–27	0.40	0.063
27–28	0.39	0.068
28–29	0.38	0.075
29–30	0.36	0.095
30–31	0.82	0.269
31–32	1.14	0.156

bus system. The variation in line reactance with the change in harmonic order is tabulated in Table 10.6. The individual harmonic voltages for reconfigured IEEE-33 bus system are presented in Fig. 10.9.

It can be observed from Fig. 10.9 that if we consider individual harmonic order, the variation in voltage will be different for all the nodes for fixed-harmonics. This result shows the dominance of harmonics voltage is different for different nodes. The individual harmonic node voltage will contribute to a change in the overall node voltage. The variation in voltage due to diverse harmonic order may result in a change in the power flow of the active distribution system as inferred from Eq. (10.21). This will restrict the PV penetration into the distribution system, which will not be the same as calculated HC in case of the planning stage. The additional harmonics, which are added to the system, may also cause various issues in the distribution system such as losses in the connected device, unwanted resonance, and heating of equipment. Any of the mentioned parameters considered as an indicator for DER integration may prematurely restrict the HC of the network. This condition may also lead to the underutilization of the available infrastructure, showcasing the importance of this research for enabling the distribution system to work on its optimal potential to meet the current and future energy demand.

10.9   Conclusion

Considering the current energy demand, DER integration into the distribution system cannot be avoided. In addition to the DER integration application of various non-linear loads are frequent. With such renewable sources and loads in the active distribution network, a wide range of harmonic content cannot be avoided. In this work effect of harmonics on network parameters are studied and following conclusions can be drawn.

i.   Line current THD was estimated for both the considered networks, according to IEEE-519.
ii.   Individual impedances are estimated corresponding to each dominant frequency of the network.
iii. The HLF analysis is used to estimate voltage for various harmonic orders. The HLF result shows that with an increase in harmonic content in the network, system parameters like the node voltage and line impedance changes. The node voltages of dominant harmonic order frequency show a higher contribution to harmonic voltage.
iv.   These changes in parameters cause the fundamental node voltage to change which further will result in the change of power flow in the network and thereby result in the change of HC.

Table 10.6 Harmonic line impedance for different harmonics of the reconfigured IEEE-33 bus system

Sl. No	From bus	To bus	R(Ω)	Fundamental X(Ω)	3rd harmonic	5th harmonic	7th harmonic	9th harmonic	11th harmonic
1.	1	2	0.0922	0.0470	0.1410	0.2350	0.3290	0.4230	0.5170
2.	2	3	0.4930	0.2512	0.7536	1.2560	1.7584	2.2608	2.7632
3.	3	4	0.3661	0.1864	0.5592	0.9320	1.3048	1.6776	2.0504
4.	4	5	0.3811	0.1941	0.5823	0.9705	1.3587	1.7469	2.1351
5.	5	6	0.8190	0.7070	2.1210	3.5350	4.9490	6.3630	7.7770
6.	6	7	0.1872	0.6188	1.8564	3.0940	4.3316	5.5692	6.8068
7.	8	9	1.0299	0.7400	2.2200	3.7000	5.1800	6.6600	8.1400
8.	10	11	0.1967	0.0651	0.1953	0.3255	0.4557	0.5859	0.7161
9.	11	12	0.3744	0.1298	0.3894	0.649	0.9086	1.1682	1.4278
10.	12	13	1.4680	1.1549	3.4647	5.7745	8.0843	10.3941	12.7039
11.	13	14	0.5416	0.7129	2.1387	3.5645	4.9903	6.4161	7.8419
12.	15	16	0.7462	0.5449	1.6347	2.7245	3.8143	4.9041	5.9939
13.	16	17	1.2889	1.7210	5.1630	8.6050	12.047	15.489	18.931
14.	17	18	0.7320	0.5739	1.7217	2.8695	4.0173	5.1651	6.3129
15.	18	33	0.5000	0.5000	1.5000	2.5000	3.5000	4.5000	5.5000
16.	2	19	0.1640	0.1565	0.4695	0.7825	1.0955	1.4085	1.7215
17.	19	20	1.5042	1.3555	4.0665	6.7775	9.4885	12.1995	14.9105
18.	20	21	0.4095	0.4784	1.4352	2.3920	3.3488	4.3056	5.2624
19.	21	22	0.7089	0.9373	2.8119	4.6865	6.5611	8.4357	10.3103
20.	12	22	2.0000	2.0000	6.0000	10.000	14.000	18.000	22.000

(continued)

Table 10.6 (continued)

Sl. No	From bus	To bus	R(Ω)	Fundamental X(Ω)	3rd harmonic	5th harmonic	7th harmonic	9th harmonic	11th harmonic
1.	1	2	0.0922	0.0470	0.1410	0.2350	0.3290	0.4230	0.5170
21.	3	23	0.4512	0.3084	0.9252	1.5420	2.1588	2.7756	3.3924
22.	23	24	0.8980	0.7091	2.1273	3.5455	4.9637	6.3816	7.8001
23.	24	25	0.8959	0.7071	2.1213	3.5355	0.9497	6.3639	7.7781
24.	6	26	0.2031	0.1034	0.3102	0.5170	0.7238	0.9306	1.1374
25.	26	27	0.2842	0.1447	0.4341	0.7235	1.0129	1.3023	1.5917
26.	27	28	1.0589	0.9338	2.8014	4.6690	6.5366	8.4042	10.272
27.	28	29	0.8043	0.7006	2.1018	3.5030	4.9042	6.3054	7.7066
28.	29	30	0.5074	0.2585	0.7755	1.2925	1.8095	2.3265	2.8435
29.	30	31	0.9745	0.9629	2.8887	4.8145	6.7403	8.6661	10.592
30.	31	32	0.3105	0.3619	1.0857	1.8095	2.5333	3.2571	3.9809
31.	8	21	2.0000	2.0000	6.0000	10.000	14.000	18.000	22.000
32.	9	15	2.0000	2.0000	6.0000	10.000	14.000	18.000	22.000

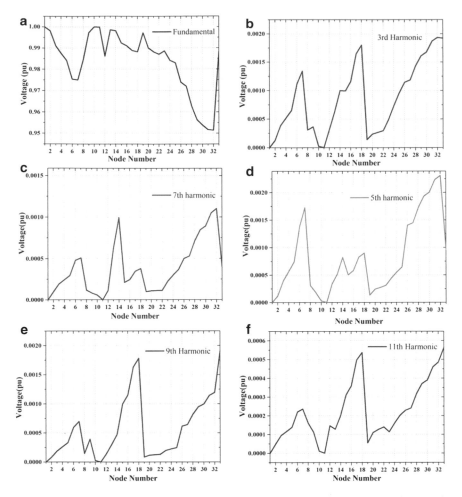

Fig. 10.9 Harmonic node voltages of reconfigured IEEE-33 bus distribution network **a** Fundamental frequency, **b** 3rd harmonic order, **c** 5th harmonic order, **d** 7th harmonic order, **e** 9th harmonic order, and **f** 11th harmonic order

In this context, Typhoon HIL real-time environment is utilized to extract real-time data to be used in the MATLAB for calculation of THD and harmonic impedances of lines. Additionally, ETAP is used for the estimation of THD in the network and results are compared with the real-time results.

References

Abdelrahman S, Milanović JV (2019) Practical approaches to assessment of harmonics along radial distribution feeders. IEEE Trans Power Deliv 34(3):1184–1192, https://doi.org/10.1109/TPWRD.2019.2901245

Abbas W, Saqib MA, Effect of nonlinear load distributions on total harmonic distortion in a power system

Ahmad J et al (2021) Performance analysis and hardware-in-the-loop (HIL) validation of single switch high voltage gain DC-DC converters for MPP tracking in solar PV system. IEEE Access 9, https://doi.org/10.1109/ACCESS.2020.3034310

Alturki M, Khodaei A, Paaso A, Bahramirad S (2018) Optimization-based distribution grid hosting capacity calculations. Appl Energy 219:350–360, https://doi.org/10.1016/J.APENERGY.2017.10.127

Alturki M, Khodaei A (2019) Increasing distribution grid hosting capacity through optimal network reconfiguration. 2018 North Am Power Symp NAPS 2018, https://doi.org/10.1109/NAPS.2018.8600645

Bagudai SK, Ray O, Samantaray SR (2019) Evaluation of control strategies within hybrid DC/AC microgrids using typhoon HIL, https://doi.org/10.1109/ICPS48983.2019.9067331

Bollen MHJ, Hassan F (2022) Integration of distributed generation in the power system. Google Books. https://books.google.co.in/books?hl=en&lr=&id=1KKN82OUXMYC&oi=fnd&pg=PA1&ots=kjdDUl3lC4&sig=_PVrM4S3HGDetCnBvW61u0H-zuA&redir_esc=y#v=onepage&q&f=false. Accessed 17 Jan 2022

Blooming T, Daniel PE, Carnovale J, Ieee PE (2007) "Application of IEEE Std 519–1992 Harmonic Limits

Brigham EO, Morrow RE (1967) The fast Fourier transform. IEEE Spectr 4(12):63–70. https://doi.org/10.1109/MSPEC.1967.5217220

Chathurangi D, Jayatunga U, Lanka S, Perera S, Agalgaonkar A (2018) Connection of solar PV to LV networks: considerations for maximum penetration level; connection of solar PV to LV networks: considerations for maximum penetration level

Ding F, Mather B (2017) On distributed PV hosting capacity estimation, sensitivity study, and improvement. IEEE Trans Sustain Energy 8(3):1010–1020, https://doi.org/10.1109/TSTE.2016.2640239

IEEE recommended practice and requirements for harmonic control in electric power systems sponsored by the transmission and distribution committee IEEE power and energy society, https://doi.org/10.1109/IEEESTD.2014.6826459

Jothibasu S, Dubey A, Santoso S (2019) Two-stage distribution circuit design framework for high levels of photovoltaic generation. IEEE Trans Power Syst 34(6):5217–5226, https://doi.org/10.1109/TPWRS.2018.2871640

Rajaram R, Sathish Kumar K, Rajasekar N (2015) Power system reconfiguration in a radial distribution network for reducing losses and to improve voltage profile using modified plant growth simulation algorithm with Distributed Generation (DG). Energy Reports 1:116–122, https://doi.org/10.1016/j.egyr.2015.03.002

Robert A et al (1997) Guide for assessing the network harmonic impedance. IEE Conference Publication, no. 438 pt 1/2, 1997, https://doi.org/10.1049/CP:19970473

Sahu SK, Ghosh D (2020) Hosting capacity enhancement in distribution system in highly trenchant photo-voltaic environment: a hardware in loop approach. IEEE Access 8:14440–14451. https://doi.org/10.1109/ACCESS.2019.2962263

Sahu SK, Ghosh D, Mohanta DK (2020) Effect of harmonics due to distributed energy resources on hosting capacity of microgrid: a hardware in loop-based assessment. Hosting Capacit Smart Power Grids, pp 47–85, https://doi.org/10.1007/978-3-030-40029-3_4

Sahu SK, Ghosh D (2021a) Operational hosting capacity-based sustainable energy management and enhancement. Int J Energy Res. https://doi.org/10.1002/ER.7317

Sahu SK, Ghosh D (2021b) Photovoltaic hosting capacity increment in an unbalanced active distribu-
     tion network. ICPEE 2021—2021 1st international conference power electronics Energy, https://
     doi.org/10.1109/ICPEE50452.2021.9358688
Santos IN, Ćuk V, Almeida PM, Bollen MHJ, Ribeiro PF (2015) Considerations on hosting
     capacity for harmonic distortions on transmission and distribution systems. Electr Power Syst
     Res 119:199–206. https://doi.org/10.1016/j.epsr.2014.09.020
Tsikalakis AG, Hatziargyriou ND (2007) Environmental benefits of distributed generation with
     and without emissions trading. Energy Policy 35(6):3395–3409. https://doi.org/10.1016/j.enpol.
     2006.11.022
Teng JH, Chang CY (2007) Backward/forward sweep-based harmonic analysis method for distribu-
     tion systems. IEEE Trans Power Deliv 22(3):1665–1672. https://doi.org/10.1109/TPWRD.2007.
     899523
Teng JH (2003) A direct approach for distribution system load flow solutions. IEEE Trans Power
     Deliv 18(3):882–887, https://doi.org/10.1109/TPWRD.2003.813818
Wazir A, Arbab N (2016) Analysis and optimization of IEEE 33 bus radial distributed system using
     optimization algorithm

Chapter 11
Non-directional Overcurrent Protection Relay Testing Using Virtual Hardware-in-the-Loop Device

Le Nam Hai Pham, Raju Wagle, Francisco Gonzalez-Longatt, and Martha Nohemi Acosta Montalvo

Abstract Protective relays are an integral part of the power grid, making it reliable and secure against abnormal conditions. *Hardware-in-the-Loop* (HIL) can be employed to test and validate digital protective relay devices in the real-time simulation comprising hardware and software before actual implementation. However, the HIL technique requires a complex and high-cost technical environment consisting of hardware devices, sensors, communication, and simulation platforms. The Virtual HIL (VHIL) Device is a new real-time simulation approach to offer the entire experience and challenges coming from the actual implementation of HIL. It allows the users to approach the possibility of learning the principles and techniques used in HIL without any concern related to using real hardware or physical devices. This chapter aims to illustrate the use of the model-based system engineering toolchains of Typhoon HIL for understanding the implementation methodology of the VHIL technique, from creating a testing model to the process of running. A modified version of the well-known three-phase radial feeder of the European MV distribution benchmark system created by CIGRE Task Force C6.04.02 is modelled and simulated for evaluating non-directional overcurrent protective relay performance through multiple short-circuit fault scenarios. The main contribution of this chapter is to systematically introduce the modelling and simulation for testing purposes of

L. N. H. Pham (✉) · F. Gonzalez-Longatt · M. N. A. Montalvo
University of South-Eastern Norway, Porsgrunn, Norway
e-mail: Le.Pham@usn.no

F. Gonzalez-Longatt
e-mail: fglongatt@fglongatt.org

M. N. A. Montalvo
e-mail: Martha.Acosta@usn.no

R. Wagle
UiT The Arctic University of Norway, Narvik, Norway
e-mail: raju.wagle@uit.no

F. Gonzalez-Longatt
Centre for Smart Grid, University of Exeter, Exeter, UK

the non-directional overcurrent protection relay in VHIL that helps power engineers evaluate the protective relay settings under more realistic conditions.

Keywords Hardware-in-the-loop · Non-directional overcurrent · Real-time simulation · Typhoon HIL · Virtual protection relay

11.1 Introduction

Power system protection plays a vital role in achieving a satisfactory level of reliability and security in power systems. Relaying power system protection is a branch of power system concerned with the principles of design and operation of equipment called protection relays or protective relays (or simply called relays). A protection relay is a device that detects abnormal power system conditions and initiates corrective action as quickly as possible to return the power system to its normal state (Horowitz and Phadke 2014).

One of the functions of a protective relay that is used in distribution networks is overcurrent protection. The protection relays used in such a function could be directional (operating for in-front events) and non-directional (will work for all) depending upon the topology of the distribution system (Aman et al. 2011). The non-directional relays are mostly used for radial distribution feeders, the most common type of worldwide distribution system. Currently, it is used in combination with the directional overcurrent protection relay according to the development of modern networks, including more than one source or multiple lines conveying electrical power in different directions.

Testing protection relay in software and hardware environment is vital in ensuring the relay setting configuration in real-world systems. *Hardware-in-the-Loop* (HIL) is a developing technology used in developing and testing complex real-time systems by providing an effective platform to help validate software systems on specially equipped test benches receiving real input data from physical devices (Review of hardware-in-the-loop 2019). The HIL technology has been used by several authors using the testing platform with different optimisation models (Hubschneider et al. 2018; Edrington et al. 2015; Kelm et al. 2022; Kezunovic et al. 2017; D'Arco et al. 2020).

Testing processes have been developed to be applied for a variety of subsystems and functionalities in the protection systems: SCADA systems (Montaña et al. 2018), communication platforms (Pazdcrin et al. 2018), overcurrent and directional overcurrent relays (Rodriguez et al. 2018), and distance protection systems as well (Celeita et al. 2018). For example, the authors in Makhzani et al. (2017) investigate the behaviour of instantaneous overcurrent protection using an OPAL-RT digital simulator through HIL testing. In Camarillo-Peñaranda et al. (2020), virtual distance protection relay testing was conducted through HIL simulation in PSCAD software. However, the HIL technique requires a complex technical environment consisting of hardware devices, sensors, communication, and simulation platforms. For instance, the authors in Celeita et al. (2018) rely on micro-controllers costing around 5000 USD for a single testing platform (Camarillo-Peñaranda et al. 2020). To allow the users to approach the possibility of learning the principles and techniques used in HIL without any concern related to using real hardware or physical devices, Virtual HIL (VHIL) Device is a new technique to offer all the experience (including challenges) coming from the real implementation of HIL.

This chapter aims to illustrate the use of model-based system engineering toolchains of Typhoon HIL to understand the implementation methodology for evaluating non-directional overcurrent protection relay performance, from creating a testing model to the process of running.

A practical and illustrative example of "How to" create a model and simulate it inside the Typhoon HIL Control Centre environment using the "Virtual HIL Device" is provided. A simple radial distribution is used for illustrative purposes through multiple short-circuit scenarios: three-phase, three-phase-to-ground, single-phase-to-ground, phase-to-phase, and two-phase-to-ground.

This chapter starts by introducing the problem and giving an overview of the overcurrent protection function of the protection relay. The workflow inside Typhoon HIL Control Centre is explained. The model creation starts with the use of HIL Schematic Editor; then, the methods of creating HIL SCADA are introduced to illustrate how to display numerical results and visualise signals. Finally, simulation results through short-circuit fault scenarios are displayed, and a discussion of the results is performed.

11.2 Problem Definition

This section is dedicated to explaining the modelling and simulating of non-directional overcurrent protection relays using a digital real-time simulation framework. Since overcurrent protection is well-known for being used in distribution systems, the authors have decided to use a straightforward radial system as a representative test system. However, all the methods explained in this chapter can be extended to more complicated topologies of distribution systems.

The authors started by considering the well-known three-phase radial feeder of the European MV distribution benchmark system created by CIGRE Task Force C6.04.02 (C.T.F. 2014). However, for the sake of simplicity, the authors decided to use a modified version of the system. In this chapter, an equivalent of the CIGRE European MV distribution benchmark system is used; it is represented by a three-bus as shown in Fig. 11.1, and it is called the "Test System" from here onwards.

The 50 Hz Test System considers two voltage levels, 110 kV representing the connection to the external grid and 20 kV as the nominal voltage of a primary distribution feeder. The Test System consists of three buses, a lumped equivalent of an extensive network called "External Grid" at a nominal voltage of 110 kV (RMS, line-to-line) at bus B0; from there, a step-down two winding power transformer "Trx"

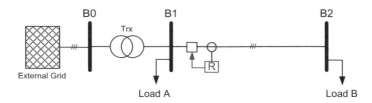

Fig. 11.1 *Test System.* Simplified equivalent model of the three-phase radial feeder of the European MV distribution benchmark system

is used to reduce the voltage to 20 kV. Buses B1 and B2 are connected throughout an overhead transmission distribution line called "Line 12". On buses B1 and B2, two lumped loads named "Load A" and "Load B" are connected to each bus.

A non-directional overcurrent protection relay is connected to the overhead distribution line "Line 12" sending terminal. The protection relay located at the beginning of the overhead line offers feeder protection. In the case of short-circuit fault situations, it must be prevented from spreading to healthy parts of the network to ensure that the power grid continues to supply energy. The main requirements for the feeder protection are as follows:

- During the short circuit, the circuit breaker nearest to the fault should open, and all other circuit breakers remain in a closed position.
- If the circuit breaker nearest to the fault fails to open, then backup protection should be provided by the adjacent protection relay ordering the opening of its circuit breaker.
- The operating time of the protection relay should be small enough to maintain the system stability without the need to trip a circuit.

The classical non-directional overcurrent protection relays use a straightforward protection principle; the protection relay senses the feeder current through the secondary winding of the *current transformer* (CT) and compares it with a predetermined threshold value. The protective relay operates if the sensed current is above the threshold and the trip signal is generated and sent to the tripping circuit of the *circuit breaker* (BK) to isolate the faulty feeder from the rest of the healthy remaining part of the system.

Depending on the technologies (analog or digital), the non-directional overcurrent protective relay may have a specific internal design; Fig. 11.2 shows a simplified block diagram of a non-directional overcurrent relay and its signal flows with the other components of the protection system.

The operating characteristics of non-directional overcurrent protection relays are defined by the operating times typically governed by a time versus current magnitude curve (*I–t*).

Considering the modern protection practice, there are three main types of time–current operating characteristics:

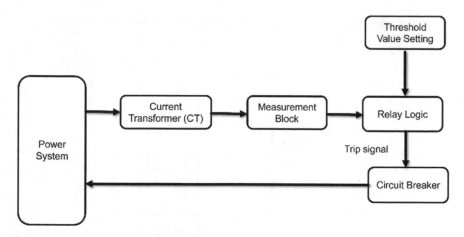

Fig. 11.2   Block diagram of overcurrent protection relay operation

- *Instantaneous.*
- *Time-dependent*: Definite time and inverse time.
- *Mixed*: Combination of the previously stated.

The *instantaneous overcurrent function* is characterised by no time delay intentionally provided. Nowadays, the operating time of the *instantaneous overcurrent function* is less than ten milliseconds and is sometimes expressed in cycles based on the power system frequency (Mehta 2005). The *instantaneous overcurrent function* has only the pickup current setting and does not have any time delay setting. The pickup current is adjustable, and the application engineer can choose various settings from a wide range. The operating characteristics of this type can be shown in Fig. 11.3.

The *time-dependent overcurrent function* characteristics have been standardised in modern times to allow protection engineers a coherent use of the protection characteristics. As the name implies, these overcurrent protection functions operate with an intentional time delay. The minimum current at which the protection relay operates (pickup current) and the delay time before the trip are both adjustable. Two types of time-dependant overcurrent functions are broken into two categories: *Define time* and *inverse time*.

The *define-time* protection relays operate with some intentional time delay and are adjustable along with the current pickup level. *Inverse-time* protection relays have an operating time depending on the magnitude of the current, generally with an inverse characteristic (the operating time of the overcurrent protection relay is smaller as the current gets more significant).

The *inverse-time* operating characteristics have evolved over time, and then standardisation allowed homologating them. Two primary standards are used worldwide,

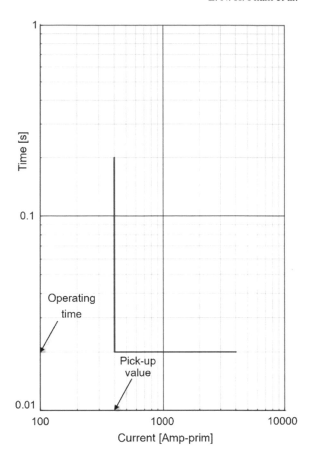

Fig. 11.3 Instantaneous overcurrent protection relay characteristics

the American standard IEEE C37.112–1996 (IEEE Standard for Inverse-Time Characteristics Equations for Overcurrent Relays 2019) and *the European IEC 60,255* (IEC 2009). The inverse-time operating characteristics according to the two standards are shown in Fig. 11.4.

Tables 11.5 and 11.6 show the equations associated with US curves and IEC curves, respectively, where *tp* is operating time in seconds, *TD* is time-dial setting, *TMS* is time multiplier setting, and *M* is applied multiples of pickup current.

The *mixed overcurrent function* consists of the different types of overcurrent protection relay elements packed into a single programmable unit. The combination of instantaneous, define-time, and inverse-time elements may be used in this type of protection relay.

ANSI/IEEE standards are used to define the protection operating characteristics in America, while IEC standards are a common practice in Europe. Therefore, it is typical to use some codes and nodes when practically referring to protection functions according to these standards.

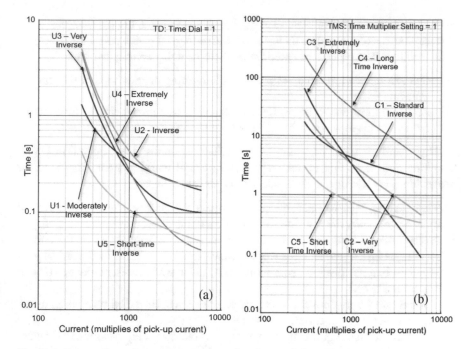

Fig. 11.4 Inverse-time overcurrent protection relay characteristic. **a** *The American standard IEEE C37.112–1996.* **b** *The European IEC 60,255*

The American standards ANSI/IEEE use numerical coding to refer to protection functionalities. The ANSI device number helps identify the features of a protective device according to *"ANSI /IEEE Standard C37.2 Standard for Electrical Power System Device Function Numbers, Acronyms, and Contact Designations"* (IEEE Standard Electrical Power System Device Function Numbers, Acronyms, and Contact Designations 2008).

When referring to overcurrent protection functions, the ANSI device number is 50 for an *instantaneous overcurrent* and *definite-time overcurrent*. In contrast, the *inverse-time overcurrent function* is denoted by the ANSI device number 51. The ANSI/IEEE C37.2 allows the use of suffixes. For instance, the suffix TD should be used (e.g., 50TD) to describe a definite-time overcurrent function.

On the other hand, the European IEC standard defines the protection functions by using logical nodes according to *"International Standard IEC 61,850-7-4 Communication networks and systems in substation"* (Schossig 2010). As per the standard, PIOC is used for an instantaneous overcurrent and PTOC for an inverse-time overcurrent. The numerical suffixes that come with PIOC and PTOC define protection elements. For instance, PIOC1 describes time overcurrent protection for ground overcurrent protection function.

The operating characteristics of the protection overcurrent relay are used to offer protection to a particular situation by predetermined settings. The instantaneous overcurrent element has only one setting, the pickup current. On the other hand, the inverse-time overcurrent element has two basic settings: the pick current and the time delay settings. The process of determining the time delay setting involves.

(1) Calculation of a time-dial or time multiplier setting in definite-time overcurrent elements.
(2) Selection in inverse-time overcurrent elements of a time–current curve from a family of curves as given in Tables 11.5 and 11.6.

These settings must be evaluated for reliability and correct operation before installing and deploying protective relays in a real-world system. The need for protection relay testing can be divided into four categories:

- **Type testing**. It is an extensive process where the quality of a newly fabricated relay or new software revision for a relay model is concerned.
- **Acceptance testing**. The protective relay is tested to prove that it is the correct model and that all the features work as they should. It consists of functional tests of inputs, outputs, displays, communication, and in some cases, predefined pickup and timing tests.
- **Commissioning testing**. It is a site-specific test to confirm all protective elements and logic settings are correct for their intended uses.
- **Maintenance testing**. It is used to ensure a protective relay continues to operate as it should.

At the same time, the protective relay testing methods and equipment become more powerful. However, they require time, effort, and expensive physical equipment to conduct tests on the protective relay.

The authors in this chapter contributed an adequate protection relay testing method that overcomes the limitations of the previous testing methods by using real-time simulation. It allows the testing model to be executed at the same rate as actual physical devices making these tests more realistic under multiple testing scenarios.

For this purpose, the testing model is created that can be used for real-time simulation by using model-based engineering toolchains of Typhoon HIL. Two overcurrent protective functions are focused on, *instantaneous overcurrent* and *inverse-time overcurrent*. The implementation methodology of creating the model and conducting non-directional overcurrent protection relay testing is described in the following sections.

11.3   Workflow in the VHIL

The authors used the modelling and simulation framework for digital real-time simulations created by Typhoon HIL.

Typhoon HIL is the technology and market leader in ultra-high-fidelity hardware-in-the-loop solutions for the design, testing, and validation of power electronics, microgrids, e-Mobility, and electric vehicle (EV) powertrain software testing, and distribution control and protection systems. Typhoon HIL platform ushered a new era of model-based testing and validation of control software and hardware with its embedded, ultra-high-fidelity real-time *Hardware-in-the-Loop* platform.

Virtual HIL or VHIL (as referred to in this chapter) is not an offline simulator. The VHIL device is a virtual machine created to emulate the computer system dedicated to real-time simulation; it brings the functionality of a physical HIL device with all "restrictions" of the real-time environment to a traditional PC.

In this chapter, the authors took advantage of the VHIL to implement and test the proposed *Test System*. The proposed *Test System* is transformed into a model in the VHIL framework that can then be used to emulate the performance of the physical device.

VHIL offers that opportunity by basically using two applications of *Typhoon HIL Control Centre*:

(i) Schematic Editor and
(ii) HIL SCADA.

To access these toolchains, Typhoon HIL Control Centre is started up.

In this book chapter, the authors used the Typhoon HIL Control Centre version 2022 SP1, as shown in Fig. 11.5a. The software initialisation window will appear immediately after double-clicking on the application icon (see Fig. 11.5b).

The initialisation process of the Typhoon HIL Control Centre can take a few seconds depending on the user's PC specification; then, the Typhoon HIL Control Centre envisagement will appear as in Fig. 11.6 and be ready for the users to proceed with the subsequent progress.

The workflow for using the Typhoon HIL modelling and simulation framework is straightforward (of course, more steps can be added depending on the complexity of the modelling and simulation framework).

The basic implementation of workflow in the VHIL can be shown in Fig. 11.7.

Considering the framework presented in Fig. 11.7, the steps are briefly explained below and then expanded during the following subsections:

Step 1: The first step is to model the simplified benchmark network and protective relay using electrical domain devices in the available library in the Typhoon HIL Schematic Editor environment. The measurement devices and signal processing devices are distributed to the model to measure electrical signals in real-time simulation and send the signal to the monitor and control platform.

Step 2: The second step is to create a control and monitor panel that displays the measured signals according to the multiple scenarios and different protection relay settings using HIL SCADA of Typhoon HIL.

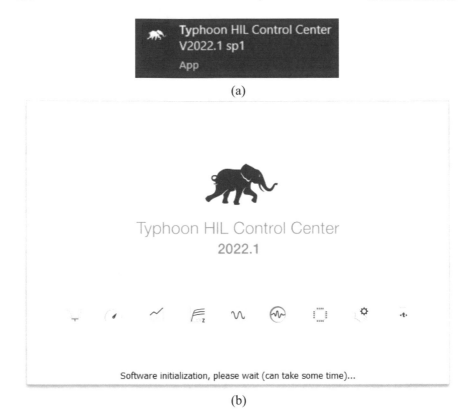

(a)

(b)

Fig. 11.5 **a** Desktop icon of Typhoon HIL Control Centre v2022.1 SP1, **b** Typhoon HIL Control Centre in software initialisation

11.3.1 Schematic Editor Workflow

To start working in *Schematic Editor Environment*, there are two alternatives for creating a new schematic. The users can use the existing access toolbar to create a new model or use the combination key "Ctrl + N" on the PC keyboard. After that, the setting window will appear and require users to define the schematic setting (see Fig. 11.8).

After finishing the setup of schematic settings, the Schematic Editor will load the *Library of Components,* and a new schematic design will be ready to start the building of the model.

To find components, the users can navigate the library tree or use the library explorer until having suitable components for the model. The model can be placed by classically dragging and dropping and/or rotating a component; the users can take advantage of the right bottom context menu when right-clicking on the component.

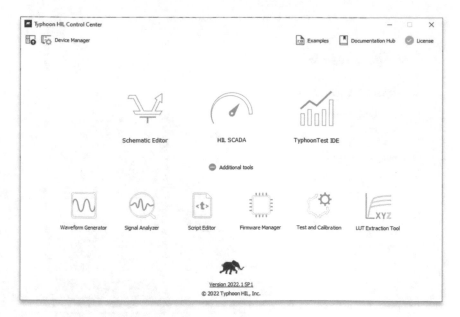

Fig. 11.6 Typhoon HIL control centre interface

Fig. 11.7 VHIL workflow

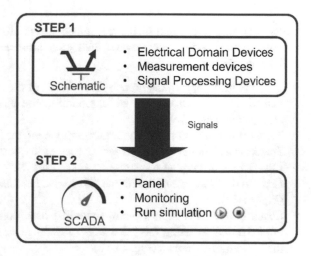

11.3.1.1 Defining the Components in the Test System

The simplified benchmark network used for the *Test System* circuit is created in the Schematic Editor. According to Fig. 11.1, the components of the *Test System* include an external grid, a two-winding power transformer ("Trx"), a non-directional overcurrent protection relay (R), three buses (B0, B1, and B2), a section of transmission line (Line 12), and three lumped three-phase loads (A and B). Additionally, short-circuit

Fig. 11.8 Window showing the setting used for a New Schematic

faults occurring at the end of the transmission line are simulated for the purpose of testing the non-directional overcurrent protection relay. These components are defined as follows:

- External grid (❶): In the branch "*Test Suite*" of the library tree, the component named "*Grid Simulator*" is used for modelling the external grid component of the test system (Fig. 11.9).
- Transformer (❷): For the transformer component, the "*Three-phase Two-Winding Transformer*" in the "*Transformer*" branch of the library is selected (Fig. 11.10).
- Line 12 (❸): For the transmission line "Line 12" component, "*Transmission Line*" in the "*PI section*" part of the "*Transmission Lines*" of the library is selected (Fig. 11.11).
- Loads (❹):Loads A and B are modelled as a constant impedance and it is implemented by using the "*Constant Impedance Load*" in section "*Loads*" of the "*Microgrid*" branch (Fig. 11.12).
- Buses (❺): For bus B0, B1, and B2 components, users can use "*Electrical port*" in the "*Ports*" branch and define a bus with three-port in and three-port out, then merge them under "*Subsystem*" (Fig. 11.13).
- Overcurrent protection relay (❻): The overcurrent protection relay device is different from the aforementioned components. It cannot be represented by individual components in the library but can be emulated by a combination of protective relay functions, current transformer, and measurement devices. This component is defined in a later section (see Sect. 3.1.3) (Fig. 11.14).

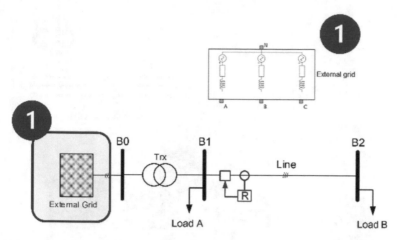

Fig. 11.9 External grid component

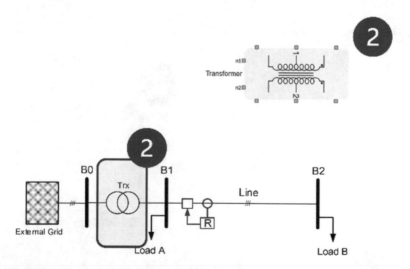

Fig. 11.10 Transformer "Trx" component

- Fault (❼): This element is created to simulate a fault caused by short-circuit fault scenarios under overcurrent protection relay tests. It can be modelled under *"Subsystem"* of *"Electrical port"*, *"Switch"*, and *"C function"* of the library. The authors used switching cases of four *"Switches"*, S1, S2, S3, and S4, corresponding to three phases A, B, and C and grounded neutral N, to simulate nominal operation and short-circuit fault scenarios (three-phase, three-phase-to-ground, single-phase-to-ground, phase-to-phase, and two-phase-to-ground) by programming scripts inside *"C function"* block named *"Control State Machine"*. These switches are labelled as *"FAN1"*, *"FBN1"*, *"FCN1"*, and *"F3P1"* for S1, S2, S3,

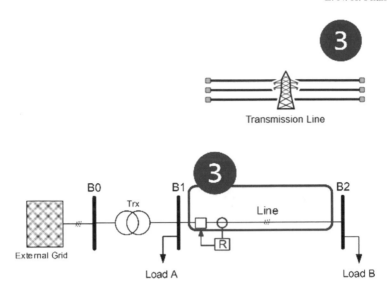

Fig. 11.11 "Line 12" component

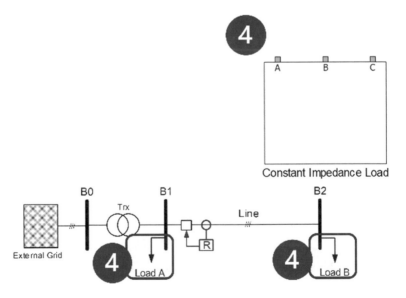

Fig. 11.12 Loads component

and S4, respectively. In the case when one of these switches is closed, a short circuit will occur. For example, four switches, S1, S2, S3, and S4, are closed for the three-phase-to-ground short-circuit case (Fig. 11.15).

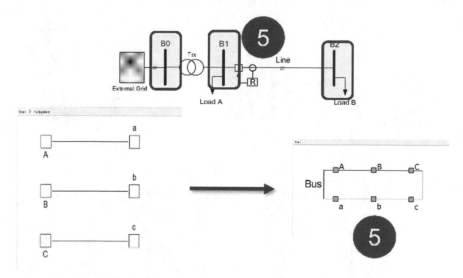

Fig. 11.13 Buses component

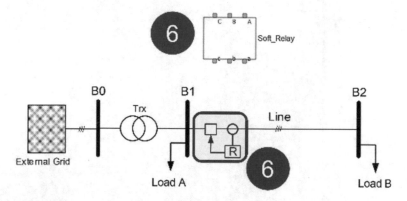

Fig. 11.14 Complete overcurrent protection relay component (R) named "*Soft Relay*"

11.3.1.2 Connecting the Component and Editing Component Data

After placing all components in the schematic environment, the components need to be connected. The users will notice that the line connecting the initial terminal is attached to the cursor and then move to the next terminal, and then perform a right-click at the ending terminal to finish the connection. The complete *Test System* model with the full connection of components can be shown in Fig. 11.16.

To edit the parameters of components, the users use double-click to open a properties window to include the appropriate data. For the external grid (❶) and the "Trx" component (❷), the user can edit the parameters according to Tables 11.1 and 11.2, respectively. To edit the transmission line (❸) and loads (

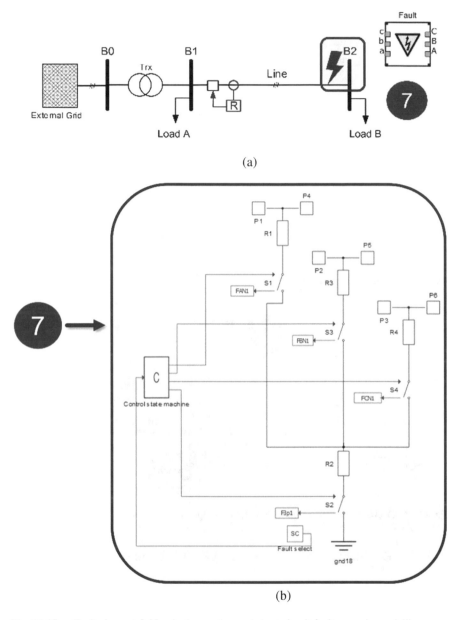

(a)

(b)

Fig. 11.15 a Fault element. **b** Nominal operating and short-circuit fault scenario modelling

Fig. 11.16 Complete *Test System* model with the full connection of components

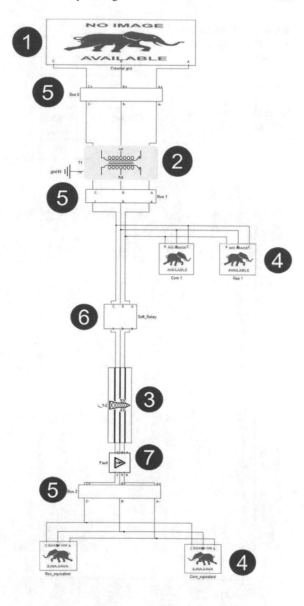

4) component, the "*Model initialisation function*" is used to define these parameters of components by adding them into the namespace in the process of compilation. The users can start the "*Model initialisation function*" by clicking on the icon shown in Fig. 11.17 on the Schematic Editor of Typhoon HIL toolbar. Then, the command panel will appear in a few seconds, depending on the users' PC and be ready for the users to perform parameter definition.

Table 11.1 External Grid equivalent network parameters

Nominal system voltage [kV]	Short-circuit power [MVA]	R/X ratio
110	5000	0.1

Table 11.2 Transformer "Trx" parameters

Connection	Primary voltage [kV]	Secondary voltage [kV]	Rated power [MVA]
3-ph Dyn1	110	20	25

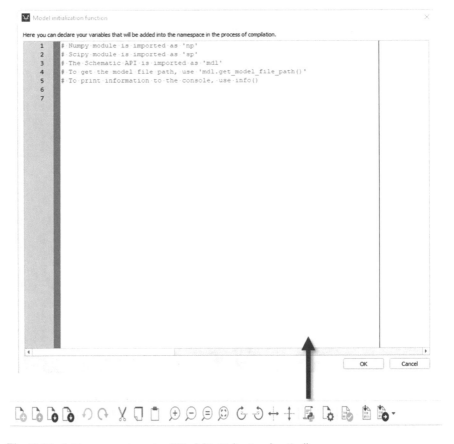

Fig. 11.17 Setting parameter using "*Model initialisation function*"

In the "*Model initialisation function*" environment, the programming language used is Python (What is Python Executive Summary 2022). The users need to apply Python scripts to define the desired parameters for the *Test System*.

For the transmission line, the authors used "*class*", a code used to construct zero and positive sequence impedance per unit length according to Table 11.3, under 3 ×

Table 11.3 Transmission line parameters with R'_{ph}, X'_{ph}, B'_{ph}, R'_0, X'_0, and B'_0 that are positive and zero sequence resistance, reactance, and susceptance values, respectively

R'_{ph} [Ω/km]	X'_{ph} [Ω/km]	B'_{ph} [μS/km]	R'_0 [Ω/km]	X'_0 [Ω/km]	B'_0 [μS/km]	Length [km]
0.501	0.716	47.493	0.817	1.598	47.493	2.82

Table 11.4 Electrical parameters considered at the loads

Load names	Apparent power [MVA]	Power factor	Load current [kA]
Load A (Commercial)	15.30	0.98 (ind)	0.43
Load A (Residential)	5.10	0.95 (ind)	0.14
Load B (Commercial)	3.48	0.97 (ind)	0.09
Load B (Residential)	1.11	0.85 (ind)	0.03

3 matrix form. As a consequence, the parameters set through the properties window will be filled in according to the nameplates of the user-set parameters. For instance, as shown in Fig. 11.18, "*class cable*" consists of common parameters of a typical cable: zero sequence and positive sequence of resistance, reactance, and capacitance; variable "*LT*" is declared as "*class cable*"; the set data for resistance parameter of the transmission line is "*LT.Rabc_matrix*" meaning resistance matrix in "*class cable*" (Table 11.3).

Similarly, for load components (Loads A and B), with the parameters shown in Table 11.4, the set data for nominal three-phase power and power factor is under matrix form, as shown in Fig. 11.19.

The buses (❺), non-directional overcurrent protective relay (❻), and fault (❼) are only connected and are not necessarily edited.

11.3.1.3 Modelling Non-directional Overcurrent Protection Relay

The non-directional overcurrent protection relay component is named "*Soft Relay*" for testing purposes. It consists of four main elements as enumerated and is shown in Fig. 11.20. ❶represents the implementation of the virtual or soft-relay, in this case, it is the combination of available non-directional overcurrent protection models as included in the Typhoon HIL Schematic library. ❷is a measurement device responsible for measuring electrical quantities such as voltage and current signals. ❸is the circuit breaker, and finally, ❹is the connected blocks for outputting the operating time of the protective relay.

The first element (❶) named "*Relay*" is created based on available functionalities of relay SEL-751 included in the Typhoon HIL Schematic library, including the ANSI

```
Transmission Line from library 'core'

Transmission line model implemented as either a PI section or
RL coupled section.
There are three ways to define the model parameters:
   - Geometry: x and y coordinates of the phase cables
   - RLC: per lenght resistance, inductance and capacitance
values of cables in the phase domain
   - Sequence: per lenght resistance, inductance and
capacitance
   values of cables in the sequence domain. Model parameters
can
   be defined this way only for three phase transmission lines

Model:              RL coupled      ▼
Number of phases:   3               ▼
Model definition:   RLC             ▼
Unit system:        metric          ▼
Length:             length[0]          km
R:                  LT.Rabc_matrix     Ω/km
L:                  LT.Labc_matrix     H/km
                    Import from Geometry

Help                          OK          Cancel
```

```python
class cable():
    def __init__(self, R0=0.817, R1=0.501, X0=1.598 , X1=0.716,
                 C0=0.1511749, C1=0.1511749, fn=50, unit='km'):
        self.R0 = R0  # ohms per self.unit [zero sequence resistance]
        self.R1 = R1  # ohms per self.unit [positive sequence resistance]
        self.X0 = X0  # ohms per self.unit [zero sequence reactance]
        self.X1 = X1  # ohms per self.unit [positive sequence reactance]
        self.C0 = C0  # nF per self.unit [zero sequence capacitance]
        self.C1 = C1  # nF per self.unit [positive sequence capacitance]
        self.fn = fn  # Hz [nominal grid frequency]
        self.unit = unit  # [distance unit used]
        self.w = 2*np.pi*self.fn  # rad/s [angular speed]

        self.Rseq_matrix = np.matrix([[self.R0, 0, 0],
                                      [0, self.R1, 0],
                                      [0, 0, self.R1]])  # ohms/unit [sequence resistance matrix]
        self.Xseq_matrix = np.matrix([[self.X0, 0, 0],
                                      [0, self.X1, 0],
                                      [0, 0, self.X1]])  # ohms/unit [sequence reactance matrix]
        self.Zseq_matrix = self.Rseq_matrix + 1j*self.Xseq_matrix  # ohms/unit [sequence impedance matrix]
        self.Lseq_matrix = self.Xseq_matrix/self.w  # H/unit [sequence inductance matrix]
        self.Cseq_matrix = np.matrix([[self.C0, 0, 0],
                                      [0, self.C1, 0],
                                      [0, 0, self.C1]])  # nF/unit [sequence capacitance matrix]

        alphax, alphay = np.cos(-120.0*np.pi/180.0), np.sin(-120.0*np.pi/180.0)
        alpha = complex(alphax, alphay)
        A = np.matrix([[1, 1,          1],
                       [1, alpha**2.0, alpha],
                       [1, alpha,      alpha**2.0]])  # sequence to phase components transf. matrix

        self.Zabc_matrix = np.dot(np.dot(A, self.Zseq_matrix), A.I)  # ohms/unit [phase impedance matrix]
        self.Rabc_matrix = self.Zabc_matrix.real  # ohms/unit [phase resistance matrix]
        self.Xabc_matrix = self.Zabc_matrix.imag  # ohms/unit [phase reactance matrix]
        self.Labc_matrix = self.Xabc_matrix/self.w  # H/unit [phase inductance matrix]
        self.Cabc_matrix = np.dot(np.dot(A, 1j*self.Cseq_matrix), A.I).imag  # nF/unit [phase capacitance matrix]

length = [2.82]
LT= cable(R0=R0, X0=X0, R1=R1,  X1=X1,C0=C0, C1=C1, fn=f, unit='km')
```

Fig. 11.18 Edit "Transmission line" parameters using "*Model initialisation function*"

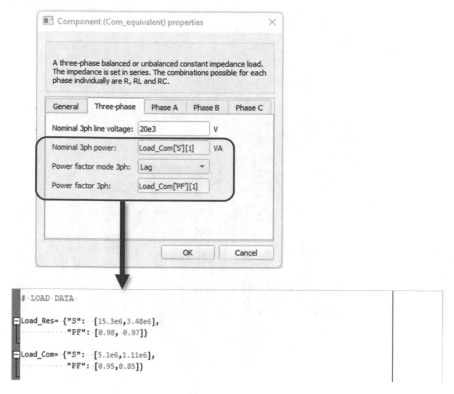

Fig. 11.19 Edit "Loads" parameters using "*Model initialisation function*"

50 and ANSI 51 elements. The inputs of this element are "*Mem_reset*", "Meas_in", "*CT*", and "*Manual_reset*".

These input signals are used by the subsystems inside the "*Relay*" as shown in Fig. 11.21. The input "*Mem_reset*" aims to send the protective functional blocks inside "*Relay*" which will reset the memory of the protective functional blocks. The input "*Meas_in*" is used to receive the measured RMS current from ❷, the input "*CT*" is intended to receive the ratio of current transformer, and the input "*Manual_reset*" is used to receive the trip signal reset of the "*Relay*". These input values can be configured in the HIL SCADA interface.

Figure 11.31 shows the subsystems inside the "*Relay*" that are "*Protective functions*", "*Trip Source combination*", "*Trip Memory reset zone*", and "*Trip Signal zone*". Each of these subsystems is explained as follows:

- *Protective functions*: In this subsystem, there are two main overcurrent protective functional blocks with the specific description and parameter settings that are available in 'ANSI protective functions' (2022), including

  - "*(50) Instantaneous Overcurrent*" (see Fig. 11.22). This block is dedicated to providing instantaneous protection against high currents. This block issues a

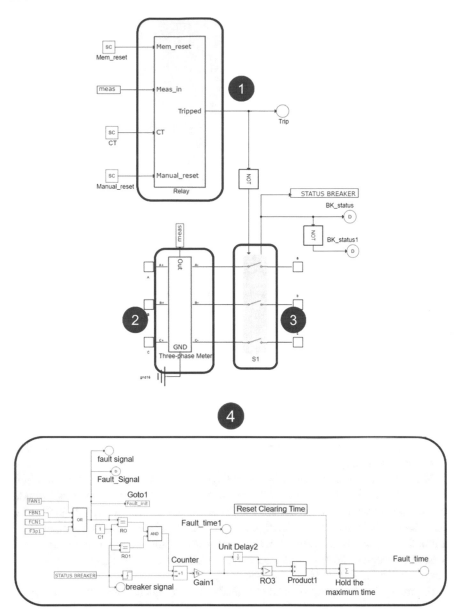

Fig. 11.20 Schematic diagram of the complete overcurrent protection scheme with four main elements, ❶for overcurrent functions, ❷for measurement device, ❸for circuit breaker, ❹and for defining operating time of protection relay

Fig. 11.21 HIL SCADA
inputs to "*Relay*" with their
nameplates

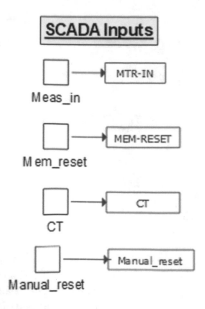

trip signal without any time delay. The inputs of this block are "*Meas_in*",
"*Mem (En/Rst)*", "*CT*", and "*OC_threshold_pu*". Two inputs, "*Meas_in*" and
"*CT*", are connected with the "*Relay*" inputs through previously declared
nameplates, the input "*Mem (En/Rst)*" is intended to reset the trip memory,
and the input "*OC_threshold_pu*" is used to receive the threshold value from
HIL SCADA interface. The output signals are "*trip_inst*", "*trip_A*", "*trip_B*",
"*trip_C*", and "*Mem*". The output "*trip_A*" is used to move the protective relay
dials towards trip position if the measured current crosses the preset threshold
value on phase A. The same purpose goes for "*trip_B*" and "*trip_C*" in phase
B and phase C, respectively. However, these outputs are not used; instead,
using the output "*trip_inst*" sends the trip signals when any currents at any
of the phases cross the defined threshold. The output "*Mem*" is used to send
the feedback of trip memory; when the tripping occurs, this output is active;
otherwise, this output is inactive.

- "*(51) AC Inverse Time Overcurrent*" (see Fig. 11.23). This block is dedicated
  to protecting against high currents based on time-dial and inverse-time overcurrent curves. There are six inputs of this block including "*Mem (En/Rst)*",
  "*Meas_in*", "*TOC_thresh*", "*TOC_curve_type*", "*TOC_timeDial*", and "*CT*".
  Two inputs, "*Meas_in*" and "*CT*", are connected with the "*Relay*" inputs
  through previously declared nameplates; the input "*Mem (En/Rst)*" is used to
  reset the trip memory; the input "*TOC_thresh*" is used to receive the threshold
  value from HIL SCADA interface; the input "*TOC_curve_type*" is used to
  define the inverse-time overcurrent curves corresponding to a family of curves
  in Tables 11.5 and 11.6, and the input "*TOC_timeDial*" is used to receive the

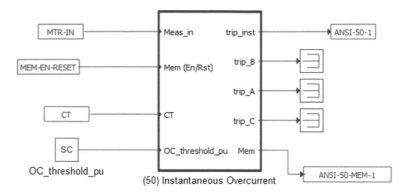

Fig. 11.22 "(50) Instantaneous Overcurrent"

TD or TMS values from HIL SCADA. The output signals are "*dial_reset*", "*time_trip*", and "*Mem*". The output "*dial_reset*" is intended for feedback on the dial states, the output "*time_trip*" is used for an inverse-time overcurrent trip command, and the output "*Mem*" is used to send the feedback of trip memory; when the tripping occurs, this output is active; otherwise, this output is inactive.

- "*Trip Source Combination*" (see Fig. 11.24): the "*OR*" block (out is True if any inputs are True) in this subsystem is dedicated to receiving the output signal "*trip_inst*" from "*(50) Instantaneous Overcurrent*" and the output signal

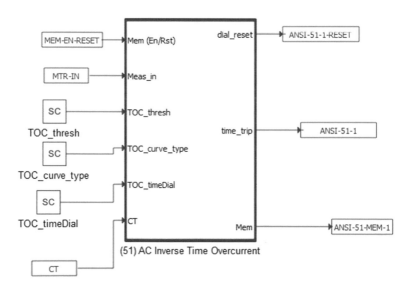

Fig. 11.23 "(51) Instantaneous Overcurrent"

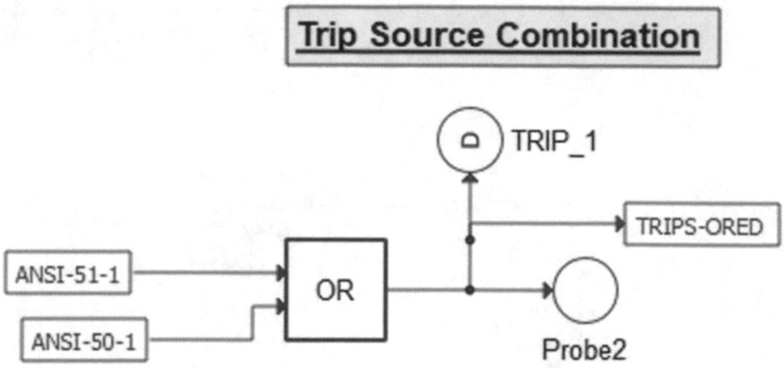

Fig. 11.24 "Trip Source Combination" zone

"*time_trip*" from "*(51) AC Inverse Time Overcurrent*". When any protective functions detect a fault, the trip is issued, which triggers a command to the output of the "*OR*" block that sends the trip signal to the HIL SCADA interface.

- "*Trip-Memory Reset Zone*": In this area, there is a combination of multiple functional blocks for memory, and trip reset of protection functional blocks as follows:

  - "*Trip Memory Combination/Memory Reset Logic*" (see Fig. 11.25). There are two logical blocks in this area, the "*NXOR*" block (out is True if all inputs are True or False) and the "*OR*" block. The "*OR*" block is dedicated to receiving the output "*Mem*" of two protective functional blocks "*(50) Instantaneous Overcurrent*" and "*(51) AC Inverse Time Overcurrent*". The "*NXOR*" block has two inputs, one is the output of the "*OR*" block, and one is "*Mem_reset*", an input of the "*Relay*" that can be configured from the HIL SCADA interface. When one of two protection functional blocks detects a fault, the output signal of "*OR*" is active; and if the "*Mem_reset*" signal is active contemporaneous via the HIL SCADA interface, the output of "*NXOR*" is activated and sends the memory reset signal to the input "*Mem (En/Rst)*" of two protection functional blocks.
  - "*Reset Source Combination*" (see Fig. 11.26). The "*AND*" block (out is True if all inputs are True) in this subsystem has two inputs, one is the output "*dial_reset*" of "*(51) AC Inverse Time Overcurrent*" and one is "*Manual_reset*", an input of the "*Relay*" that can be configured from the HIL SCADA interface. The output of the "*AND*" block is "*Resets-Relay*" that is dedicated to resetting the trip state of protective functional blocks.
  - "*Trip-Reset Logic/Control*" (see Fig. 11.27). This block is dedicated to issuing the trip signal to the HIL SCADA interface. The inputs of this block are "*Trip*", "*Reset*", and "*Min-Time*". The input "*Trip*" is used to receive the trip signal from two protective functional blocks and is the output of "*Trip Source Combination*" the input "*Reset*" is used to receive the trip-reset signal and is the

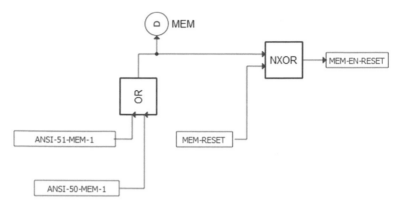

Fig. 11.25 "Trip Memory Combination/Memory Reset Logic" zone

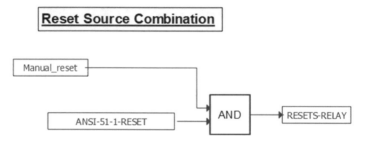

Fig. 11.26 "Reset Source Combination" zone

output of "*Reset Source Combination*", the input "*Min-Time*" is to define the time delay signal for needed backup protection (this input is set at 0 because no backup protection in this section). The output of this block is a trip signal; if the protective relay is active, this output is enabled, otherwise, in case the protective relay is not in the trip position, or there is a trip-reset signal, this output is inactive.

- "*Trip Signal zone*" (see Fig. 11.28): This area consists of two "*SR Flip Flop*" blocks which implement the functionality of the SR Flip Flop sequential logic (Failed 2022). S input can be viewed as a "*Set input*" and R as a "*Reset input*". If S is active, the "*Flip Flop*" will store the value 1. If R is active, the "*Flip Flop*" will store the value 0. A particular case is when both S and R are active. This is an invalid case, and the state of the "*Flip Flop*" will be -1. This area is intended to identify the status of the "*Relay*" with two protective functional blocks, whether it is at the trip position or at the reset state. For the inputs of these blocks, S input is "*trip_inst*" of "*(50) Instantaneous Overcurrent*", "*time_trip*" of "*(51) AC*

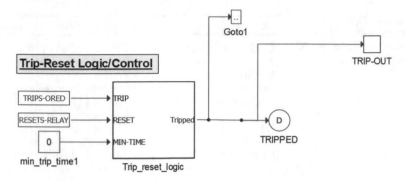

Fig. 11.27 "Trip-Reset Logic/Control" zone

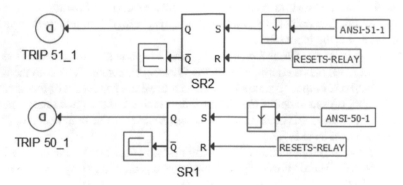

Fig. 11.28 *"Trip Signal"* zone

Inverse Time Overcurrent", and "Resets-Relay" of "Reset Source Combination". The outputs of these "SR Flip Flop" blocks are named "TRIP 51_1" and "TRIP 50_1" to determine which functional block operates.

Table 11.5 Equations associated with US curves following *the American standard IEEE C37.112–1996* (IEEE Standard for Inverse-Time Characteristics Equations for Overcurrent Relays 2019)

Curve type	Relay operating time	Equations
U1 (Moderately Inverse)	$t_p = TD \times \left(0.0226 + \frac{0.0104}{M^{0.02}-1}\right)$	(1)
U2 (Inverse)	$t_p = TD \times \left(0.18 + \frac{5.95}{M^2-1}\right)$	(2)
U3 (Very Inverse)	$t_p = TD \times \left(0.0963 + \frac{3.88}{M^2-1}\right)$	(3)
U4 (Extremely Inverse)	$t_p = TD \times \left(0.0352 + \frac{5.67}{M^2-1}\right)$	(4)
U5 (Short-Time Inverse)	$t_p = TD \times \left(0.0262 + \frac{0.00342}{M^{0.02}-1}\right)$	(5)

Table 11.6 Equations
associated with IEC curves
following *the European IEC
60,255* (IEC 2009)

Curve type	Relay operating time	Equations
C1 (Standard Inverse)	$t_p = TMS \times \left(\frac{0.14}{M^{0.02}-1} \right)$	(6)
C2 (Very Inverse)	$t_p = TMS \times \left(\frac{13.5}{M-1} \right)$	(7)
C3 (Extremely Inverse)	$t_p = TMS \times \left(\frac{80}{M^2-1} \right)$	(8)
C4 (Long-Time Inverse)	$t_p = TMS \times \left(\frac{120}{M-1} \right)$	(9)
C5 (Short-Time Inverse)	$t_p = TMS \times \left(\frac{0.05}{M^{0.04}-1} \right)$	(10)

The second element (❷) is represented by the *"Three-phase meter"* in the Schematic Editor library to measure RMS values of current flowing through the protective relay. The output of this element is named *"meas"*, including RMS values of current that is the input of ❶.

"Triple-Pole Single-Throw Contactor" is used to define the circuit breaker as the third element (❸) that is connected with the *"Three-phase meter"*. This element has one input and one output. The input is used to receive the trip signal of the *"Relay"*, and the output is used to define the status of the circuit breaker in two states, open or closed. The output of ❸is nameplated as *"Status Breaker"*. The connection of ❷and ❸can be depicted as in Fig. 11.29.

Finally, ❹is the combination of logic and mathematical blocks that are available in the Schematic Editor library for calculating the protection relay operating time (see Fig. 11.30).

The inputs are the short-circuit cases tagged with nameplates from the *"Fault"* element of the *Test System* (see Sect. 3.1.1) and the name *"Status Breaker"* from the third element *"Triple Pole Single Throw Contactor"* of *"Soft_Relay"*.

When a short-circuit fault occurs (tagged by the nameplate *"Fault_Signal"*) and the status of the circuit breaker (labelled by the name *"Status Breaker"*) is in the open state, the time counter is enabled. The counter is deactivated when the short-circuit fault is removed; the removal time depends on the output of two overcurrent protection functional blocks, *"trip_inst"* of *"(50) Instantaneous Overcurrent"* and *"time_trip"* of *"(51) AC Inverse Time Overcurrent"*. The protection relay's operating time is maintained when the counter stops working and when short-circuit fault is cleared.

The algorithm for protection relay's operating time is provided to illustrate precisely the construction method of the associative blocks of ❹.

```
Algorithm for the relay operating time calculations
1        Inputs:
             Ts = Execution rate
             SB = Status of Circuit Breaker
             SBC = Circuit Breaker change position state
             en = Encounter
             t = Operating time of relay
             SB close = 1; SB open = 0
```

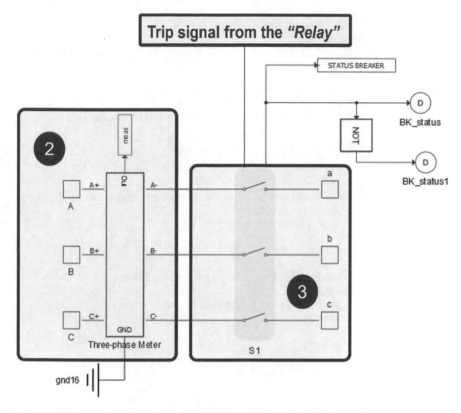

Fig. 11.29 *"Three-phase meter"* connected with *"Triple-Pole Single-Throw Contactor"*

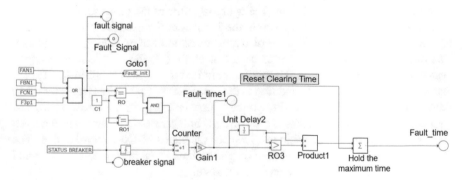

Fig. 11.30 Protection relay's operating time calculation zone

```
             Fault on = 1; Fault off = 0
             Ts = 20e-06
2            if (Fault = 1) and (SB = 1) [Condition to enable the counter]
3                    en = 1;
4            else
5                    en = 0;
             end if
6            if SB (t-1) = SB (t-1) [Identify state change of SB]
7                    SBC = 0;
8            else
9                    SBC = 1;
             end if
10           count = 0;
11           if en = 1 and SBC = 1 [Condition to enable the counter]
12                   count = count + 1;
13                   t_faultON = count * Ts;
14           else
15                   count = 0;
             end
                 Maintain the relay operating time
16           Max = 0;
17           if t_faultON(t-1) > t_faultON(t)
18                   c = 1;
19           else
20                   c = 0;
             end
21           Max = Max + t_faultON(t-1)*c
22           Outputs:
                 The operating time of protection relay
```

The algorithm includes two sections: (a) calculations of the protection relay's operating time and (b) maintaining the operating time of the protection relay.

In the first section of the algorithm, the inputs are the execution rate, the status of the circuit breaker, state change of circuit breaker, state of fault, and encounter. The discrete execution rate is set as default according to the sample time assigned system components of 20 μs. This section includes three logical conditions with "*If*" statements. The first logical condition is the condition to enable the encounter when there is a short-circuit fault, and the circuit breaker is still in the closed state. The short-circuit fault is happening, and the circuit breaker needs to switch to the open state; this is the second condition to identify the status of the circuit breaker. The third condition is to count the timer of operating time when the counter is active and the state of the circuit breaker is in an open state.

The second section of the algorithm includes one logical condition to stop the time counter from maintaining the operating time of the protection relay. If the short-circuit fault is cleared, the counter is deactivated, stopping the timer. The logical condition is the product of counter time and previous time 1.0 s if greater than the product of counter time (is forced to stop when the fault is cleared) and real-time, the counter time or the operating time of protective relay is held that will be displayed in HIL SCADA interface (Fig. 11.31).

Fig. 11.31 The structure of "*Relay*" element in non-directional overcurrent protection relay component

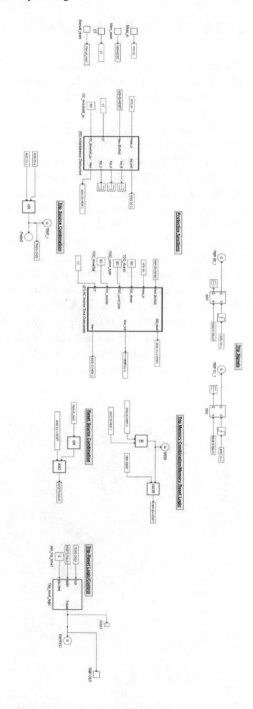

11.3.2   SCADA HIL Workflow

The model of the *Test System* is built using the Typhoon HIL schematic, and then the model is compiled to translate the computer model created in Schematic Editor into a programming language code that can be used in the over elements of the VHIL toolchain.

The users use two buttons in the toolbar to compile the model, "*Compile schematic*" (Compile the current schematic model) and "*Compile and load model in HIL SCADA*" (Compile the current schematic model and load it into HIL SCADA). The alternative way is using the Schematic Editor menu "*Model*" and selecting the option "*Compile schematic*" or "*Compile and load model in HIL SCADA*" as needed.

The HIL SCADA interface can be depicted in Fig. 11.32. The users need to select the option "*Load Model in the Virtual Device*" to load created model in the schematic to HIL SCADA. This option emulated the real HIL 40x/60 × device, but no external I/O to real devices is possible. Then, the HIL SCADA editor is ready to use, and the graphical environment is prepared to create a specific interface with the real-time model. To create a new panel, the users need to press the "*Create new Panel*" at the command toolbar. The blank panel appears and is ready for the users to place available widgets from the library, as shown in Fig. 11.33.

An efficient monitoring dashboard requires users to customise the widget for the panel depending on the arrangement and the quantities that users want to display and control. The panel includes two main functions: *result displaying* and *adjusting the desired values*.

Fig. 11.32 Loading the created schematic model into the virtual HIL device

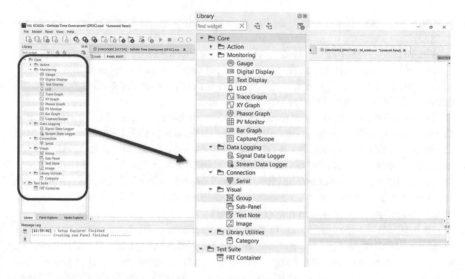

Fig. 11.33 Working panel in HIL SCADA with available widgets library

In this chapter, the complete monitoring and controlling panel was created using HIL SCADA, and its implementation is depicted in Fig. 11.34.

The widgets in the *"Monitoring"* and *"Action"* core of the widget library are intended to create the monitoring and controlling panel. These widgets are adjusted by performing a double-click on the widget to open the properties window. In the window settings, the users can do navigation commands that require widgets to

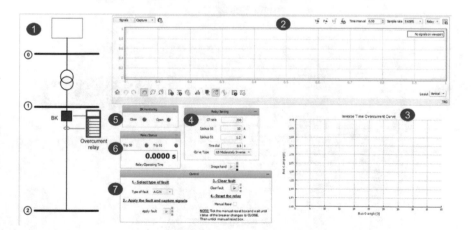

Fig. 11.34 Complete monitoring and controlling panel created by HIL SCADA, ❶for the visual image, ❷for capturing voltage and current waveforms, ❸for inverse-time overcurrent curves visual graph, ❹for protection relay settings dashboard, ❺for circuit breaker's status monitoring, ❻for protective relay's status monitoring, and ❼for control dashboard

perform requests of the users, including displaying desired values or signals for monitoring functions or adjusting parameters.

The first element (❶) is used to visualise protective relay tests through images. To create this element, the users need to select the "*Image*" widget in the "Visual" of the HIL SCADA widget library; then, the users perform a right-click to define the direct path of the desired image (see Fig. 11.35).

The visual image used in the panel is *Test System* based on Fig. 11.1. Additionally, the authors used the image in Scalable Vector Graphics (SVG) format to reconstruct the image under two cases, nominal operating and short-circuit fault cases. For the nominal operating case, the circuit breaker (named BK in the image) is highlighted in red colour, which indicates the closed state. For short-circuit fault cases, the circuit breaker is in green colour, which shows the opened state, and the fault icon appears at the end of the transmission line (Lines 1–2).

To perform the image format in the two aforementioned conditions, the authors used the "*Macro*" widget in "*Action*" of the HIL SCADA widget library; then, the users need to do a right-click to open the properties window and start to parse SVG image as XML (Extensible Markup Language). The "*Fault Signal*" and "*Status Breaker*" from the fourth element (❹) of "*Soft Relay*" (see Sect. 3.1.3) are the input data to execute "BK" and fault icon via the "xml.etree.ElementTree" module ('xml.etree.ElementTree 2022) with associated commands (see Fig. 11.36).

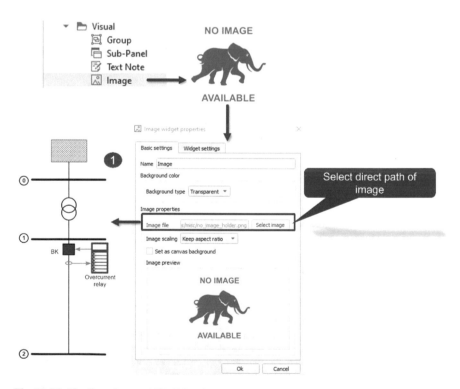

Fig. 11.35 The first element (❶) "Visual Image"

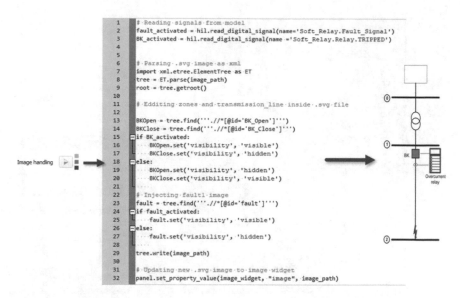

```
1   # ·Reading·signals·from·model
2   fault_activated = hil.read_digital_signal(name='Soft_Relay.Fault_Signal')
3   BK_activated = hil.read_digital_signal(name ='Soft_Relay.Relay.TRIPPED')
4
5
6   # ·Parsing· .svg·image·as·xml
7   import xml.etree.ElementTree as ET
8   tree = ET.parse(image_path)
9   root = tree.getroot()
10
11  # ·Editing·zones·and·transmission_line·inside· .svg·file
12
13  BKOpen = tree.find('''.//*[@id='BK_Open']''')
14  BKClose = tree.find('''.//*[@id='BK_Close']''')
15  if BK_activated:
16      BKOpen.set('visibility', 'visible')
17      BKClose.set('visibility', 'hidden')
18  else:
19      BKOpen.set('visibility', 'hidden')
20      BKClose.set('visibility', 'visible')
21
22  # ·Injecting·fault1·image
23  fault = tree.find('''.//*[@id='fault']''')
24  if fault_activated:
25      fault.set('visibility', 'visible')
26  else:
27      fault.set('visibility', 'hidden')
28
29  tree.write(image_path)
30
31  # ·Updating·new· .svg·image·to·image·widget
32  panel.set_property_value(image_widget, "image", image_path)
```

Fig. 11.36 Define "Visual Image" in specific circumstances, normal operation, and fault cases

The second element (❷) (see Fig. 11.37) is "*Capture/Scope*" in "*Monitoring*" of the HIL SCADA library. The users can right-click to select "*Switch to embedded mode*", then select "*Signal*" to capture the desired signals to observe during the real-time simulation.

The third element (❸) (see Fig. 11.38) is created to visualise inverse-time over-current curves during the test. The users can select "*XY graph*" in "*Monitoring*", then right-click to open the properties window. At this window, the authors have set the X-axis as multiples of the pickup current and the Y-axis as the protective relay's operating time based on the equation of the curves' family. In this programming part, the authors used the input of the RMS values of the three-phase current measured from the three-phase meter (see Sect. 3.1.1); CT and time-dial values, along with the parameters of curve equations, are set in the fourth element (❹), "*Relay setting*".

The fourth element (❹) (see Fig. 11.39) is the "*Relay setting*" dashboard, which is a group of widgets in "*Action*", including "*Text Box*", for importing the desired values of protection relay settings. They are CT ratio, ANSI 50/51 pickup current, time-dial, and inverse-time overcurrent curves. The users use the "*Text Box*" widget to define the values of CT ratio, ANSI 50/51 pickup current, and time dial. The "*Combo Box*" widget is used to select inverse-time overcurrent curves. These settings can be defined by double-clicking on the widget and using the Python command in the "*Macro code*" section. They are the inputs of the overcurrent protection functional block created in Schematic Editor previously (see Sect. 3.1.2).

The fifth element (❺) (see Fig. 11.40) is the "*BK monitoring*" dashboard, including the "*LED*" widget in "*Monitoring*". The LED expression is defined inside the widget properties according to the state of BK of the *Test System*; the LED is red when BK is closed and green when BK is opened.

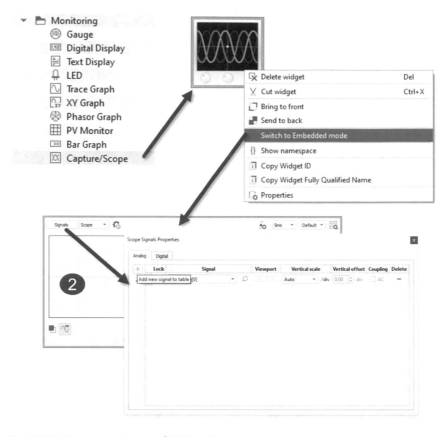

Fig. 11.37 The second element (❷) "Scope"

The sixth element (❻) (see Fig. 11.41) is "*Relay status*", including "*LED*" and "*Digital Display*" widgets. The "*LED*" widget is used to represent the trip signal coming from which overcurrent functional blocks, "*(50) Instantaneous Overcurrent*" or "*(51) AC Inverse Time Overcurrent*". The "*Digital Display*" widget represents the operating time of the protective relay. The users can use right-click on the widget and select the desired signal.

The seventh element (❼) is the "*Control*" dashboard created to perform multiple actions, "*Fault selection*", "*Apply Fault*", "*Clear Fault*", and "*Reset Relay*" (see Fig. 11.42.

For the "*Fault selection*", the "*Combo Box*" widget in "*Action*" of the HIL SCADA library is used to perform this action. The users can define the name of short-circuit cases in "*Available values*" of widget properties, then, according to declared types of fault in the Fault element of the test system (see Sect. 3.1.1). The nominal operating case is named "*NO FAULT*". At the same time, short-circuit fault can be of different types: three-phase, three-phase-to-ground, single-phase-to-ground, phase-to-phase,

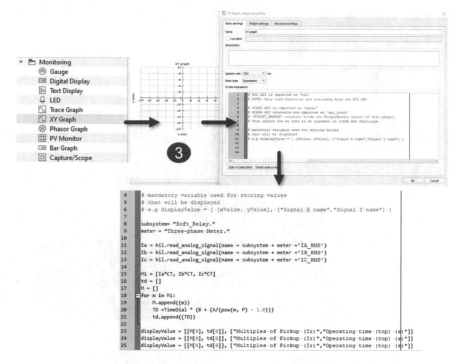

Fig. 11.38 The third element (❸) "*XY graph*"

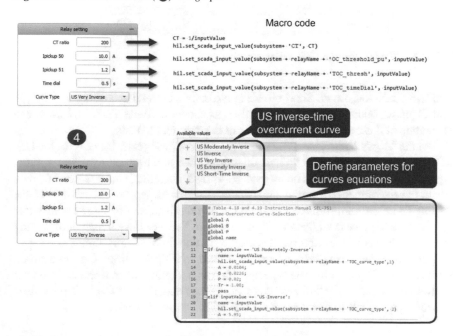

Fig. 11.39 The fourth element (❹) "*Relay setting*" dashboard

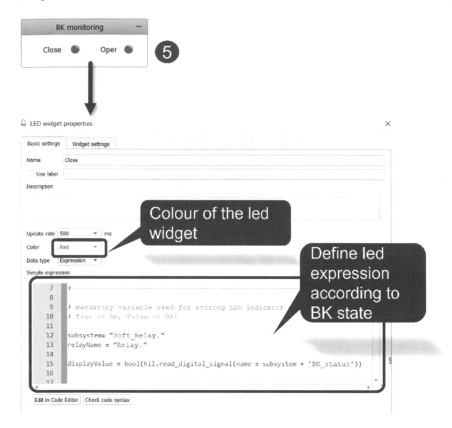

Fig. 11.40 The fifth element (⑤) "*BK monitoring*"

and two-phase-to-ground. Short circuit is typically identified according to the *A*, *B*, and *C* phases and the grounded neutral *N*, as shown in Table 11.7. The users can select one of these short circuits to conduct the appropriate tests.

For the "*Apply Fault*", the users select the "*Macro*" widget in "*Action*" of the HIL SCADA library. This button defines the sample rate and time interval for captured signal in "*Capture/Scope*". The users need to right-click the "*Capture/Scope*", then select "*Copy Widget ID*" and use this ID number for Macro widget settings of the "*Apply Fault*" button. Additionally, the types and starting times of short-circuit cases can be defined in these widgets via Python scripts. For the convenience of observing short-circuit events, the authors selected the starting time of short circuit at 0.1 s.

For "*Clear Fault*", the "*Macro*" widget is used. In widget settings, the short-circuit fault is cleared by defining the type specified as a nominal operating condition via Python scripts inside the "*Macro*" code. For "*Reset Relay*", the "*Check Box*" is used to send reset signals to overcurrent functional blocks in "*Relay*".

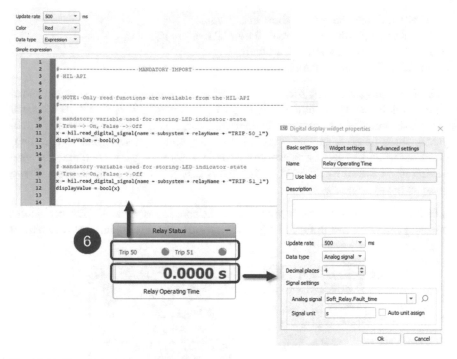

Fig. 11.41 The sixth element (**6**) *"Relay status"*

11.4 How To Use The Created Panel

The monitoring and controlling panels were created inside HIL SCADA that can be used for non-directional overcurrent protective relay testing.

The users need to run the real-time simulation by pressing the *"Start simulation"* button on the toolbar (after the HIL Schematic has been appropriately compiled).

Simulation using the VHIL will start, and time will be presented at the Control bar.

The circuit breaker monitoring dashboard has two LED indicators displaying the status of the circuit breaker; when the circuit breaker is closed, the LED named *"Close"* is lit in red colour, as shown in Fig. 11.43.

To test the non-directional overcurrent functions, the protection relay settings configuration, including CT ratio, ANSI 50/51 pickup current, time-dial, and inverse operating time curves, can be predetermined at the *"Relay setting"* dashboard as shown in Fig. 11.44.

The procedure of using created panel includes four steps shown in the *"Control"* dashboard in Fig. 11.45 as follows:

STEP 1: Select the type of short-circuit fault.
STEP 2: Apply the short-circuit fault and capture signals.
STEP 3: Clear the short-circuit fault.
STEP 4: Reset the non-directional overcurrent protective relay.

In **STEP 1**, there are many scenarios, including nominal operating cases and short-circuit fault cases. The nominal operating case is named "*NO FAULT*". At the same time, short-circuit fault cases can be defined by many types of short circuits: three-phase, three-phase-to-ground, single-phase-to-ground, phase-to-phase, and two-phase-to-ground, in each case, according to the *A*, *B*, and *C* phases and the grounded neutral *N* as shown in Table 11.7. The users can select one of these scenarios to conduct the appropriate test.

STEP 2 is to press the button "*Apply fault*". The consequences in SCADA panel after pressing can be shown in Fig. 11.46. The captured signal is automatically activated

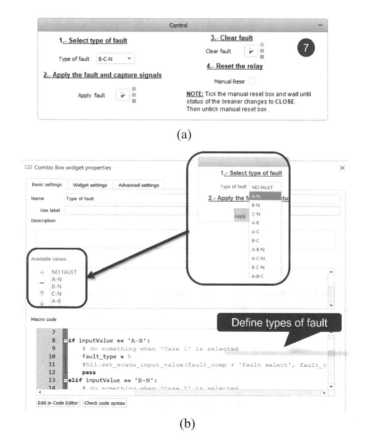

(a)

(b)

Fig. 11.42 **a** "*Control*" dashboard; **b** "*Fault selection*" action; **c** "*Apply Fault*" action; **d** "*Clear Fault*" action; **e** "*Reset Relay*" action

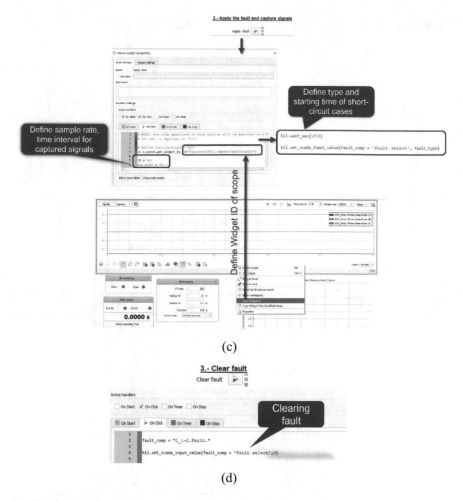

(c)

(d)

Fig. 11.42 (continued)

by the condition predefined in "*Capture/Scope*" (see Sect. 3.2). In the visual image, the BK is lit in green colour, and the fault icon appears simultaneously. At the "*BK monitoring*" dashboard, the LED expressing the open state of the circuit breaker is turned on. At the "*Relay Status*" dashboard, the LED indicates which overcurrent function is active, and the operating time of the protective relay is shown. The inverse-time overcurrent curve is plotted through the XY-graph widget according to preset values in the "*Relay setting*" dashboard, as shown in Fig. 11.47.

STEP 3 and **STEP 4** are performed simultaneously for the purpose of returning the protective relay to its initial state to conduct further tests. The users need to click the button "*Clear Fault*" first and then tick the box "*Manual Reset*". The clicking of the button "*Clear Fault*" clears the fault in the test system. Ticking the "*Manual*

Fig. 11.42 (continued)

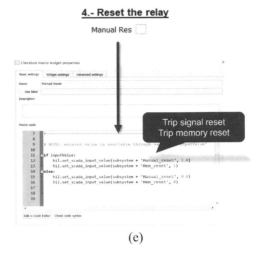

4.- Reset the relay

(e)

Table 11.7 Nameplates corresponding with short-circuit fault cases

Name	Short-circuit fault cases
A-N	Single-line-to-ground
B-N	
C-N	
A-B	Phase-to-phase
A-C	
B-C	
A-B-N	Two-phase-to-ground
A-C-N	
B-C-N	
A-B-C	Three-phase
A-B-C-N	Three-phase-to-ground

Fig. 11.43 The red colour of the LED in "*Close*" indicates the ready state to perform tests

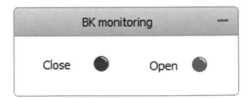

Reset" box returns the test system to the initial state. All necessary calculations and observations are computed and stored before running another test. The users need to untick the "*Manual Reset*" box before implementing further tests.

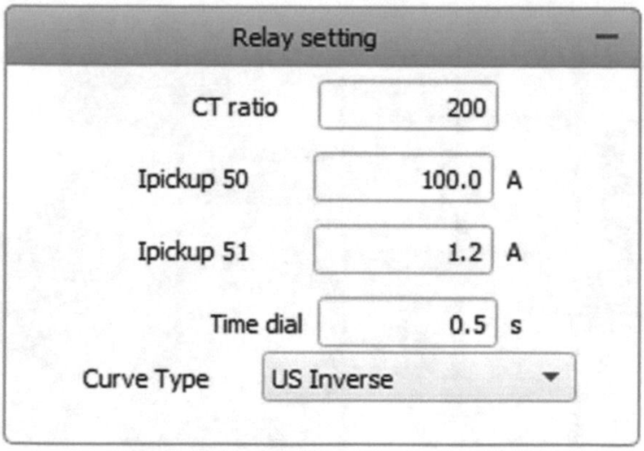

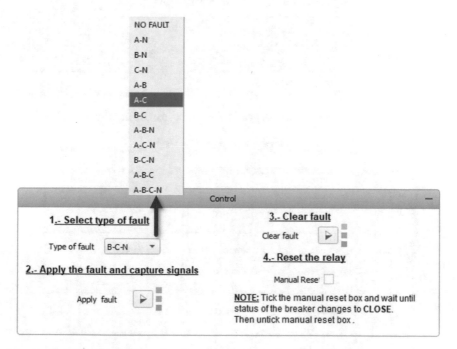

Fig. 11.44 The red colour of the LED in "Close" indicates ready to perform tests

Fig. 11.45 Procedure of using *"Control"* dashboard

11.5 Results and Discussions

This section presents the results of simulating different short-circuit fault scenarios at the end of the test model's overhead distribution line (Lines 1–2). The main objective of this section is to observe the behaviour of the overcurrent protection relay through model-based toolchains of Typhoon HIL as the following assumptions:

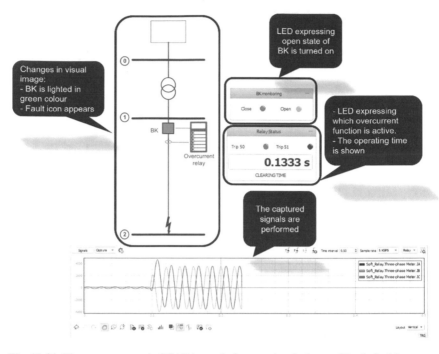

Fig. 11.46 The consequences in SCADA panel after pressing the button *"Apply fault"*

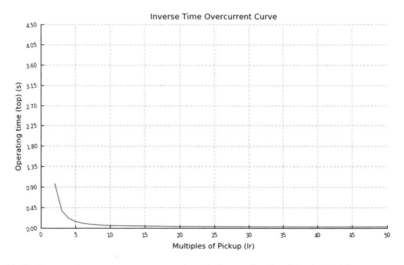

Fig. 11.47 Inverse-time overcurrent curve plotted after performing *"Apply Fault"*

- The overcurrent protection functions ANSI 50 and ANSI 51 are considered. Time-dial, ANSI 50/51 pickup current, and curve characteristics are known as protection relay settings. The operating time of the protection relay in two cases, instantaneous ANSI 50 and different inverse overcurrent time curves of ANSI 51, is presented.
- The current waveforms under multiple short-circuit cases, three-phase, three-phase-to-ground, single-phase-to-ground, phase-to-phase, and two-phase-to-ground, are presented using Typhoon HIL.

For these objectives, the authors provided a proposed set of predetermined settings for non-directional overcurrent functions for testing the operating time of the protection relay.

11.5.1 Predetermined Protective Relay Settings

The set of predetermined settings consists of CT ratio and values of pickup current ANSI 50/51. This section illustrates the proposed methodology of predetermined setting selection.

For CT ratio, the accurate CT ratio reflects secondary values of current through the protection relay and directly influences protection relay's operation whether to send a trip signal or not. The methodology to select the CT ratio for the overcurrent protection relay is based on the nominal primary operating current according to Failed (2000). Therefore, to calculate the CT ratio, the total load current up to the point of the protection relay should be considered. The total load current through the protection relay is the sum of load B current which is given in Table 11.4 with a value of 120A. The users can retrieve this value from the waveform of the nominal operating current by selecting the *"NO FAULT"* option in **STEP 1** (see Sect. 11.4), as shown in Fig. 11.48.

The authors used the CT ratio for protection relay due to the 125% nominal operating current as an overload situation. As a consequence, the current transformer

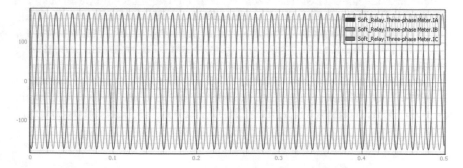

Fig. 11.48 Waveforms of three-phase current in nominal operating case

CT ratio of 200 (the nearest value above the overload current) is selected and used for all the simulation scenarios.

The pickup current is the operation condition of the overcurrent protective relay upon the current through it exceeds that value. The selection of threshold values depends on the specific system in the particular situations corresponding to the stated objective of protection engineers. The value of pickup current (ANSI 50 and ANSI 51) proposed by the authors is twice the value of the nominal operating current through the protection relay of 240 A on the primary side and 1.2 A on the secondary side, according to the ratio of CT.

There are two specific testing cases with particular settings according to the main purpose of this chapter is testing overcurrent functions, ANSI 50 and 51, as follows:

- **Case I, Instantaneous**: For testing the operating time of the instantaneous over-current function (ANSI 50), the pickup current of the ANSI 50 function is 1.2A.
- **Case II, Inverse Time**: The pickup current settings in inverse-time overcurrent testing (ANSI 51) is 1.2A and the time dial, TD, is 0.5.

11.5.2 Short-Circuit Study Validation

The RMS values of short-circuit scenarios performed by using the VHIL Device in the *Test System* are compared with the results from the DIgSILENT PowerFactory modelling. DIgSILENT PowerFactory is the widespread and leading power system analysis software, which covers the full range of functionality from standard features to novel, sophisticated, and advanced applications (PowerFactory—DIgSILENT 2022) (Fig. 11.49).

The authors used SP2 DIgSILENT PowerFactory version 2022 to model *Test System* and to perform the implementation of short-circuit calculation. The *"Complete"* method in calculating short circuit is applied.

To compare the RMS short-circuit values obtained from VHIL and DIgSILENT PowerFactory, a summary of RMS values of short-circuit current (*Ik*) for all scenarios is presented in Table 11.8. The RMS value from DIgSILENT PowerFactory is obtained by RMS simulation. The RMS value of short-circuit current from the VHIL device is extracted from the waveform as shown in Fig. 11.50. The waveform in Fig. 11.50 is obtained from the steady-state waveform of short-circuit current when the settings in **CASE II** are applied to the protective relay.

The maximum relative error between RMS short-circuit current in all scenarios is 1.65%. This indicates a significant similarity between the RMS values of the short-circuit cases retrieved from the VHIL Device and DIgSILENT PowerFactory.

Fig. 11.49 Modelling *Test System* in DIgSILENT PowerFactory

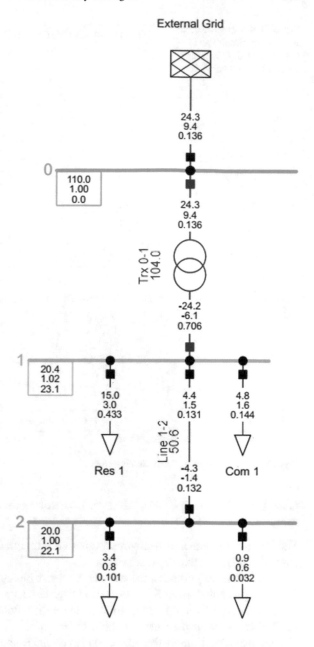

11.5.3 Testing Non-directional Overcurrent Protective Relay Using VHIL Device

Considering the aforementioned settings of the non-directional overcurrent protection relay, the results of two testing cases are presented in this section.

Table 11.8 Comparison of RMS short-circuit values between VHIL Device and DIgSILENT PowerFactory

Short-circuit case	VHIL Device Ik [kA]	DIgSILENT PowerFactory Ik [kA]	Per cent error [%]
A-N	2.38	2.42	1.65
B-N	2.38	2.42	1.65
C-N	2.38	2.42	1.65
A-B	2.28	2.28	0.00
A-C	2.28	2.28	0.00
B-C	2.28	2.28	0.00
A-B-N	2.53	2.55	0.78
A-C-N	2.53	2.55	0.78
B-C-N	2.53	2.55	0.78
A-B-C	2.68	2.64	1.51
A-B-C-N	2.68	2.64	1.51

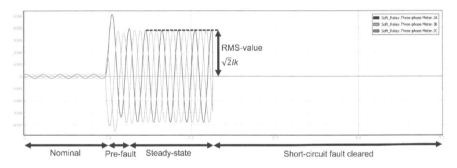

Fig. 11.50 State period of short-circuit event

Case I: Testing the operating time of the instantaneous overcurrent protective relay

The idea of instantaneous overcurrent function is no intentional time delay when the current gets higher than a preset value.

For illustration purposes, the waveforms and tripping sequences for three-phase faults are presented using SCADA HIL in Fig. 11.51.

A summary of the operating time of the entire simulation considering several short-circuit cases is presented in Table 11.9.

The operating time of the protection relay under short-circuit fault scenarios is extremely small (less than 10 ms), showing the concept of the overcurrent function testing that is consistent with the idea of instantaneous overcurrent function ANSI 50.

The results shown in Table 11.9 give the excellent performance of the non-directional overcurrent protective relay implemented in the real-time simulation.

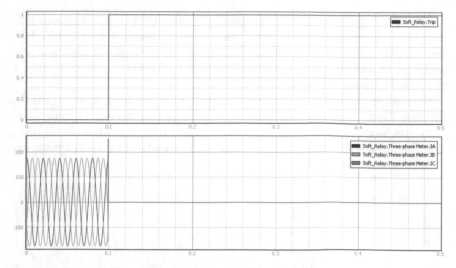

Fig. 11.51 Trip sequence and waveforms of three-phase current considering a three-phase short circuit in instantaneous overcurrent testing. The short-circuit event starts at 0.1 s, and the trip operates instantly at 0.1 s (from 0—inactive to 1—active)

Table 11.9 The operating time of protection relay in instantaneous overcurrent testing

Short-circuit cases	Operating time (ms)	Cycles[a]
A-N	0.200	0.010
B-N	0.200	0.010
C-N	0.200	0.010
A-B	0.080	0.004
A-C	0.080	0.004
B-C	0.080	0.004
A-B-N	0.060	0.003
A-C-N	0.060	0.003
B-C-N	0.060	0.003
A-B-C	0.060	0.003
A-B-C-N	0.060	0.003

[a]One cycle is equal to 20 ms (ms) at 50 Hz

Case II: Testing the operating time of the inverse-time overcurrent protective relay

For the inverse time overcurrent relay, the operating time is changed according to the case of short-circuiting and inverse-time overcurrent curves. For illustration purposes, the waveforms of three-phase current and tripping sequences for multiple short-circuit fault scenarios are presented using SCADA HIL in Fig. 11.52.

The operating time of the protective relay in this section can be retrieved from Eq. (1) to Eq. (5) based on the RMS values of short-circuit current derived from DIgSILENT PowerFactory in Table 11.8. A summary of the protection relay's operating time of the entire simulation, considering several short-circuit cases, is presented in Tables 11.10 and 11.11.

Considering the aforementioned fault scenarios, it is noticed from the results that the short circuit is cleared after delay time depending on the type of short-circuit cases and inverse-time overcurrent curve. As seen from the above results, there is a significant similarity between the case study simulated in VHIL Device and equation-based calculation.

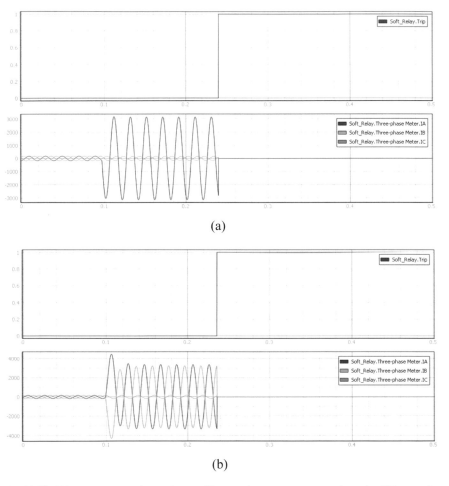

Fig. 11.52 Trip sequences and waveforms of inverse time overcurrent testing using U1 curve for short-circuit cases: **a** single-line-to-ground (A-N), **b** phase-to-phase (A-B), **c** two-phase-to-ground (A-B-N), and **d** three-phase (A-B-C)

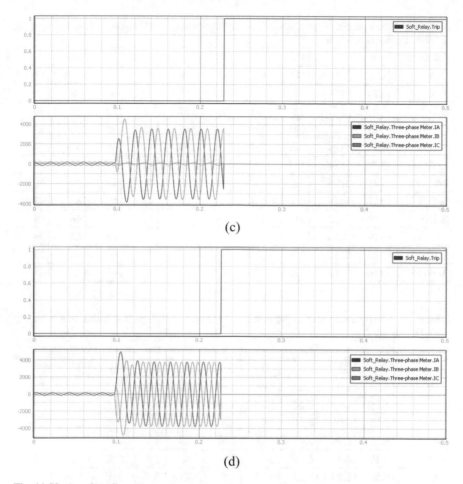

(c)

(d)

Fig. 11.52 (continued)

The most significant error or discrepancy between equation-based calculation and VHIL simulation is 6% for a single-line-to-ground short-circuit fault as shown in Fig. 11.53. The difference between equation-based calculation and VHIL simulation may be due to the differences (assumptions made during modelling, errors in measurement, and signal processing) between the real object and its assumed model. However, considering the overall discrepancy, this error can be assumed at an acceptable level. This allows for comparing the operation results with theoretically expected ones since both models are comparable.

Table 11.10 The operating time of protection relay in milliseconds. (1) Equation-based calculation, (2) VHIL Device simulation

Curve	U1		U2		U3		U4		U5	
Cases	(1)	(2)	(1)	(2)	(1)	(2)	(1)	(2)	(1)	(2)
A-N	125	121	124	119	70	67	49	50	48	49
B-N	125	121	124	119	70	67	49	50	48	49
C-N	125	121	124	119	70	67	49	50	48	49
A-B	122	124	121	123	68	69	47	49	47	50
A-C	122	124	121	123	68	69	47	49	47	50
B-C	122	124	121	123	68	69	47	49	47	50
A-B-N	119	118	115	116	64	65	42	43	46	48
A-C-N	119	118	115	116	64	65	42	43	46	48
B-C-N	119	118	115	116	64	65	42	43	46	48
A-B-C	116	117	114	115	63	64	40	41	46	47
A-B-C-N	116	117	114	115	63	64	40	41	46	47

Table 11.11 The operating time of protection relay in cycles. (1) Equation-based calculation, (2) VHIL Device simulation

Curve	U1		U2		U3		U4		U5	
Cases	(1)	(2)	(1)	(2)	(1)	(2)	(1)	(2)	(1)	(2)
A-N	6.25	6.05	6.20	5.95	3.50	3.35	2.45	2.50	2.40	2.45
B-N	6.25	6.05	6.20	5.95	3.50	3.35	2.45	2.50	2.40	2.45
C-N	6.25	6.05	6.20	5.95	3.50	3.35	2.45	2.50	2.40	2.45
A-B	6.10	6.20	6.05	6.15	3.40	3.45	2.35	2.45	2.35	2.5
A-C	6.10	6.20	6.05	6.15	3.40	3.45	2.35	2.45	2.35	2.5
B-C	6.10	6.20	6.05	6.15	3.40	3.45	2.35	2.45	2.35	2.5
A-B-N	5.95	5.90	5.75	5.80	3.20	3.25	2.10	2.15	2.30	2.40
A-C-N	5.95	5.90	5.75	5.80	3.20	3.25	2.10	2.15	2.30	2.40
B-C-N	5.95	5.90	5.75	5.80	3.20	3.25	2.10	2.15	2.30	2.40
A-B-C	5.80	5.85	5.70	5.75	3.15	3.20	2.00	2.05	2.30	2.35
A-B-C-N	5.80	5.85	5.70	5.75	3.15	3.20	2.00	2.05	2.30	2.35

11.6 Conclusions and Recommendations

This chapter implemented a simplified version of the three-phase radial feeder, MW CIGRE European's distribution benchmark system, using a real-time simulation platform.

The implementation of a non-directional overcurrent protection relay using the Virtual HIL Device from model-based toolchains of Typhoon HIL is presented in

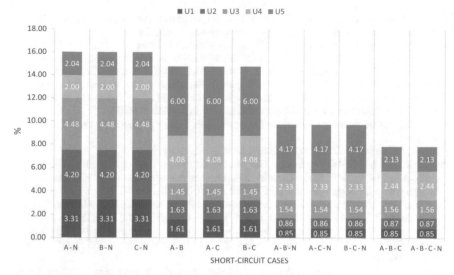

Fig. 11.53 Per cent error of operating time of protection relay when comparing values from VHIL Device with Equation-based calculation in short-circuit cases

this chapter. The so-called "*Soft Relay*" can be used to test and validate protection settings without the need to use real hardware or any other physical devices.

The test system was used to test the overcurrent protection functions of a protection relay. Instantaneous overcurrent and inverse-time overcurrent functionalities of a non-directional protection relay were tested considering predetermined setting configuration.

Presenting the methodology for creating and testing a model, including a protection relay to be used in real-time simulation, allows the scientific community to use this chapter as a starting point for many other possible protection schemes.

References

Aman MM, Khan MQA, Qazi SA (2011) Digital directional and non-directional over current relays: modelling and performance analysis. NED Univ J Res 8(2):70–85

'ANSI protective functions'. https://www.typhoon-hil.com/documentation/typhoon-hil-software-manual/References/ansi_protective_functions.html. Accessed 11 Jul 2022

Camarillo-Peñaranda JR, Aredes M, Ramos G (2020) Hardware-in-the-loop testing of virtual distance protection relay. In: 2020 IEEE/IAS 56th industrial and commercial power systems technical conference (I&CPS), Jun. 2020, pp 1–6, https://doi.org/10.1109/ICPS48389.2020.9176775

Celeita D, Flores A, Ramos G, Pohl M (2018) Design of virtual distance protection for offline transmission line relay testing. In: 2018 IEEE 38th central America and Panama convention (CONCAPAN XXXVIII), Nov. 2018, pp 1–6, https://doi.org/10.1109/CONCAPAN.2018.8596426

C. T. F. C6. 04.02, Benchmark systems for network integration of renewable and distributed energy resources. International council on large electric systems Paris, France

D'Arco S, Duong TD, Are Suul J (2020) 'P-HiL evaluation of virtual inertia support to the nordic power system by an HVDC terminal. In: 2020 IEEE PES innovative smart grid technologies Europe (ISGT-Europe), pp 176–180.https://doi.org/10.1109/ISGT-Europe47291.2020.9248905

Edrington CS, Steurer M, Langston J, El-Mezyani T, Schoder K (2015) Role of power hardware in the loop in modeling and simulation for experimentation in power and energy systems. Proc IEEE 103(12):2401–2409, https://doi.org/10.1109/JPROC.2015.2460676

Fonti P (2000) Current transformers: how to specify them. Schneider Electr Cah Tech Merlin Gerin, no. 194

Horowitz SH, Phadke AG (2014) Power system relaying. Wiley

HubschneiderS et al (2018) Requirements for power hardware-in-the-loop emulation of distribution grid challenges. In: 2018 53rd international universities power engineering conference (UPEC), Glasgow, Sep. 2018, pp 1–6. https://doi.org/10.1109/UPEC.2018.8541851

IEEE standard for inverse-time characteristics equations for overcurrent relays. IEEE Std C37112–2018 Revis. IEEE Std C37112–1996, pp 1–25, https://doi.org/10.1109/IEEESTD.2019.8635630

IEEE standard electrical power system device function numbers, acronyms, and contact designations. IEEE Std C372–2008 Revis IEEE Std C372–1996, pp 1–48, https://doi.org/10.1109/IEEESTD.2008.4639522

KelmP et al (2022) Hardware-in-the-loop validation of an energy management system for LV distribution networks with renewable energy sources. Energies 15(7):2561, https://doi.org/10.3390/en15072561

Kezunovic M, Esmaeilian A, Manimaran G, Mehrizi-Sani A (2017) The use of system in the loop, hardware in the loop, and co-modeling of cyber-physical systems in developing and evaluating new smart grid solutions, https://doi.org/10.24251/HICSS.2017.385

Montaña DAM, Rodriguez DFC, Ivan Clavijo Rey D, Ramos G (2018) Hardware and software integration as a realist SCADA environment to test protective relaying control. IEEE Trans Ind Appl 54(2):1208–1217, https://doi.org/10.1109/TIA.2017.2780051

Makhzani AS, Zarghami M, Falahati B, Vaziri M (2017) Hardware-in-the-loop testing of protection relays in distribution feeders with high penetration of DGs. In: 2017 North American power symposium (NAPS), Sep. 2017, pp 1–6, https://doi.org/10.1109/NAPS.2017.8107191

Mehta VK, Mehta R (2005) Principles of power system: including generation, transmission, distribution, switchgear and protection : for B.E/B.Tech., AMIE and Other Engineering Examinations. S. Chand Publishing

Pazdcrin AV, Samovlenko VO, Tashchilin VA, Chusovitin PV, Dymshakov AV, Ivanov YV (2018) Platform for testing Iec 61850 control systems using real-time simulator. In: 2018 international youth scientific and technical conference relay protection and automation (RPA), Sep. 2018, pp 1–14. https://doi.org/10.1109/RPA.2018.8537188

'PowerFactory—DIgSILENT'. https://www.digsilent.de/en/powerfactory.html. Accessed 6 Jun 2022

Rodriguez DFC, Osorio JDP, Ramos G (2018) Virtual relay design for feeder protection testing with online simulation. IEEE Trans Ind Appl 54(1):143–149, https://doi.org/10.1109/TIA.2017.2741918

Review of hardware-in-the-loop—a hundred years progress in the pseudo-real testing', E+E Scientific Journal, Aug. 09, 2019. https://epluse.ceec.bg/review-of-hardware-in-the-loop-a-hundred-years-progress-in-the-pseudo-real-testing/. Accessed 7 Aug 2022

S. IEC, 'Measuring relays and protection equipment-Part 151: Functional requirements of over/under current protection', IEC 60255–151 (2009)

Schossig T (2010) Testing in IEC 61850—advanced topics and extended possibilities, pp 61–61, Jan. 2010, https://doi.org/10.1049/cp.2010.0233

'SR Flip Flop'. https://www.typhoon-hil.com/documentation/typhoon-hil-software-manual/References/sr_flip_flop.html. Accessed 29 Jul 2022

'What is Python? Executive Summary', Python.org. https://www.python.org/doc/essays/blurb/. Accessed 16 Jan 2022

'xml.etree.ElementTree—The ElementTree XML API—Python 3.10.5 documentation'. https://docs.python.org/3/library/xml.etree.elementtree.html. Accessed 30 Jul 2022

Chapter 12
Directional Overcurrent Relay Protection System Implementation on 8-bus System Using Typhoon HIL

Juan David Hernández Santafé, Felipe Antonio Gómez Olaya, and Paulo Manuel De Oliveira De Jesús

Abstract The coordination of directional overcurrent relays (DOCR) is a complex problem. Generally, optimization problems are posed using simplified positive sequence network models. The 8-bus transmission system proposed by A. S. Braga and J. Tomé Saravia in 1996 has been widely used as a standard test case on DOCR to get optimal solutions. However, detailed three-phase network models are required in order to analyze the quality of the resulting relay settings obtained from optimization procedures. In this chapter, the 8-bus transmission system was implemented in Typhoon-HIL (Hardware in the loop) in order to evaluate protection system performance, considering a detailed three-phase model for generators, transformers and lines, pre-fault conditions, load currents, current transformers (CT) saturation and the effect of transient topologies. Protection engineers will benefit from the proposed implementation to evaluate the quality of any directional overcurrent coordination solution, optimized or not, under more realistic conditions.

Keywords Coordination of protective relays · Core · Pre-fault conditions · Directional overcurrent protections · Real time software · Simulation couplings · Transient currents · Transient topologies

12.1 Introduction

The use of directional overcurrent protections is widespread in the electric power system industry. The coordination of directional overcurrent protections (DOCP) is a challenging problem for protection engineers. For many years, attempts have been made to find the optimal settings for DOCP systems aiming to assure high levels of reliability and speed. In Urdaneta et al. (1988), the first optimization model based on linear programming for finding the DOCP settings is presented. This was

J. D. H. Santafé (✉) · F. A. G. Olaya · P. M. De Oliveira De Jesús
Department of Electrical and Electronical Engineering, University of the Andes, South America, Colombia
e-mail: jd.hernandezs@uniandes.edu.co

© The Author(s), under exclusive license to Springer Nature Singapore Pte Ltd. 2023   339
S. M. Tripathi and F. M. Gonzalez-Longatt (eds.), *Real-Time Simulation and Hardware-in-the-Loop Testing Using Typhoon HIL*, Transactions on Computer Systems and Networks, https://doi.org/10.1007/978-981-99-0224-8_12

followed by several authors, who developed different optimization models that allow obtaining the relay pick-up current settings and time delay constants for minimum overall operation time subject to selectivity and sensitivity constraints (Hussain et al. 2013; Birla et al. 2005; Raza et al. 2014; Singh and Panigrahi 2014; Noghabi et al. 2009; Amraee 2012; Mahari 2013; Albasri et al. 2015; Damchi et al. 2018).

Existing network representations used to get optimal coordination settings are generally based on static positive-sequence models. For instance, the 8-bus transmission system proposed by A. S. Braga and J. Tome Saravia in 1996 has been widely used as a standard test case on DOCR to get optimal solutions (Braga and Saraiva 1996). However, as this test system is structured as a positive sequence network model, some important aspects such as the synchronous generator model, transient and sub-transient fault currents per phase, dynamic topologies generated by non-simultaneous actuation of a circuit breaker, current transformers saturation and the effect of unbalanced loads (Urdaneta et al. 1995; Mahboubkhah et al. 2020; Sorrentino and Rodríguez 2020) must be also considered to evaluate the quality of any relay coordination solution in real-world conditions. For this reason, it is necessary to implement existing test systems in real-time simulators considering detailed network models whose time-variant currents affect the response of the DOCP system. Literature shows some contributions to this objective. In Martín (2016), Almas et al. (2012), Hussin et al. (2016), detailed network models were developed in MATLAB/Simulink to evaluate the response of directional overcurrent relays with ANSI (American National Standards Institute) 67 function.

This chapter presents the implementation of the 8-bus Braga's test system (Braga and Saraiva 1996) under realistic conditions using the soft-real-time Typhoon HIL platform (Typhoon-HIL 2020). The system comprises 2 generators, 2 power transformers, 8 busbars and 7 transmission lines with 14 directional overcurrent relays with ANSI 67 function. Three-phase fault current waveforms are determined per-phase at all line ends considering the transient and sub-transient components, dynamic topologies generated by non-simultaneous actuation of circuit breakers, load currents and current transformers saturation.

To properly evaluate the performance of the protection system for a given coordination solution (time dials and pick-up currents settings), it is necessary to insert faults at different locations of the line. As the set of faults is large in a system with 8 busbars and 7 transmission lines, modifying fault positions in the schematic editor and observing the results in Typhoon SCADA HIL(Supervisory Control And Data Acquisition) is cumbersome. The Typhoon HIL solvers have a rather limited memory, mainly for those components that need signal processing for their operation. Even if the different processing cores that Typhoon HIL devices have been used, the contactors that can be used to perform faults at the same time are very few (Typhoon-HIL 2020).

To overcome these limitations, for each inserted fault, a methodology was developed in the Script editor of Typhoon HIL to calculate the relay operation times using the exact IEC (International Electromechanical Commission) formula [IEC 60,255–151:2009, 19]. Operation times for several fault locations are obtained in an automated way using Typhoon HIL API and Typhoon HIL Test. The main advantage is that added calculations or procedures internally can be directly evaluated, without the need to export data and/or use other software for such evaluation.

As a result, this tool allows protection engineers testing in real-world conditions a DOCP coordination solution for the Braga's 8-bus test system (Braga and Saraiva 1996), optimized or not, considering several fault locations in terms of overall speed, tripping sequence and the ability of the directional overcurrent relays to detect the corresponding fault contributions.

The organization of the chapter is as follows. Theorical framework section presents the theory behind directional overcurrent protection systems. Methodology section discussed the developed methods to implement the 8-bust test system in Typhoon HIL. Likewise, simulation details are presented in the results and validation section and finally conclusions are drawn in the conclusions and recommendations chapter.

12.2 Theorical Framework

In this section some theoretical definitions are presented, such as the overcurrent and directional protection functions, as well as the definition of directional blocking.

12.2.1 Directional Protection Function

Directionality of the power flow is measured from the phases of the voltages and currents on the line. Table 12.1 shows the current and voltage pairs used for power flow determination. There are several types of polarizations used both in theory and in practice, but the most used is the phase-phase quadrature in which it is determined whether the polarization of the flow is positive ("Forward") or negative ("Backward") as will be seen below:

Thus, taking the voltage polarization angle as λ_1 and current operation angle as λ_2, and $\psi = \lambda_2 - \lambda_2$, it is satisfied that the torque angle is proportional to $T = \lambda_1 \lambda_2 \sin(\psi)$. This value will be positive if $0 < \psi < \pi$ and negative when $\pi < \psi < 2\pi$. Considering the above, the operation area can be created considering the Relay

Table 12.1 Bias voltages and currents for the three phases of the system

Phase	Bias current [A]	Bias voltage [V]
A	I_a	V_{bc}
B	I_b	V_{ac}
C	I_c	V_{ab}

Characteristic Angle (RCA) to determine directionality as shown in Eqs. (12.1) and (12.2).

"Forward" if:

$$-90° < \lambda_1 + \text{RCA} - \lambda_2 < 90° \tag{12.1}$$

"Backward" if:

$$90° < \lambda_1 + \text{RCA} - \lambda_2 < -90° \tag{12.2}$$

Note: All relays used during this investigation have an RCA angle of 45°.

12.2.2 Overcurrent Protection Function

Directional overcurrent relays operate based on inverse time curves. Fundamental property of this type of relay is that they operate in a time inversely proportional to the fault current, according to an available set of characteristic curves given from the variation of the parameters used to calculate relay operation times.

The function that determines the relay "r" operation time can be seen in Eqs. (12.3) and (12.4) considering current transients. It presents constants that change depending on different curves given by the American and European standards (ANSI, IEEE (Institute of Electrical and Electronics Engineers), UK (United Kingdom) and IEC), which offer different values to the constants of the function as shown in Table 12.2.

$$T_r(G) = TMS\left(\frac{\alpha_{r1}}{\frac{G(t)}{I_{pk}}^{\alpha_{r2}} - 1} + L\right) \tag{12.3}$$

$$\int_0^{T_r} \frac{1}{T(G)} dt = 1 \tag{12.4}$$

Where:

$T_r(G)$:   Relay operation time
$G(t)$:   Fault current signal.
$\alpha_{r1} \& \alpha_{r2}$:   Characteristic slope of relay
L: Constant
TMS: Time multiplier, different for ANSI−IEEE UK−and IEC
I_{rms} = RMS current
I_{pk} = Minimum drive current "Pick up"

Table 12.2 shows the above parameters used for different standards.

Table 12.2 Constant values for ANSI/IEEE and IEC standards (IEEE Standard for Inverse-Time Characteristics Equations for Overcurrent Relays 2019)

Curve description	Standard	α_{r1}	α_{r2}	L
Standard inverse	IEC	0,14	0,02	0
Very inverse	IEC	13,5	1	0
Extremely inverse	IEC	80	2	0
Long time inverse	UK	120	1	0
Rectifier	UK	45,900	5,6	0
Moderately inverse	IEEE	0,0515	0,02	0,114
Very inverse	IEEE	19,61	2	0,491
Extremely inverse	IEEE	28,2	2	0,1217
Inverse	US	5,95	2	0,18
Short time inverse	US	0,16,758	0,02	0,11,858

12.2.3 Directionality Blocking

Figure 12.1 shows a three-node system with a fault at line "b". As can be seen there, the directions of the power flow when the externality occurs are shown in red. From this image it is possible to see the pairs of primary relays ("3", "4") with their respective backups ("1", "6"), which will always have a positive polarization and will operate in coordination given the input settings. However, the relays ("2", "5") also see a high current that will surely be above the minimum pickup current required to trip their associated breaker, but because they observe a polarity opposite to the direction of the center of the line they protect, they will not be able to operate at any time. This property increases the selectivity of a protection system, and it is essential to take it into account when modeling directional overcurrent relays.

12.3 Methodology

This section will present the parameters of the system to be implemented and the details of system implementation. It is necessary to clarify that due to the limited contactor capacity that can be used in the software, it was not possible to implement the fourteen existing relays in the system simultaneously. For this reason, 7 different schematic files (.tse) with their respective HIL API files (.py) were created. In each of these files the same system is presented, but with the fault located in each of the different 7 lines and with their respective primary and backup relays. The project files and a step-by-step video tutorial can be found in https://github.com/pmdeolive iradejesus/DOCP-TyphoonHIL.git.

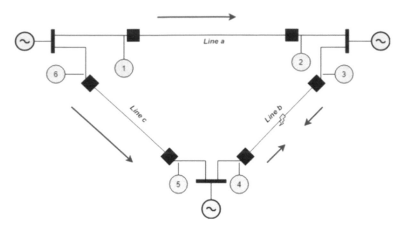

Fig. 12.1 Three-node system, used to account for directionality blocking

12.3.1   8-Bus Test System Implementation

12.3.1.1   System Parameters

The case study used during the development of this chapter was obtained from Braga and Saraiva (1996) and has electrical devices such as transformers, generators, transmission lines and circuit breakers operated by directional overcurrent relays. Figure 12.2 shows the topology of the 8-node system, followed by tables with the parameters of each of the elements mentioned above (Tables 12.3, 12.4, 12.5 and 12.6).

12.3.1.2   Overall Simulation Parameters

There are several models of Typhoon HIL devices (402, 404, 602, 603 and 604), each with distinct characteristics. For the present project Typhoon HIL 402 was used, which has 4 cores, 16 analogy inputs/outputs, 32 digital inputs/outputs, an analogy voltage range of ± 10 V and a resolution of 16 bits. Moreover, the simulation step and method must be chosen to define the accuracy of calculations and simulation time. The candidate methods are "Exact", "Trapezoidal", "Euler" or "Bilinear", Euler" or "Bilinear". On the other hand, the "Hardware configuration id" box is of significant importance because it allows choosing the HIL 402's own limit configurations, such as the number of System Processing Cores (SPC), "Machine solvers", signal generators, contactors, and others.

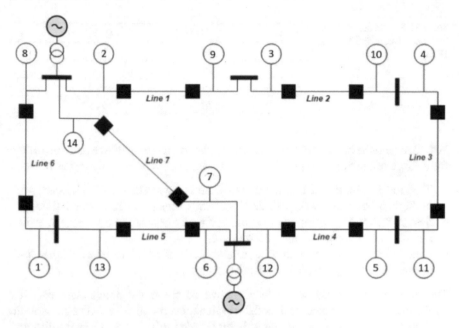

Fig. 12.2 Case study, 8 nodes system (Braga and Saraiva 1996)

Table 12.3 8 nodes system, generator data (Braga and Saraiva 1996)

Node	S [MVA]	Vp [KV]	x [%]
7–1	150	10	15
8–6	150	10	15

Table 12.4 8 nodes system, line data (Braga and Saraiva 1996)

Node	Node	R [Ω/km]	X [Ω/km]	Y [S/km]	Length [km]
1	2	0,04	0,5	0	10
1	3	0,057	0,714	0	7
3	4	0,05	0,563	0	8
4	5	0,05	0,45	0	10
5	6	0,045	0,409	0	11
2	6	0,044	0,5	0	9
1	6	0,05	0,5	0	10

Table 12.5 8 nodes system, transformers data (Braga and Saraiva 1996)

Node	S [MVA]	Vp [KV]	Vs [KV]	x [%]
7–1	150	10	150	4
8–6	150	10	150	4

Table 12.6 8 nodes system, load data (Braga and Saraiva 1996)

Node	P [MW]	Q [MVAr]
2	40	20
3	60	40
4	70	40
5	70	50

To understand the concept of SPC used in "Typhoon" it is essential to make use of the following explanation: The processing of a system in Typhoon is divided as follows:

- Typhoon FPGA (Field Programmable Gate Arrays) solver: FPGA board with multiple processing cores, optimized for time-domain electrical simulations.
- System CPU: Processing application, indirectly controlled by the user, which serves as an assistance to the FPGA board.
- User CPU: Processing application, directly controlled by the user, which is used to execute signal processing models.

The different processing levels listed above compose the architecture used by Typhoon to process the electrical models created for subsequent real-time simulation.

Figure 12.3 shows the processing structure made by the FPGA where the different cores generated in the system are processed. This means that it is the user who decides how many cores should be used for processing in the system. However, this decision is limited because the larger the modelled system is, the more cores are needed. In Typhoon, the term SPC is used to refer to the cores and its use is essential if the modelled system has many non-linear and time-varying elements such as:

- Signal generator: Blocks that generate different waveforms.
- LUT (Look-Up-table): Block used to simulate non-linear elements such as battery or solar panels.

Fig. 12.3 Basic processing architecture used in Typhoon (Typhoon-HIL 2020)

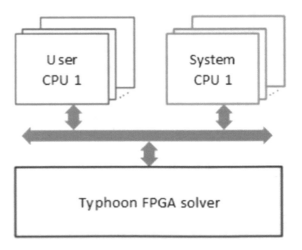

Table 12.7 Chosen hardware settings and circuit solver settings this project

Hardware settings	
Device	HIL 402
Configuration id	2
Circuit solver settings	
Discretization method	Bilinear
Simulation step	10e-6

- Machine solvers: Blocks of synchronous machines that have incorporated the electromagnetic part that composes these machines.
- PMW (Pulse Width Modulation) modulator: Pulse-width-modulation pulse modulations.
- Switches: Ideal and non-ideal contactors.

In general, SPCs are used because the use of non-linear elements increases the size of the state matrices considerably and memory overflows are generated if the correct division in SPCs of the circuit to be simulated is not made.

Considering the enormous number of ideal contactors that were foreseen to be necessary for the implementation of the system and that every hardware configuration id has the same ideal contactors per SPC, the id with the largest number of SPC was chosen, i.e., 2, with 4 SPC. Besides, a balance between computation time and accuracy of results was bearing in mind to determine the simulation step and method.

Table 12.7 shows the chosen configuration for hardware settings and circuit solver settings for this project. The other settings were left as default.

12.3.1.3   Relay Modelling in Schematic Editor

This section will present the development of the directional function of the relays in the schematic editor of the real-time simulation tool "Typhoon HIL" and the capture of necessary signals for the overcurrent function. The overcurrent function will be discussed in detail in Sect. 12.3.2.

Relay block implementation

This section will explain the internal creation of the relay block, i.e., it will show how the directionality and overcurrent measurements are performed so that the device can open a circuit breaker in case of fault conditions are met.

In Fig. 12.4, the internal topological layout of the relay is presented, where the rate transition function, the directionality and overcurrent logic block (ANSI 67) and trip output post-processing are located. These elements will be explained below:

- Rate transition

The first step is to sample the current and voltage data entering the protection device. This is necessary since the real relays cannot take all the existing data and it is

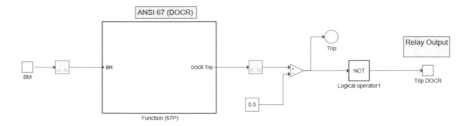

Fig. 12.4 The relay's internal topology has three zones: ANSI 67 Function, Transition Rate and Output (Trip)

Fig. 12.5 "Rate Transition" element in the Typhoon library

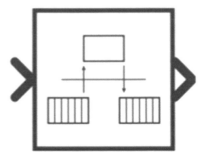

necessary to take samples every certain time of the input signal. Therefore, a block called "Rate Transition" provided by Typhoon (Fig. 12.5) was used, which allows a change in the time step during the simulation of the signals, which simulates a data sampling at rates lower than those at which the system is working.

The system has a simulation time of 10e-6, and the relays have a simulation time of 5e-4. The relay's internal step time can be varied to smaller values from the initialization window provided by the schematic editor, but it is necessary that its value be a multiple of the system's step time. It is for this reason that the value of 5e-4 was taken, to be able to take 32 samples/cycle = 5.208888e-4, which is used by real relays.

- ANSI 67 block

This block is divided into six different data processing zones, which allows us to determine directionality and overcurrent given the input voltage and current signal data. Each of these is explained below:

- **Basic data**: In this area the data of voltages and phase currents entering the relay are acquired (Fig. 12.6).
- **RMS Values**: In this area, RMS values of instantaneous currents are obtained (Fig. 12.7).
- **Phase measurement**: In this area the angle values of input voltages and currents are obtained. This is done by means of the phase difference of an ideal sin(wt)

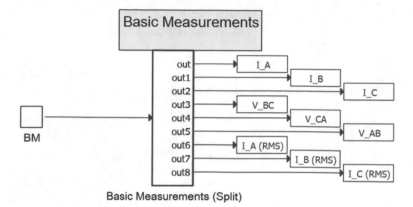

Basic Measurements (Split)

Fig. 12.6   Area: basic data

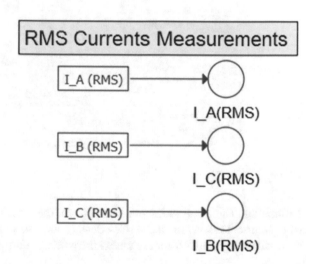

Fig. 12.7   Area: RMS current measurements

sine signal with the input signal by means of a block inside the tool library called "Phase Diff" (Fig. 12.8).

- **HIL API inputs**: In this area the values that can be configured during simulation in the HIL API interface, or in SCADA HIL if that is necessary, are obtained, as will be seen later (Fig. 12.9).
- **Directionality function**: In this area there is a block that internally allows the development of code in C# language, with which the angle input data are obtained, and the relay polarization is checked (Fig. 12.11). Figure 12.10 shows the flowchart associated with the algorithm developed to find the polarity with which the relay observes the power flow.

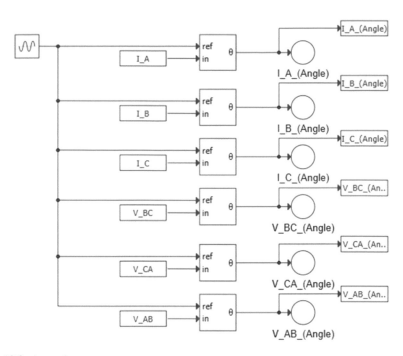

Fig. 12.8 Area: phase measurements

- **Directional Blocking**: This zone is intended to perform the blocking property of directional overcurrent relays. In case a relay detects negative polarity, it will never operate whatever the fault current is unless the polarity returns to positive (Fig. 12.12).

12.3.1.4   System Modelling

To model the 8-bus case study in Typhoon, elements provided by tool library and the input parameters described in system parameters were used. The devices used during the development of the case study were the following:

- Three-phase transformer with impedance parameters at rated power, frequency, RMS line-to-neutral voltage, positive short-circuit sequence, Y-Y connection, active losses and linear or non-linear core model (Fig. 12.13).
- An ideal three-phase source with nominal frequency, RMS line-to-neutral voltage and phase input parameters. Series impedances are added to account transient fault contribution (Fig. 12.14).

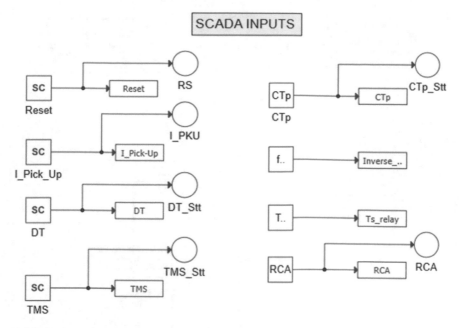

Fig. 12.9 Area: HIL API Inputs

- Constant impedance load which, depending on its power factor, either in forward or reverse, is made up of series or parallel resistance circuits with inductances and capacitances respectively. Among its input parameters are the line-to-line voltage, connection type, rated power, and power factor (Fig. 12.15).
- PI model transmission lines with input parameters that can be entered in three ways, by means of geometry, sequence data, and coupled RL. These types of lines are modelled as resistors in series with inductances and in parallel with capacitance at the ends, which are fed by voltage sources controlled by the input parameters entered. Transmission line geometry is assumed symmetrical (Fig. 12.16).
- Because of PI section lines were placed instead of RL (resistance-inductance) and the circuit is too large, it was necessary to place simulation couplers, which are needed by the software to divide the system into subsystems, allowing parallel processing of the cores or SPCs that were discussed in the theoretical framework of this document. These simulation couplings do not change in any way the properties of the lines (mutual or self-impedances), they are called couplings because they have two ideal voltage sources that help to perform the division discussed above. The coupling used was a three-phase transmission line coupling, as can be seen in Fig. 12.17.

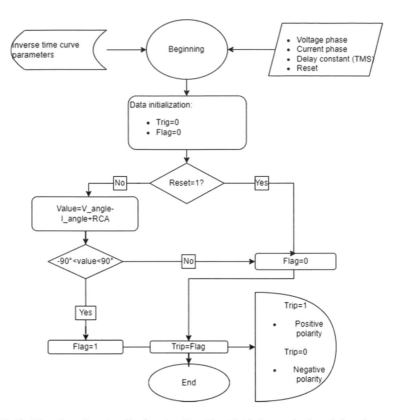

Fig. 12.10 Flowchart directionality function *Note* Fig. 12.10 shows a logic switch at the output of the function block, which enables or disables the directionality from the device mask

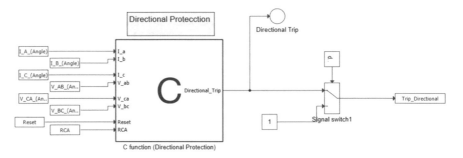

Fig. 12.11 Area: directionality function *Note* Fig. 12.11 shows a logic switch at the output of the function block, which enables or disables the directionality from the device mask

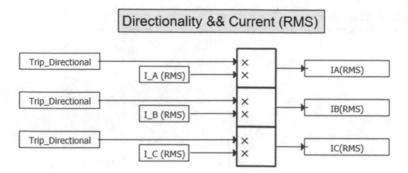

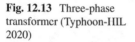

Fig. 12.12 Area: Directional locking

Fig. 12.13 Three-phase transformer (Typhoon-HIL 2020)

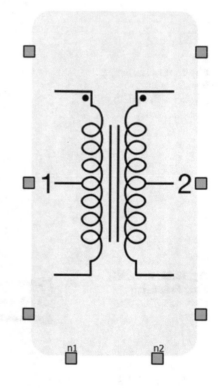

The use of simulation couplings together with lines with PI configuration, completely changes the properties of the system, since as the system has contactors, topological problems are generated when one of these opens, varying the voltages, in magnitude and phase of the system. Other possible problems when placing couplings in the system are:

Fig. 12.14 Three-phase
generator (Typhoon-HIL
2020)

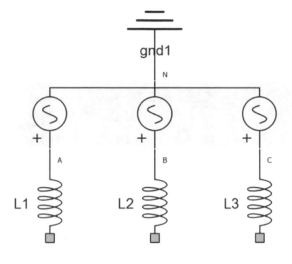

Fig. 12.15 Constant
impedance load
(Typhoon-HIL 2020)

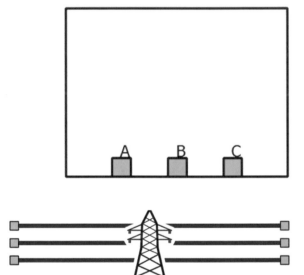

Fig. 12.16 PI model
transmission line
(Typhoon-HIL 2020)

1. The couplings enter inductances to the system, which must be considered at the time of making the system.
2. Tune coupling parameters is not an easy challenge because off there is not an exact rule to do this to have the system as close as possible to the original one. It is necessary to choose correct inductance value, if it is necessary to get embedded inductors or capacitors and choose the correct value of "Embedded ratio", which is configured to be calculated automatically. However, it can be set manually and if this value is set incorrectly, it can cause erroneous data to be read into the

Fig. 12.17 Three-phase
transmission line coupling
(Typhoon-HIL 2020)

Core Coupling 1

system. For example, it can cause one side of the coupling to read ten times less current than the other side.

3. Due to the use of contactors in the system, a topological error arises when they open, varying the magnitude and phase of the voltages in the system nodes. This produces discrepancies when the relays act. The problem when opening a contactor is that the line attached to it is with an "incorrect" voltage, since its value should be zero. However, when the contactor is opened, the simulation couplings cause a parasitic voltage in these lines that generate current changes on the adjacent side of the coupling, causing the other nodes of the system to change the voltages and therefore discrepancies are generated when the modelled relays act.

With the aim of solve problem 2, it was necessary to read the Typhoon documentation and perform various tests, where it was found that the most similar results to those expected were obtained without embedded inductors or capacitors, which means that it is not necessary to set the TLM (Transmission Line Model) radius (Fig. 12.18).

On the other hand, the optimum inductance of the couplings is to select as coupling inductance the value of the inductance of the nearby line.

Furthermore, the solution found to solve problem 3, was to make use of capacitances on both sides of the couplings. These function as voltage dischargers when a switch is open and maintain a constant voltage when they are closed, or the system is reset. In this way, stray voltages, and their effects on the other nodes in the system are avoided.

- Final measurements block, relay, and switch

Figure 12.19 shows how a relay block is implemented in the case study where it will be seen that each relay has its respective switch and measurement block, in which

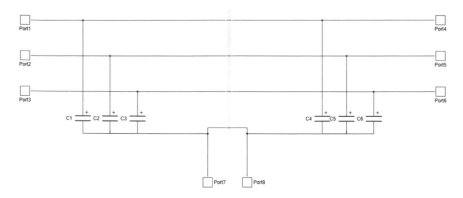

Fig. 12.18 Simulation coupling and its capacitances on both sides, to avoid stray voltages on the line. stray voltages on the line

the instantaneous phase currents and voltages are measured and are sent to HIL API by using the probes available in the simulation tool as shown in Fig. 12.20.

12.3.1.5 Modelled System with Defined Cores

Using what was shown during this section, the complete circuit was assembled, and the 7 different schematics were created for when the fault occurs in each of the 7 lines. Figures 12.21, 12.22, 12.23, 12.24, 12.25, 12.26 and 12.27 show the divisions generated by the couplings for each of the modelled schematics and their respective primary and backup relays. It is important to remember that the circuit is separated into four different cores and that they allude to the maximum allowed by the HIL 402 device.

It is important to stress out that when dividing the circuit into cores, the "Ground" points must be separated and because the circuit must have this same point in common, they must be represented by "tags" joined in the same coupling. If the above is not developed, topological problems will arise in the network that will not allow to compile the system because it will not know which is the correct division of the "SPC's".

12.3.1.6 Current Transformers Saturation

Current transformer saturation was included in the model. The main requirement of a current transformer is the ability to accurately reflect the waveform of the primary current. However, the principle of a traditional electromagnetic transformer is realized based on the core coupling between the primary and secondary windings. Due to the hysteresis of the CT core and its nonlinear characteristics, the secondary current during the fault may contain large harmonics and decaying DC components that can

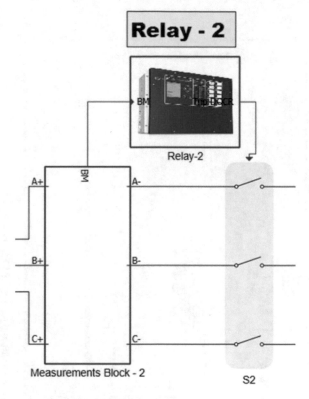

Fig. 12.19 Relay, measurement, and switch system

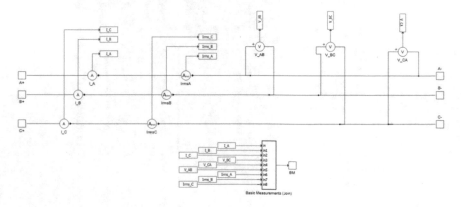

Fig. 12.20 Measurements block

358 J. D. H. Santafé et al.

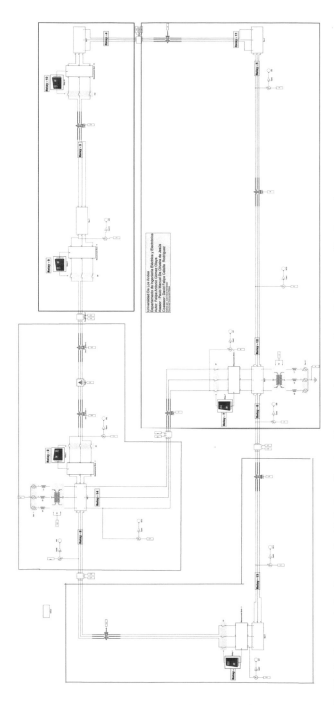

Fig. 12.21 Case study implemented with fault on line 1 and its respective core splits and primary and backup relays

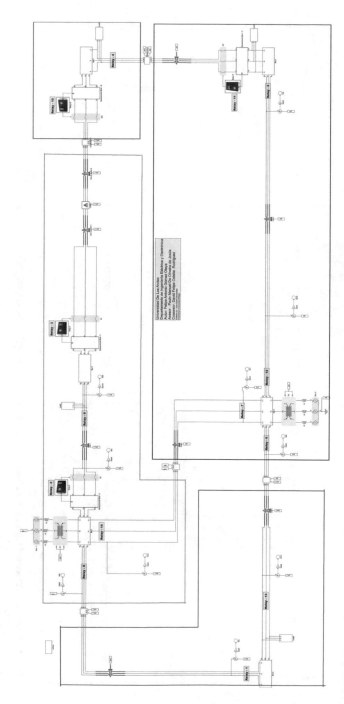

Fig. 12.22 Case study implemented with fault on line 2 and its respective core splits and primary and backup relays

360

J. D. H. Santafé et al.

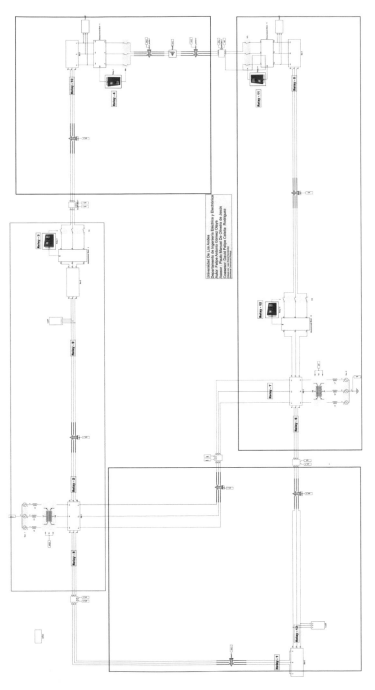

Fig. 12.23 Case study implemented with fault on line 3 and its respective core splits and primary and backup relays

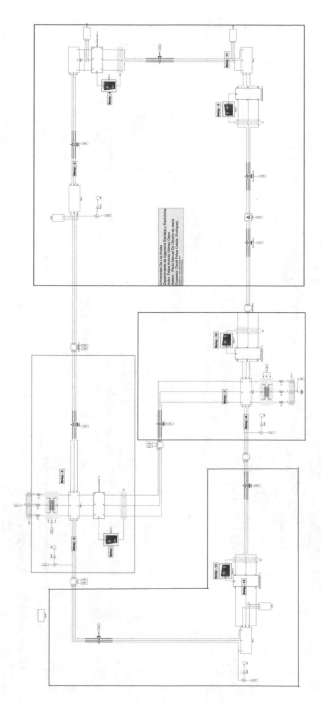

Fig. 12.24 Case study implemented with fault on line 4 and its respective core splits and primary and backup relays

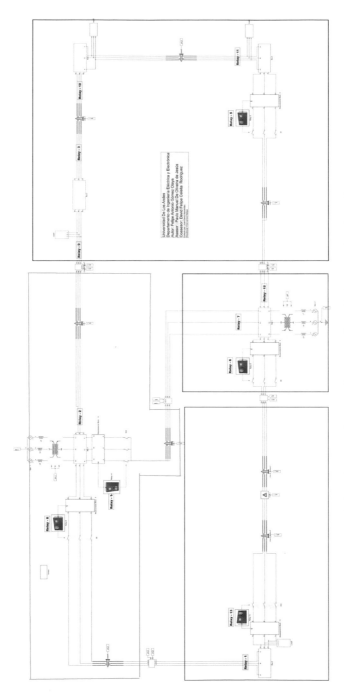

Fig. 12.25 Case study implemented with fault on line 5 and its respective core splits and primary and backup relays

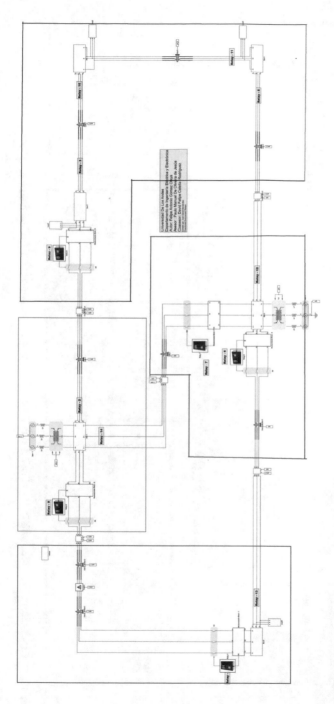

Fig. 12.26 Case study implemented with fault on line 6 and its respective core splits and primary and backup relays

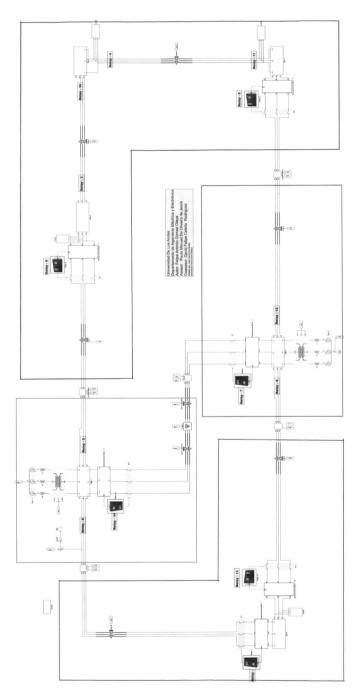

Fig. 12.27 Case study implemented with fault on line 7 and its respective core splits and primary and backup relays

cause the CT to have severe saturation. Consequently, the distortion of the secondary current may cause protection malfunction. For this reason, the current signal in the secondary of the current transformer can experience severe distortion due to saturation and affects the performance of the protection relay (Fernando and Quiñónez 2021).

We used the magnetization curves for different primary-secondary current transformer ratios included in Gers and Holmes (2004). These curves determine the magnetization current and voltage, which, if noted, present some breakpoints that cause the current not to increase significantly after these points. The ratio chosen in this case is 200:5.

12.3.1.7   Transient Topologies

The implemented system considers the transient topologies of the system given at the difference in the tripping time between the two primary relays as shown in this section.

Power system model

A power system with a protection scheme based on directional overcurrent relays can be described with 7 transmission lines (in this study they are presented as π model), 8 transmission nodes, 2 transformers, and 2 generation nodes. Each of the system's k lines is protected by two relays function 67 phases with their respective switch, giving a total number of relays equal to twice the number of lines in the system. The number of relays is 14. 20 is the number of relay pairs main backup. Each fault position h in a faulted line is given per unit line from 5 to 95% ($0.05 \leq h \leq 0.95$).

Protection system model

In a power system with a protection scheme based on directional overcurrent relays there are two types of relays: primary and backup. The primary relays are both associated to the two circuit breakers at the ends of the faulted line. On the other hand, backup relays are those linked to the breakers connected to the ends of the next transmission lines and should operate in if the primary relays do not work properly. In addition, it is generally considered that when there is a fault on a transmission line, both primary relays trip simultaneously to clear the fault. However, this never happens because the relay that senses a higher fault current, usually the one that is closer to the fault, trip first. This results in a transient state of the system between the operation time of the first primary relay to trip, which is called relay "q", until the final operation of the other primary relay, which is called relay "i". Likewise, backup relays of main relay "q" are called relays "p" and backup relays of relay "i" are called relays "j". This information can be seen more clearly in Fig. 12.28.

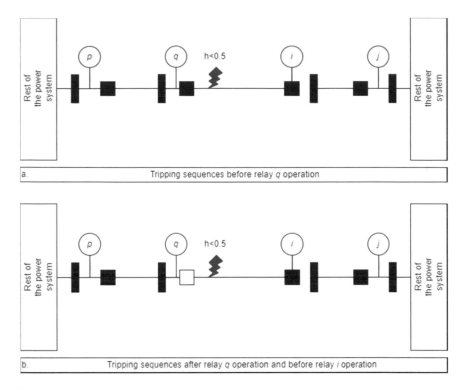

Fig. 12.28 Trigger sequences considering transient topologies

Operation times

Considering only IEC standardized curves and the power system model presented, the operation time of a relay r given by a fault h on a line k (T_{rkh}), can be recalculated for relays q and p with any drawback as can be seen in Eqs. (12.5) and (12.6).

$$T(G)_{rkh} = \frac{\alpha_{r1} D_r}{\frac{G(t)_{rkh}^{\alpha_{r2}}}{P_r} - 1} \tag{12.5}$$

$$\int_0^{T_{rkh}} \frac{1}{T(G)_{rkh}} dt = 1 \tag{12.6}$$

It is important to highlight that the shown integral is a summation that depends on the sampling time of the signal $G(t)$, which is the phase "A" rms current fault seen by the corresponding relay along the time. RMS values are calculated from the signal without any filter. In this case this sampling time is 50 ms. $G(t)$ signal for an example fault can be seen in Fig. 12.29.

However, in order to calculate the operation times of relays i and j, it is necessary to consider the behaviour of the system from relay q operation time until relay i

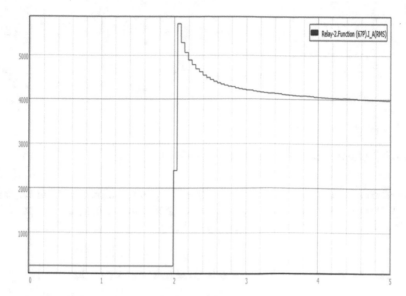

Fig. 12.29 G(t) signal for a free earth fault at 5% of line 1

operation time or its backup relay j operation time in case i is blocked ($T_{qkh} \leq t \leq T^*_{ikh}$ or $T_{qkh} \leq t \leq T^*_{jkh}$). For understanding the computation of the operation times, it is necessary to imagine that a ground fault occurs in a location of the line, at that time, relay q trip first. However, while this is going on, relays i and j are calculating operation times $\left(T_{(i-j)kh}\right)$ in the interval $0 \leq t \leq T_{qkh}$. When relay q is operation, the current seen by i and j increases, so, operation times $\left(T'_{(i-j)kh}\right)$ (denoting transients with (')) are smaller. As a result, effective transient operation times $\left(T^*_{(i-j)kh}\right)$ lies between above-mentioned values. It can be seen graphically in Fig. 12.30 and mathematically in Eqs. (12.7) and (12.8) as (Sorrentino 2020).

Fig. 12.30 Second interval operation times of relays i or j considering transient topologies

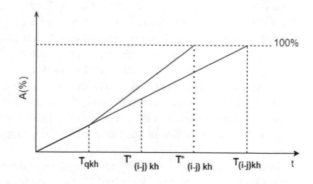

$$T_{ikh}^* = T_{qkh} + T_{ikh}' \left(1 - \frac{T_{qkh}}{T_{ikh}} \right) \tag{12.7}$$

$$T_{jkh}^* = T_{qkh} + T_{jkh}' \left(1 - \frac{T_{qkh}}{T_{jkh}} \right) \tag{12.8}$$

12.3.2 Overcurrent Function of Relays

This section will describe the programming and implementation of overcurrent function of relays whose operation times are calculated considering transient period of the fault currents following Eqs. (12.5) and (12.6).

12.3.2.1 Typhoon HIL Environments Used in Implementation

Relays implementations were performed in the Typhoon HIL Script editor. This tool provides full test-automation capabilities. It allows you to write, open, and execute various automated testing scripts written in Python, using appropriate Typhoon API libraries. In this case, three types of API libraries will be used, which are the HIL API, schematic editor API and Typhoon Test.

HIL API are collection of functions that allow users to real-time control the HIL simulation process from the Python scripts. Functions are divided into four groups: Functions for controlling and initializing the simulation process (function for loading model, starting/stopping simulation…) Functions for setting and changing the state of the simulation (various sct_ functions) Functions for getting information from the simulation (get_ functions including Capture functionality) Various utility functions (Typhoon-HIL 2020).

On the other hand, Schematic API provides a set of functions/methods to manipulate existing schematic models (tse files) and create new ones from scratch programmatically. This is most used for creating scripts for testing and automating repetitive tasks, but its use is not restricted for these use cases only (Typhoon-HIL 2020).

Finally, TyphoonTest is a testing framework that provides high-level functions to test power electronics and power systems equipment. It also provides integration with pytest and Allure framework and is so useful to automate tests. TyphoonTest API is meant to be easier to use and abstract away from HIL-specific details, meaning clearer tests that convey more effectively the intent and that could be used, in the future, for other targets than HIL simulation. Traditional HIL API and TyphoonTest API can be used together in the same test code without problems (Typhoon-HIL 2020).

As mentioned at the beginning of the system implementation section, a.py file was created with a function for each of the lines, added to a main.py file from where the inputs and outputs of each of these functions will be managed.

12.3.2.2   Algorithm Used to Implement Overcurrent Function

Algorithm

1.   Inputs:

    Power system schematics
    [one for each of the seven lines with their main and back-up relays]
    $\alpha r1$, $\alpha r2 \in \alpha R$, $Dr \in D$, $Pr \in P$, $\forall r \in R$ [Relay settings data]
    Relay-pairs data
    Simulation parameters [signal capture time, sampling rate, fault time]
    Working directory path

2.   for k=1 to K do [n_k lines are analyzed]

3.   for f=1 to n_f do [n_f faults are simulated]

4.   h=0.05+((i-1)*(0.95 - 0.05)/(n_f- 1)) [Fault location h is assigned]

5.   for r=1 to n_r do [n_r RMS signals are capturing considering parameters]

6.   Calculate T_{rkh} without considering transient state [Operation times Equations (5) and (6)]

7.   Identify θ_{rkh} [Tripping angle without considering transient state]

8.   Find the faster primary relay

9.   end for

10.   for r=1 to n_{br} do [n_{br} (back up relays number) RMS signals are capturing considering simulation parameters]

11.   Calculate T_{rkh} considering transient state [Operation times Equations (5) and (6)]]

12.   Identify θ_{rkh} [Tripping angle considering transient state]

13.   end for

14.   end for

15.   end for

16. Outputs:

    Relay operation times [Equations (5) and (6))]
    Tripping angle

Input data are grouped in five blocks: (a) power system schematic (complete power system implemented in Typhoon HIL schematic editor); (b) settings of relays (relay curve types, pickup and time dial settings); (c) relay pairs data (vector of relay-pairs associated to each transmission line; (d) simulation parameters (signal capture time, sampling rate, fault time); (e) working directory path (it is necessary to upload the recompiled model in each iteration). The algorithm has four loops to calculate operation times and tripping angle. The first one assigns the faulted line k, therefore, the schematic editor associated to k line. The second one assigns the counter f to analyze n_f fault locations h. These fault positions are calculated uniformly from the number of faults "n_f" faults that the user wants to perform and obtain the operation times of the primary and backup relays of each line. The third loop traverses the relays vector to compute each operation time with relay q block, and four one does the same but considering relay q operation. Results are reported in row 16.

It is important to highlight that to simulate the change of position of the faults, what was done was to divide the transmission lines into two sections, having the fault connected in the center of these, and to vary the distances of the sections of the line. On the other hand, it is important to highlight that the calculation of the distances of the lines is made equidistantly from 5% of the line to 95%. This is because if the fault is closer to the nodes, the software presents numerical problems.

12.4   Results

This section presents the results of simulating different failures in different lines of the base system. The goal is to observe the behavior of the tool when simulating a large system and to do so according to the following assumptions:

1. Only the 67-phase function is considered. Relay time dial and pickup settings as well as curve characteristics are known parameters.
2. Three-phase short-circuit current waveforms are determined using Typhoon HIL. Short circuit current magnitude is considered variant over time considering the loads.
3. Pre-fault conditions are considered.
4. Conventional current transformer saturation is considered.
5. Circuit breaker operation times are not included in the model.
6. Relay time operation considering transient network configurations is estimated according to IEC (2009).

It is important to clarify that, since the implemented system has many nodes, the computational cost increases and the tool is limited in developing the solution of the differential equations in exact time or called Robust Real Time. Therefore, it is necessary to increase the time between each step of solution of the equations and generates a delay in the simulation time although there is a synchronization in the step time, receiving the name of Soft Real Time.

12.4.1 Short Circuit Study Validation

A short circuit study was first performed to compare the steady-state fault currents obtained with the simulated system and those found in Rajput and Pandya (2016). These results allow verifying the veracity of the system simulated in the tool. It is important to remember that, in Typhoon HIL, loads, pre-fault conditions and CT saturation are considered and this impact the results.

The current magnitude results and the error results are presented with respect to those of the literature in Table 12.8. In this table R1 refers to a primary relay and R2 to a backup relay.

As can be seen from the above results, the maximum relative error found is 5.5%. This indicates a great similarity between the case study simulated in the Typhoon tool and the results obtained in Rajput and Pandya (2016). This allows for comparing the relay operation results with the theoretically expected ones since both models are comparable.

Table 12.8 Fault currents obtained at Typhoon and relative errors of fault currents found at Typhoon and in the literature (Rajput and Pandya 2016)

Line	R1	R2	ETAP [24]		Typhoon HIL		Error	
			R1(kA)	R2(kA)	R1(A)	R2(A)	R1(%)	R2(%)
6	1	6	2,7	2,7	2591,71	2589,86	4,01	4,08
1	2	1	5,38	0,81	5454,96	792,13	1,39	2,21
1	2	7	5,4	1,54	5454,96	1514,71	1,02	1,64
2	3	2	3,34	3,34	3440,05	3441,16	3,00	3,03
3	4	3	2,24	2,24	2290,03	2273,19	2,23	1,48
4	5	4	1,36	1,36	1307,89	1308,41	3,83	3,79
5	6	5	4,99	0,42	5036,77	402,12	0,94	4,26
5	6	14	4,99	1,54	5036,77	1477,27	0,94	4,07
7	7	5	4,26	0,42	4442,1	417,38	4,27	0,62
7	7	13	4,26	0,81	4442,1	816,69	4,27	0,83
6	8	7	4,99	1,54	5022,87	1470,12	0,66	4,54
6	8	9	4,99	0,42	5022,87	396,89	0,66	5,50
1	9	10	1,45	1,45	1431,93	1432,44	1,25	1,21
2	10	11	2,34	2,34	2321,49	2320,45	0,79	0,84
3	11	12	3,49	3,49	3599,3	3600,71	3,13	3,17
4	12	13	5,38	0,81	5471,23	795,39	1,70	1,80
4	12	14	5,38	1,54	5471,23	1533,59	1,70	0,42
5	13	8	2,5	2,5	2413,32	2411,56	3,47	3,54
7	14	1	4,26	0,81	4430,08	813,67	3,99	0,45
7	14	9	4,26	0,42	4430,08	419,32	3,99	0,16

12.4.2 Example of Relay Tripping Sequence

Figures 12.31, 12.32, 12.33 and 12.34 show, for illustration purposes, the results considering an inserted three-phase solid faults located at 95% with coordination solution provided by Sorrentino (2020). Waveforms and tripping sequences are presented using SCADA HIL. The aim of these results of instantaneous currents and relay logic outputs is to appreciate the proper behavior of the relays involved during the externality.

In Figs. 12.31 and 12.32, transients do not affect the operation of the relay. As can be seen, the instant the fault is initiated a transient arises in relay "2" which is cleared at approximately 1.1 [s], and due to the opening of its associated breaker the impedance at the point of fault changes, generating a transient current seen by relay "9" being cleared at 1.2 [s] from the start of the fault. As can be seen, the effect of current variation is not only seen by the primary devices, but the backup devices are also subject to these abrupt variations and at approximately 1.25 [s] relay "10" 39

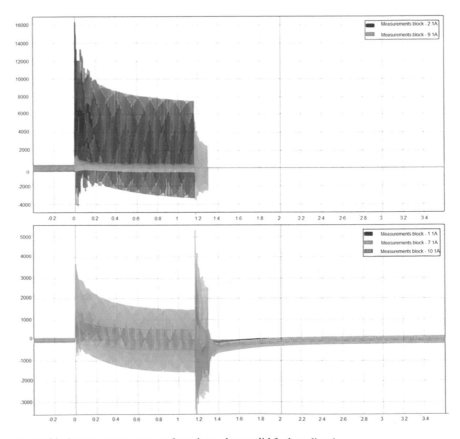

Fig. 12.31 Instantaneous currents for a three-phase solid fault on line 1

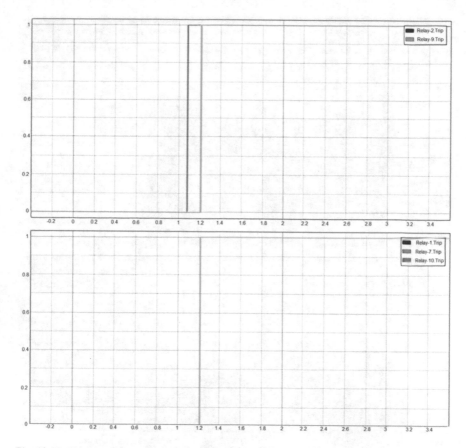

Fig. 12.32 Trip signal for a three-phase solid fault on line 1

operates. This really shows the good performance of the system implemented in the tool, but on the other hand suggests a big challenge when placing the relay input settings so that selectivity problems do not occur as seen with the backup relay in this case, where it should not operate because the primaries did it first.

In Figs. 12.33 and 12.34, at the onset of the fault, relay "10" that is closest to the fault will operate first and generate the transient currents seen in "3". Unlike the previous example, the selectivity in this example is optimal since the backup relays did not trip due to the clearing done by their primary pairs.

12.4.3 Resulting Operation Times

This section presents the resulting DOCP operation times in the 8-bus test system implemented in Typhoon considering one fault per line (located at 50% of line

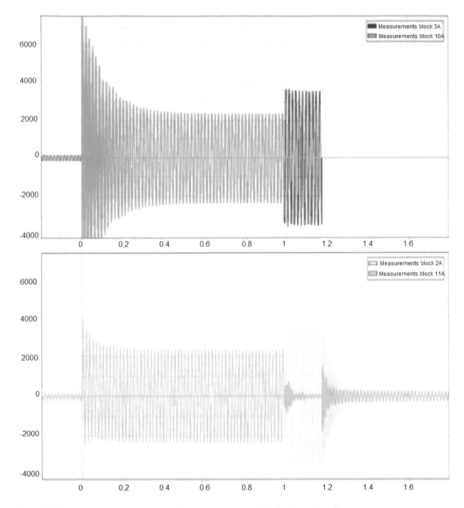

Fig. 12.33 Instantaneous currents for a three-phase solid fault on line 2

length) for the coordination solution provided by the optimization method presented in Sorrentino (2020).

Relay operation times for all primary and backup relays shown in Table 10 were computed using the exact IEC formula (Eqs. 12.5 and 12.6) [IEC 60,255–151:2009, 19] accounting three-phase fault current waveforms at all line ends considering stable and transient components, dynamic topologies generated by non-simultaneous actuation of a circuit breakers, load currents and current transformers saturation.

Operation times associated with primary relays i and q and, backup relay p and j (see Table 12.9) are written in cells shaded in red, green, yellow, and blue, respectively. Each row of Table 10 corresponds to the tripping sequence for a three-phase

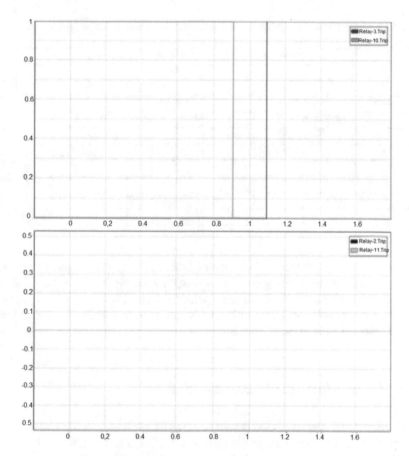

Fig. 12.34 Trip signal for a three-phase solid fault on line 2

solid fault located at 50% of line length from line 1 to line 7 and each column corresponds to a relay enumerated as shown in Fig. 12.2.

Consider a fault at 50% of the line 1, between relay 2 and 9. The faster primary relay "q" is 9 with an operation time of 0.83 s (green color). The slowest primary relay "i" is 2 with 1.31 s (red color). Backup relays "j" of "i" are 1 and 7 with operation time of 1.76 s and 1.69 s (blue color), respectively. Backup relay "p" of "q" is 10 with 1.38 s (yellow color).

In this case, when a fault is inserted in line 1, relays 1, 2, 7, 9, and 10 are sensitive and the separation times in relay pairs 2–1, 2–7 and 9–10 are 0.45 s, 0.38 s and 0.55 s. The coordination solution provided by Sorrentino (2020) consider an allowable coordination interval of 0.3 s. In this case, when a fault is inserted in line 1, relays pairs 2–1, 2–10 and 9–10 are selective.

Table 12.9 Operation times considering transients, saturation and loads in Typhoon (exact IEC formula)

Fault	Operation times with parameters obtained in [14] considering transients													
50%	Relay													
/	1	2	3	4	5	6	7	8	9	10	11	12	13	14
Line	Times [s]													
1	1,76	1,31					1,69		0,83	1,38				
2		1,59	1,18							1,04	1,51			
3			1,49	1,02							1,19	1,66		
4				1,37	0,82							1,32	1,89	1,71
5					2,59	1,13	1,49						1,04	1,87
6	0,98					1,49	1,91	1,09	2,81					
7	NT				2,98		1,22		NT				2,89	1,17

If we consider a fault in line 7 (last row of Table 12.9), we can observe that backup relays 1 and 9 are not tripping (NT) since they are unable to detect the fault, and therefore there are no selectivity in relay pairs 14–1 and 14–9.

The implementation in Typhoon is able to get operation times for several faults located between 5 and 95% of each line.

12.5 Conclusions and Recommendations

In this chapter, an interconnected power system of eight nodes, seven 150 kV transmission lines and 14 relays originally proposed by Braga and Saraiva in (1996) was implemented under a real-time simulation platform (Typhoon HIL) for the verification of the behavior of a protection system based on directional overcurrent relays (DOCR).

The simulation was performed in soft real-time. This allows the platform to determine the sequence of operation, the fault current detection capability and the operation time of the directional overcurrent relays under solid three-phase to ground fault conditions.

The developed methodology to obtain the operation times considering the transient period of several fault currents in the Script Editor can be very useful for the protection engineers since it would allow them to review the performance of a DOCP scheme configuration considering as many faults as they consider necessary, making a more accurate calculation of the operation times of the relays and get results automatically.

Protection engineers will benefit of the proposed implementation to evaluate the quality of any directional overcurrent protection scheme, optimized or not, under realistic conditions.

We recommend the use of the proposed implementation with the real relay in Hardware in the loop simulation. Additional improvements can be included in the future such as detailed modelling of synchronous generators and the uncertainty associated with the angle of insertion of the faults. Finally, it is recommended to implement a filter in the fault current signals used for the calculation of the RMS values which in turn will be used in the calculation of the operating times.

Acknowledgements The authors are grateful to Professor Elmer Sorrentino for the valuable comments and suggestions.

References

Albasri FA, Alroomi AR, Talaq JH (2015) Optimal coordination of directional overcurrent relays using biogeography-based optimization algorithms. IEEE Trans Power Delivery 30(4):1810–1820, https://doi.org/10.1109/TPWRD.2015.2406114

Almas MS, Leelaruji R, Vanfretti L (2012) Over-current relay model implementation for real time simulation amp; Hardware-in-the-Loop (HIL). IECON 2012—38th annual conference on IEEE industrial electronics society, pp 4789–4796, https://doi.org/10.1109/iecon.2012.6389585

Amraee T (2012) Coordination of directional overcurrent relays using seeker algorithm. IEEE Trans Power Delivery 27(3):1415–1422, https://doi.org/10.1109/tpwrd.2012.2190107

Braga AS, Tome Saraiva J (1996) Coordination of overcurrent directional relays in meshed networks using the Simplex method 3:1535–1538, https://doi.org/10.1109/MELCON.1996.551243

Birla D, Maheshwari RP, Gupta HO (2005) Time-overcurrent relay coordination: a review. Int J Emerg Electr Power Syst 2(2)

Damchi Y, Dolatabadi M, Mashhadi HR, Sadeh J (2018) MILP approach for optimal coordination of directional overcurrent relays in interconnected power systems. Electric Power Syst Res 158:267–274, 0378–7796, https://doi.org/10.1016/j.epsr.2018.01.015. dirección: http://www.sciencedirect.com/science/article/pii/S0378779618300233

Fernando B, Quiñónez C (2021) Estudio de saturación de transformadores de corriente : análisis y simulación Current transformer saturation study : analysis and simulation. Brazilian Appl Sci Rev, 1222–1243, https://doi.org/10.34115/basrv5n2-045

Gers JM, Holmes EJ (2004) Protection of electricity and distribution networks, ép. 47, second edition. IET y Power energy

Hussain MH, Rahim SR, Musirin IJ (2013) Optimal overcurrent relay coordination: a review. Procedia Eng 53:332–336, https://doi.org/10.1016/j.proeng.2013.02.043

Hussin NH, Idris MH, Amirruddin M, Ahmad MS, Ismail MA, Abdullah FS, Mukhta NM (2016) Modeling and simulation of inverse time overcurrent relay using Matlab/Simulink. 2016 IEEE international conference on automatic control and intelligent systems (I2CACIS), pp 40–44, https://doi.org/10.1109/i2cacis.2016.7885286

IEC 60255-151:2009 Measuring relays and protection equipment—Part 151: Functional requirements for over/under current protection (2009)

IEEE Standard for Inverse-Time Characteristics Equations for Overcurrent Relays. IEEE Std C37.112–2018 (Revision of IEEE Std C37.112–1996), pp 1–25. https://doi.org/10.1109/IEEESTD.2019.8635630

Martín SS, Fernández MB (2016) Model and performance simulation for overcurrent relay and fault-circuit-breaker using Simulink. Int J Electr Eng Educ 43(1):80–91, https://doi.org/10.7227/ijeee.43.1.8

Mahari A, Seyedi H (2013) An analytic approach for optimal coordination of overcurrent relays. IET Gener Trans Distrib 7(7): 674–680, https://doi.org/10.1049/iet-gtd.2012.0721

Mahboubkhah A, Talavat V, Beiraghi M (2020) Considering transient state in interconnected networks during fault for coordination of directional overcurrent relays. Electric Power Syst Res 186:106 413, issn: 0378–7796. https://doi.org/10.1016/j.epsr.2020.106413. dirección: http://www.sciencedirect.com/science/article/pii/S0378779620302194

Noghabi AS, Sadeh J, Mashhadi HR (2009) Considering different network topologies in optimal overcurrent relay coordination using a hybrid GA. IEEE Trans Power Delivery 24(4):1857–1863, https://doi.org/10.1109/TPWRD.2009.2029057

Rajput V, Pandya K (2016) On 8-bus test system for solving challenges in relay coordination. 2016 IEEE 6th international conference on power systems (ICPS), pp 1–5, https://doi.org/10.1109/icpes.2016.7584009

Raza SA, Mahmood T, Bukhari SB (2014) Optimum overcurrent relay coordination: a review. The Nucleus 51(1):37–49

Singh M, Panigrahi BK (2014) Minimization of operating time gap between primary relays at near and far ends in overcurrent relay coordination, 1–6, https://doi.org/10.1109/NAPS.2014.6965354

Sorrentino E, Rodríguez (2020) A novel and simpler way to include transient configurations in optimal coordination of directional overcurrent protections. Electric Power Syst Res 180:106–127, https://doi.org/10.1016/j.epsr.2019.106127

Typhoon-HIL (2020) Typhoon hardware-in-the-loop testing solutions, dirección: https://www.typhoon-hil.com/. Accessed 2020

Urdaneta AJ, Perez LG, Restrepo H, Sanchez J, Fajardo J (1995) Consideration of the transient configurations in the optimal coordination problem of directional overcurrent relays, 169–173, https://doi.org/10.1109/ICCDCS.1995.499138

Urdaneta AJ, Nadira R, Jimenez LP (1988) Optimal coordination of directional overcurrent relays in interconnected power systems. IEEE Trans Power Delivery 3(3):903–911, https://doi.org/10.1109/61.193867

Chapter 13
Distance Protection Relay Testing Using Virtual Hardware-in-the-Loop Device

Le Nam Hai Pham and Francisco Gonzalez-Longatt

Abstract The complexity of modern power system phenomena challenges power system protection testing to obtain the required adequacy of the testing environment before actual implementation. Consequently, a virtual hardware-in-the-loop (HIL) testing platform for line protection is developed and tested in this chapter. The distance protective relay is tested using a real-time HIL simulation validated by theory-based calculation and DIgSILENT PowerFactory software with three-phase and single-line-to-ground short-circuit fault cases. A methodology of creating virtual HIL distance protection relay based on Typhoon HIL (framework for the testing real-time embedded system) is proposed to allow protection engineers to access the performance of the protective relay without concerns related to hardware or physical device.

Keywords Distance protection testing · Typhoon HIL · Virtual hardware-in-the-loop

13.1   Introduction

Modern protection relays are becoming more advanced, with the number of built-in features and capabilities increasing. Most protection relays would not only have one protective relaying function, but they would also have several. The protective relay functions are highly dependent on the software/firmware of the protective relay and

L. N. H. Pham (✉) · F. Gonzalez-Longatt
University of South-Eastern Norway, Porsgrunn, Norway
e-mail: Le.Pham@usn.no

F. Gonzalez-Longatt
e-mail: fglongatt@fglongatt.org

F. Gonzalez-Longatt
Centre of Smart Grid, University of Exeter, Exeter, UK

© The Author(s), under exclusive license to Springer Nature Singapore Pte Ltd. 2023   379
S. M. Tripathi and F. M. Gonzalez-Longatt (eds.), *Real-Time Simulation and Hardware-in-the-Loop Testing Using Typhoon HIL*, Transactions on Computer Systems and Networks, https://doi.org/10.1007/978-981-99-0224-8_13

only be limited by the number of inputs and outputs the physical protection relay has.

At the same time, the protection relay testing methods and equipment are becoming more powerful. The need for protection relay testing can be divided into four categories:

- **Type testing**. It is an extensive process where the quality of a newly fabricated relay or new software revision for a relay model is concerned.
- **Acceptance testing**. The protective relay is tested to prove that it is the correct model and that all the features are working as they should. It consists of functional tests of inputs, outputs, displays, communication, and in some cases, pre-defined pickup and timing tests.
- **Commissioning testing**. It is a site-specific test to confirm all protective elements and logic settings are correct for their intended uses.
- **Maintenance testing**. It is used to ensure that a protective relay continues to operate as it should.

These types of testing protection relaying functions play a vital role in ensuring protection systems' secure and reliable operation. Many researchers have investigated testing platforms for a variety of subsystems in the protection system; ranging from SCADA systems (Montaña et al. 2018), communication platforms (Pazdcrin et al. 2018), non-directional overcurrent and directional overcurrent relays (Rodriguez et al. 2018), and distance protection relays (Camarillo-Peñaranda et al. 2020).

However, as modern systems grow in complexity, particularly in software, these critical tests of protection relay are easier said than done. *Hardware-in-the-loop* (HIL) is the solution for solving the difficulties of testing complex systems by creating a scalable test system ensuring comprehensive test coverage. A HIL distance function testing platform for feeder protection was shown in Camarillo-Pcñaranda et al. (2020); it uses PSCAD software and the virtual relay function with a commercially available microcontroller. The authors in Camarillo-Peñaranda et al. (2020) emphasised the needed realistic and flexible platforms for testing protective relay functions.

This chapter proposes a methodology for creating virtual HIL distance protection relay based on Typhoon HIL (software for the testing real-time embedded system); it allows protection engineers to access the protective relay's performance without related concerns to hardware or physical devices. An example of a distance protection relay connecting two grids with fault injection from Distance protection relay with false tripping prevention (2022) is performed. This chapter explains the distance protection function testing platform for an illustrative test system using model-based engineering toolchains of Typhoon HIL, Schematic Editor and HIL SCADA. The methodology of creating and modelling distance protection relay, the logic of the protection relay, and communication between two toolchains through multiple tests is presented. Validation of the proposed approach in Typhoon HIL is conducted considering two different approach methods: theory-based calculation and built-in model DIgSILENT PowerFactory with real protective relay SEL-411L (Latif et al. 2020). Through this chapter, readers can grasp the concept of distance protection relays through testing cases of accessing protective relay performance in different methods.

This chapter begins by introducing the problem and giving an overview of the distance protection function. The protective relay and testing system model creations are introduced in the following section; then, the proposed methodology is presented. The results of the proposed distance protective relay with the validation are performed. Finally, the conclusions close this chapter.

13.2   Problem Definition

13.2.1   *Distance Protection Function*

In the design of reliable and secure transmission systems, the *distance protection function* is denoted in the USA as ANSI 21 according to the ANSI Standard Device Number (ANSI/IEEE Standard C37.2) (IEEE Standard Electrical Power System Device Function Numbers, Acronyms, and Contact Designations 2008) and in Europe as PDIS according to *The European IEC 61,850–2* (Dede et al. 2014).

Distance protection is a method of protecting transmission lines from faults in the power system. It measures the impedance between the protective relay location and the point where the short circuit is located and compares it with the pre-determined value (Anderson 1999). If the measured impedance ($|Z_{meas}|$) is less than a pre-defined set value ($|Z_{set}|$), the distance protective relay operates by sending a trip signal to the circuit breaker to isolate the faulty section. Since the total transmission line impedance ($|Z_{line}|$) is directly proportional to line length, the impedance of fault location, therefore, can be determined directly.

The distance protection function provides the protection for transmission lines under multiple faults as follows: (i) Short-circuit faults, (ii) Failed broken conductor and voltage transformer fuse, (iii) Dead line charging, and (iv) Power swing.

Distance protection schemes are commonly employed to provide primary or main protection and backup protection for AC transmission and distribution lines against three-phase, phase-to-phase, and phase-to-ground short circuits (Erezzaghi and Crossley 2003).

In an overall scheme of distance protection, it is necessary to provide a number of relays to obtain the required discrimination. Therefore, modern practice is to adopt the method of protection applied in three zones.

A number of distance protection relays are used in association with timing relays so that the power system is divided into a number of zones with varying tripping times associated with each zone.

The first zone tripping, which is instantaneous, is normally set to 80% to 90% of the protected section. Zone 2 protection with a time delay sufficient for circuit breaker operating time and discriminating time margin covers the remaining 20% portion of the protected section plus 25% to 40% of the next section. In the distance

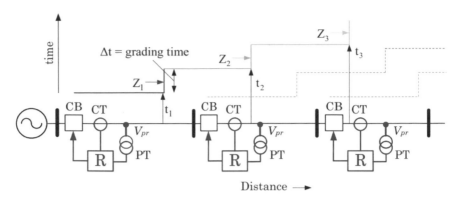

Fig. 13.1 Distance protection principle, graded protection zone

protection relay scheme, zone 2 also provides backup protection for the protective relay in the next section. Zone 3, with still more time delay, offers complete backup protection for all faults at all locations (Horowitz and Phadke 2014) (Fig. 13.1).

The advantage of distance protection is that it does not require a communication channel, making its application more inexpensive than more complex and sensible protection schemes such as differential protection. Additionally, it can be applied to a line with bidirectional power flows, so this protection scheme is more suitable for transmission and some distribution networks.

It is worth mentioning that distance protection is more selective than non-directional and directional overcurrent protections (Anderson 1999). However, distance protection also has some drawbacks. For instance, it does not cover the total length of the line, making it necessary to set various protection zones to protect the entire line and guarantee proper coordination between lines.

Depending upon the distance protection characteristics, there are the following types of distance protection relay:

- Reactance relay (X).
- Impedance relay $(|Z|)$.
- Mho relay $(|Y|)$.

The reactance (X) type protection relay is preferred for protecting short-line sections as they are not affected by the arc resistance (R_f), and more percentage of the line can be protected at high speed. The impedance $(|Z|)$ type protection relay is suitable for phase fault protection for lines of moderate length. The arc affects the impedance relay more than the reactance relay. Mho type $(|Y|)$ protection relay is best suited for long lines and particularly where a severe synchronising power surge may occur. The mho protective relay is reliable because it combines both the directional and distance measuring functions in one unit.

Distance protection plays a vital role in protecting the bulk power industry's transmission lines, especially in highly meshed systems. Due to this vital role, it is crucial to appropriately conduct acceptance testing to guarantee the correct operation of entire functional modules of the distance protection relay. Therefore, virtual real-time simulation techniques are needed to conduct this type of distance protection relay test.

13.2.2   Fault Impedance "Seen" by the Distance Protective Relay

An exciting term used in the protection environment is the word "seen". The impedance seen by a distance protection relay is calculated as the ratio between the voltage and current measurements at the installation location. During a short circuit fault, the impedance "seen", or "measured" by the distance protection relay is different from the normal condition; consequently, the distance relay operates in response to changes in the ratio of measured current and voltage.

The distance protection relay measures the impedance at the secondary of the measurement transformers, and the impedance measurement depends on the connection and the ratio of the voltage transformer (VT) and current transformers (CT).

According to Anderson (1999), the phases voltages and line currents measured and injected into a distance protection relay are line-to-neutral bus voltages and line currents obtained at the secondary side of the measurement's transformer can be expressed as follows:

$$V_{pr} = \frac{V_{pb}}{N_{vt}} \tag{13.1}$$

$$I_{pr} = \frac{I_{pl}}{N_{ct}} \tag{13.2}$$

where V_{pr} represents the voltage of the secondary side of the transformer; V_{pb} is the voltage of the primary side of the transformer; N_{vt} is the ratio of VT; I_{pr} represents the current of the secondary side of the transformer; I_{pl} is the current of the primary side of the transformer; N_{ct} is the ratio of VT.

Now, consider that the protective relay calculates the ratio of secondary voltage and secondary line current applied to it. The ratio (measured impedance $|Z_{meas}|$) is given by the following equation:

$$|Z_{meas}| = \frac{V_{pr}}{I_{pr}} = N_f \cdot m \cdot Z_l \tag{13.3}$$

where N_f is the ratio of impedance "seen" by the protective relay; m is the distance from the bus to the short-circuit location in km, Z_l is the positive-sequence impedance of the line in ohms/km.

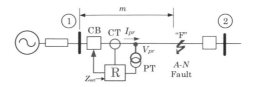

Fig. 13.2 Single line diagram showing a single-line-to-ground short-circuit at phase A, point "F"

In this chapter, the authors decided to consider two types of short-circuit conditions for the *Test System*. The short-circuits are used for testing purposes of the distance protection relay; specifically, three-phase short-circuits and single-line-to-ground short-circuits.

A three-phase short-circuit is a symmetrical fault that is not statistically the most frequent short-circuit in the transmission system but is typically the most severe. Since it is a symmetrical terminal condition and the system is assumed to be balanced, as a consequence, the magnitude of the voltage in all phases is equal ($|V_{an}| = |V_{bn}| = |V_{cn}|$), and the same principle applies to currents ($|I_a| = |I_b| = |I_c|$).

A single-line-to-ground short-circuit is an unsymmetrical fault and is statistically the most common fault in the power grid. A single line diagram showing a single-line-to-ground short-circuit can be shown in Fig. 13.2. Considering the theory of symmetrical component, the equivalent sequence networks are connected in series for this kind of short-circuit. The current only flows in the faulted phase and is connected to the ground ($|I_a| = I_f$ and $|I_b| = |I_c|$=0)., while the current in the other two phases is zero.

According to Saadat (1999), the three-phase short-circuit current, and the single-line-to-ground short-circuit current can be calculated by Eqs. (13.4) and (13.5), respectively.

$$I_{3f} = \frac{V_f}{Z_f + Z_{s1}} \qquad (13.4)$$

$$I_{1f} = \frac{V_f}{Z_f + \frac{1}{3}(Z_{s0} + Z_{s1} + Z_{s2})} \qquad (13.5)$$

where I_{3f} is 3-phase short-circuit current; I_{1f} is single-line-to-ground short-circuit current; V_f is normal phase voltage at the short-circuit location; Z_f is the short-circuit fault impedance; Z_{s0}, Z_{s1}, Z_{s2} is the zero sequence, positive sequence and negative sequence network impedance, respectively.

It is noted that for a single-line-to-ground short-circuit, a distance protective relay sees a combination of line impedance and ground impedance. Compensation factors allow a protection relay to factor out the portion of the ground impedance seen at the protection relay location.

In order to accomplish the goal of measuring the distance to the short-circuit location in terms of positive sequence, some forms of compensation factors are used. These factors depend on the type of distance protection relay from different manufacturers with different formulas, compensation type names, and symbols. For instance, some of the most used factors are symbolised as K_N, K_0, K_E or K_G, which are also referred to as earth-return compensation factor, ground-return compensation factor, and neutral impedance correction factor, respectively.

Taking the earth-return compensation factor K_N (also known as residual compensation factor) into a practical common practice in the protection engineering world. In this case, the factor is used to modify the measured current, and the impedance seen by the distance protection relay is expressed as:

$$Z_{relay} = \frac{V_{pl}}{I_{pl} + K_N I_n} \qquad (13.6)$$

$$K_N = \frac{\frac{Z_1}{Z_0} - 1}{3} \qquad (13.7)$$

where, Z_{relay} is the fault impedance "seen" by the distance protection relay; V_{pr} is the protection relay voltage; I_{pr} is the protection relay current; I_n is the residual current; Z_0 is the zero-sequence of the transmission line; Z_1 is the positive sequence of the transmission line. These compensation factors can be illustrated specifically in Sorrentino (2014). However, these factors are not considered in this chapter.

Conventional formulas used in this chapter

The conversion between the real and imaginary parts of the complex number is used in this chapter. Specifically, the fault impedance (Z_{relay}) "seen" by the protection relay can be written as:

$$Z_{relay} = Z\angle\theta° = R + jX \qquad (13.8)$$

where, Z is the magnitude of the fault impedance, θ is the phase angle, R is the resistance component, and X is the reactive component.

R and X can be calculated as follows:

$$R = Z \cos\theta \qquad (13.9)$$

$$X = Z \sin\theta \qquad (13.10)$$

13.2.3 Test System

For distance protection relay testing, the authors have decided to use a straightforward *Test System*; it consists of a transmission system interconnecting two simplified power

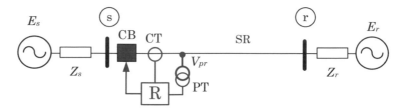

Fig. 13.3 *Test System.* The single-line diagram of the 230 kV transmission system with the distance protective relay ("R") is located at the sending terminal ("s")

Table 13.1 Definition and values of the test system

Parameter	Value	Unit
Rated system voltage	110	kV
Frequency	50	Hz
Length of transmission line	100	Km
Line positive sequence impedance	$77.564\angle74.574\ °$	Ω
Line zero sequence impedance	$198.148\angle74.914\ °$	Ω
Sending side impedance	$2.651\angle3.994\ °$	Ω
Receiving side impedance	$77.564\angle74.574\ °$	Ω
CT ratio	500/1	A
VT ratio	230,000/200	V

Table 13.2 Secondary sequence and series impedance

Parameter	Value	Unit
Line positive sequence impedance, Z_1	$20.164\angle74.574\ °$	Ω
Line zero sequence impedance, Z_0	$51.518\angle74.914\ °$	Ω
Sending side impedance, Z_s	$0.689\angle3/994\ °$	Ω
Receiving side impedance, Z_r	$2.778\angle1.7339\ °$	Ω

systems represented by a model of a constant voltage source behind the impedance, and is shown in Fig. 13.3. The external grids on both ends (s for sender and r for receiving) have a three-phase voltage source an inductive impedance.

Both voltage sources (E_r and E_s) have the same system voltage with no phase difference, and there are no loads connected; therefore, no current flows between the sending and receiving end at nominal operation. Thus, the previously mentioned condition represents no load condition in the system previous to the short circuit. The numerical values and definitions of the *Test System* parameters are shown in Table 13.1, and the impedances of transmission line, and voltage sources are shown in Table 13.2. Between the grid on the left side and the transmission line "SR", a distance protection relay is located at the sending terminal and is responsible for transmission line protection (primary protection).

13.3 Virtual HIL Distance Protection Relay

This section presented the main aspects of the modelling and simulation of distance protection relays using a digital real-time simulation framework via model-based toolchains of Typhoon HIL (two critical modelling and simulation environments, Schematic Editor and HIL SCADA).

As distance protection is well-known for being used in transmission systems, the authors have decided to use a straightforward test system described in the previous section. However, all the methods explained in this chapter can be extended to more complicated topologies of distribution systems.

13.3.1 Schematic Editor

The actual implementation of the test system shown in Fig. 13.3 inside the Typhon HIL environment has been slightly modified to consider the features required by the real-time simulation.

The *Test System* consists of two external grids (❶-left side and ❶-right side) connected to a single transmission line, but it has been divided into two sections of transmission lines connected in series (❷) and a circuit breaker (❻) and two customisable fault points (❸). This *Test System* is designed to emulate a transmission system with multiple infeed.

The transmission system is assumed to be working at a single voltage of 230 kV, 50 Hz and the single line diagram is shown in Fig. 13.3, and the schematic implementation based on Typhoon HIL Schematic Editor is accordingly modelled and can be depicted in Fig. 13.4.

For testing purposes of distance protection relay, two customisable faults are located at the transmission line; for the sake of simplicity, short-circuit cases are in the middle and at the end of the transmission line.

The *Test System* model consists of the following components: (❶) External grids, (❷) Transmission line, (❸) Fault, (❹) Circuit Breaker, (❺) Meter, (❻) Distance protective relay. The components from (❶) to (❺) can be defined by available

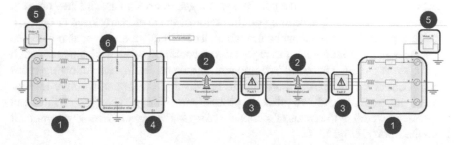

Fig. 13.4 *Test System*: Complete model of 230 kV transmission line system from Schematic Editor

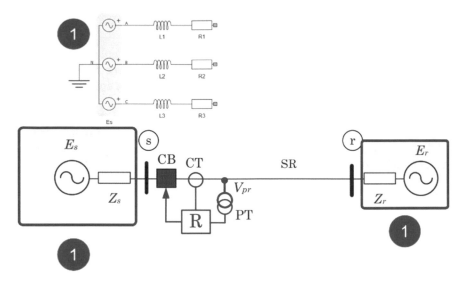

Fig. 13.5 *"External grid"* component (❶)

elements in the library of Schematic Editor that can be given in detail in Component library (2022). However, distance protection relay (❻) is the complex combination of multiple functional blocks in the library.

External grids (❶): these components can be defined by *"Three-phase Voltage Source"* (*"Source"* library) connected with "Inductor" and "Resistor" (*"Passive Component"* library) (Fig. 13.5).

Transmission line (❷): the transmission line (Line SR) is created by two components, *"Transmission line"* (*"PI Section"* of the *"Transmission Lines"* library) in series. Each component has the same zero sequence, positive sequence impedance according to Table 13.1; however, the length of each component is half of the transmission line, Line SR (Fig. 13.6).

Faults (❸): This component consists of a set of switches and a control block named "Control State Machine". The control state machine is used to define the type of short-circuit to be simulated; the control system generates signals, and they are sent to the four switches and, depending on the status of the switches (closed or open), a specific type of short circuit can be modelled. The "Fault" component also includes resistances in the configuration to replicate the condition of arc resistance for each type of short circuit. The "fault" component allows modelling the following types of short circuits: three-phase short-circuit, line-to-line short-circuit, single-line-to-ground short-circuit, and two-phase-to-ground short-circuit. The type of fault can be defined by coding the status of switches (*"Single Pole Single Throw"* component from the library) (Fig. 13.7).

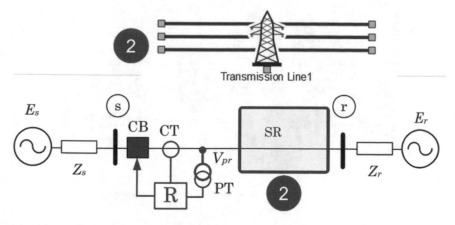

Fig. 13.6 Transmission line "Line SR" component (➋)

Circuit Breaker (➍): An ideal circuit breaker can be simulated with this component, the name in Typhoon HIL schematic library is *"Triple Pole Single Throw Contact"*, and it is created from the *"Ideal Contactors"* branch of the library (Fig. 13.8).

Meters (➎): This component is used to generate the system frequency based on a single phase PLL from a voltage measurement. It can be created by *"Single phase PLL"* from the *"Signal Processing"* branch; see the implementation details in Fig. 13.9.

Distance protective relay (➏): The distance protective relay is modelled by combining multiple available functional blocks. This component consists of three inputs and four outputs. Three inputs (A+, B+, C+) and three outputs (A-, B-, C-) are used to connect three phases A, B, and C, to other 3-phase elements. The output named *"switch signal"* is intended to send the trip signal to the circuit breaker (➍) (Fig. 13.10).

Inside the distance protection relay, there are two main functional areas as shown in Fig. 13.11: (➊) Measurements of line current (I_a, I_b and I_c) and line-to-neutral voltage on the transmission line (V_{an}, V_{bn} and V_{cn}), and (➋) Logic of the distance protection relay. These main functional areas can be illustrated as follows:

(➊) Measurements of current and voltage on transmission line

This functional area consists of a *"current measurement"* block and a *"voltage measurement"* block from the Schematic library as main elements. These blocks are used to measure current and voltage on the transmission line. The testing model does not have CT and VT. Therefore, to represent a CT and VT, a gain must be used to transfer the primary quantities and obtain the secondary current and voltage. The CT ratio and VT ratio are set according to Table 13.1. The measured current (with the tagged name *"IA"*) and measured voltage (with the tagged name *"VAn"*) are the inputs of blocks in (➋) logic of the distance protection relay.

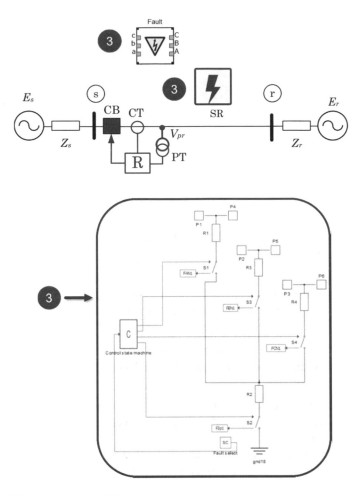

Fig. 13.7 *"Fault"* component (❸)

(❷) logic of the distance protection relay

This functional area is dedicated to receiving the measurement values from ❶and outputting the trip signal according to pre-determined values from HIL SCADA interface if the fault occurs.

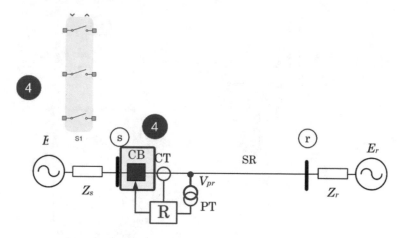

Fig. 13.8 Circuit breaker (CB) component (❹)

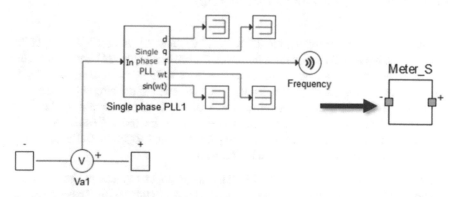

Fig. 13.9 Meter component (❺)

This area consists of the connection of three main functional blocks. They are: "*Fault detection*", "*Trip logic*", and "*CODO*" (closing-opening-difference-operator) algorithm. These blocks are the combination of subsystems that can be illustrated as follows:

• "*Fault detection*"

This block is dedicated to detecting the fault and determining the protection zone. There are two protection zones, zone 1 and zone 2, set in this block with two corresponding trip signal outputs (Fig. 13.12).

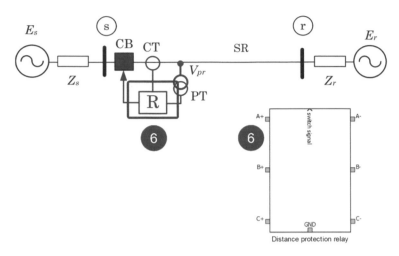

Fig. 13.10 "Distance protection relay" component (❻)

This block has three inputs, including *"I"*, *"V"*, and *"contactor"*. The input *"I"* and *V"* is the measured current that is connected from the inputs *"IA"* and *"VAn"* from the functional area, *"Measurements of current and voltage on transmission line"* (❶). The input *"contactor"* is intended to define the state of the circuit breaker.

To detect the fault and determine which protection zone operates, there are three main functional components connected inside the block *"Fault Detector"*, including: ❶*"RX Calculation"*, ❷*"Holding Inputs"*, and ❸*"Zone Determination"*. These components can be illustrated as follows:

- ❶*"RX Calculation"* (see Fig. 13.13). The instantaneous values of the measured current (*I*) and voltage (*V*) of line SR are the inputs to this block. These measurements are transformed into RMS values by using *"RMS"* block from the *"Signal Processing"* of the Schematic library. The magnitude of the impedance (*Z*) seen by the distance protection relay can be calculated according to RMS values of voltage (*Vrms*) and RMS values of current (*Irms*) by using Ohm's laws (*Z* = *Vrms/Irms*).

The phase angle difference (*θ*) between voltage phase angle and current phase angle is calculated and converted into electrical degrees unit by using the *"Phase Difference"* block from *"Signal Processing"* of the Schematic library.

These values are used to calculate the two components of the impedance, real component *R* (resistance) and imaginary component *X* (reactive) according to the Eqs. (13.8) and (13.9).

- ❷*"Holding Input"* (see Fig. 13.14). The holding input block is a function implemented in the *"C function"* of the Schematic library. This function is designed to keep the output values of the resistance (*R*) and reactance (*X*) in case the circuit breaker is open due to a trip signal.

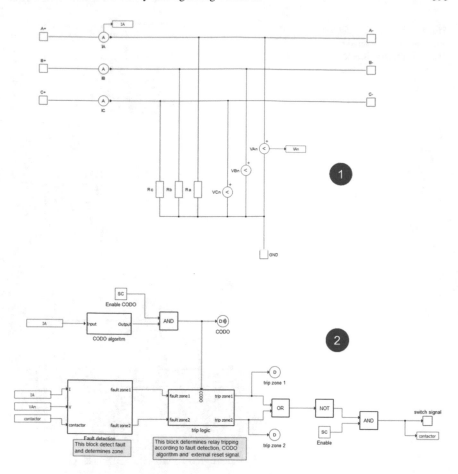

Fig. 13.11 Two functional areas of distance protective relay

- ❸"*Zone Determination*". (see Fig. 13.15). *R* and *X* values from "*RX Calculator*" are input to the zone determination, including zone 1 protection and zone 2 protection blocks. Other inputs of this block are the magnitude, angle and zone reach of the protection relay from HIL SCADA interface.

Both zone protection (zone 1 and zone 2) blocks are created by the "*C function*". They contain a function with a simple if-statement that is coded in block properties. If the calculated values are less than the setting, the function generates an output signal called "fault" equal to 1; otherwise, it will send a signal 0 (normal or not fault condition).

After determining which protection zones operate, the output of the operated block is activated to send the trip signal. However, the time of sending this signal is delayed, and this delay time is different according to which protection zone operates. The delay time for zone 1 is 50 ms, while the delay time for zone 2 is 0.3 s. To gain

Fig. 13.12 *"Fault Detector"* block and details inside the *"Fault Detector"* block

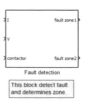

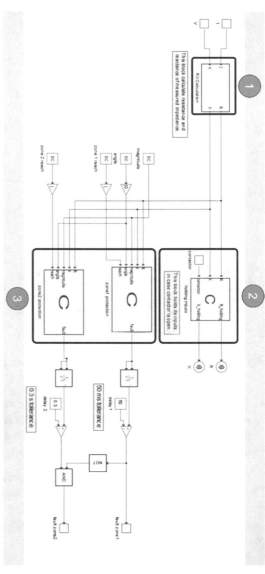

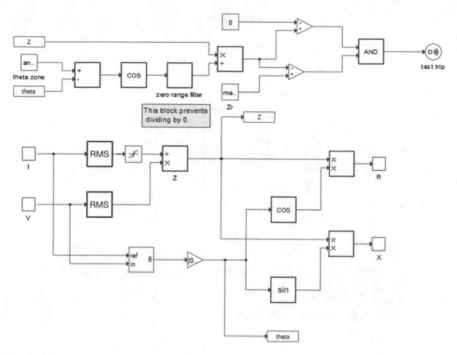

Fig. 13.13 Inside "*RX Calculation*" component ①

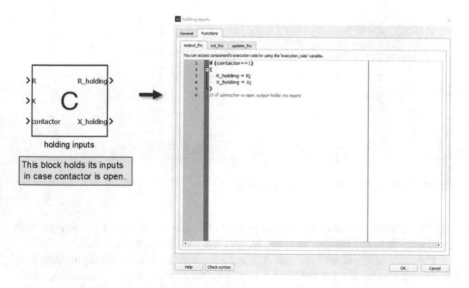

Fig. 13.14 "*Holding Input*" component ②

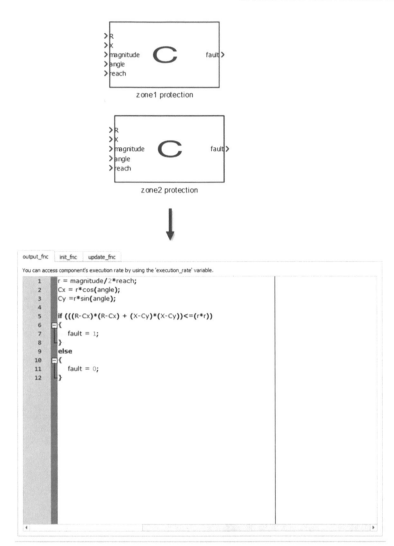

Fig. 13.15 "*Zone Determination*" component ③

this delay time, the outputs of both protection zone blocks need to connect with multiple functional blocks. This connection can be shown in Fig. 13.16.

The output from zone 1 goes to an integrator which multiplies the input value with the execution rate at each simulation step. This value is input to a comparator which compares the time-delay component with a value of 50 ms. Even if the protective relay is set instantaneously, it is impossible to trip simultaneously when a fault is selected. If the value from the integrator is greater than the time delay, the output will be set at 1. If the value is less, the output is set at 0 and will not send a signal

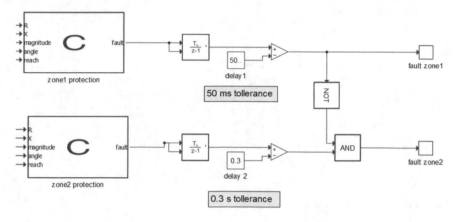

Fig. 13.16 The output signal of zone protection blocks are delayed with the corresponding time to the protection zone

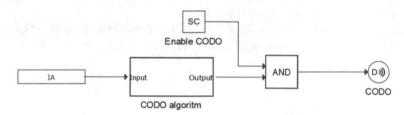

Fig. 13.17 CODO algorithm

of fault. The zone 2 protection block has the same approach as the zone 1 protection block but has a different time-delay value of 0.3 s which is a typical time delay for zone 2 protection. This time delay includes the operating time of zone 1 and a time interval for giving time to zone 1 to clear the short circuit. For detecting a fault from zone 2, there must be an output of 1 from the comparator and an inverted value to 1 from zone 1.

- *CODO algorithm*

The closing-opening-difference-operator) (CODO) algorithm' calculates the fault filtering signal and transforms it into a non-linear signal by applying mathematic operations, such as addition, subtraction, maximum and minimum (Distance protection relay with false tripping prevention 2022). The input of this block is the measured current of the transmission line (tagged name "*IA*") seen by the distance protection relay. The output of this block is active through the signal "*enabled CODO*" from the HIL SCADA interface and block "*AND*" as shown in Fig. 13.17.

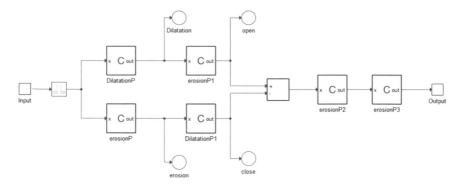

Fig. 13.18 Inside CODO algorithm

The purpose of this function is to filter the false trip signal due to the power swings event. The change in frequency will influence the impedance such that the relay protection zones may detect it as a fault.

The output signal of CODO algorithm block is formed by using the Dilatation, Erosion, Opening and Closing algorithm, as shown in Fig. 13.18.

```
Delaytion
  1.   Inputs:
       a = 40 [constant]
       x = RMS current of transmission line
       m[0]=x
  2.   for (i = a-2) and (i > 0)do:
  3.        i = i-1;
  4.        m[i+1] = m[i];
  6.   for (i = 0) and (i <= a-1) do
  7.        i = i+1;
  8.        y[i]=m[i]+0.1;
  9.   n=y[0];
  10.  for (i = 0) and (i <= a-1) do
```

```
11.          i = i+1;
12.          if y[i]>n do [find the maximum of
13. n]
                n=y[i]
14.   Out:
         f=n
```

Erosion
```
1.    Inputs:
      a = 40 [constant]
      x = RMS current of transmission line
2.    m[0]=x
3.    for (i = a-2) and (i > 0) []
4.          i = i-1;
5.          m[i+1] = m[i];
6.    for (i - 0) and (i <= a-1) []
7.          i = i+1;
8.          y[i]=m[i]-0.1;
9.    n=y[0];
10.   for (i = 0) and (i <= a-1) do
11.          i = i+1;
12.          if y[i]<n do [find the minimum of
13. n]
                n=y[i]
14.   Out:
         g=n
```

Opening:

Out=f⊖g

Closing:

Out=f⊖g

The dilation algorithm and erosion algorithm use three loops. The first one assigns the size of the matrix called "m" due to the pre-set constant number as 40. For the dilation algorithm, the second loop is to increase the step size of 0.1 for the measured RMS value of current, as opposed to the erosion algorithm. The third loop in the dilation algorithm is used to figure out the maximum input values, and for the erosion algorithm, this loop is dedicated to finding the minimum input values.

The dilation algorithm (usually represented by ⊕) is dedicated to the structure of the measured current element for expanding the tolerances contained in the input current according to the change of frequency. As opposed to the dilation algorithm, the erosion algorithm (usually represented by ⊖) is used to erode away the tolerances of the measured current caused by the change of frequency.

The opening and closing algorithms are the combinations of dilatation and erosion. The opening algorithm is used to restore or recover the original measured current to the maximum possible extent. The closing algorithm is used to smoother the contour of the distorted current and fuse back the narrow breaks and long thin gulfs. Closing is also used to get rid of the small holes of the obtained current.

- *Trip logic*

The *"Trip logic"* block is dedicated to determining the trip signal according to the *"Fault detection ", "CODO algorithm"*. This block has three inputs, *"fault zone 1"*, *"fault zone 2"*, and *"CODO"*. The inputs *"fault zone 1"* and *"fault zone 2"* are the output signals of zone protection blocks in *"Fault detection"*. The input *"CODO"* is the output signal from CODO algorithm that is used to ensure the protective relay not trip when there is a false fault (see Fig. 19a).

Inside this block, there are two main blocks, *"trip logic zone 1"* and *"trip logic zone 2"* (see Fig. 19b). These blocks are intended to send the trip signal or the reset signal to the protective relay.

The output of the trip logic block is *"trip zone 1"* and *"trip zone 2"* from *"trip logic zone 1"* and *"trip logic zone 2"*. These outputs are combined with the signal *"enable"* from HIL SCADA to send the signal *"trip"* or *"not trip"* to the circuit breaker, as shown in Fig. 13.20.

13.3.2   *HIL SCADA*

The HIL SCADA panel offers the essential user-interface elements (widgets) to monitor and interact with the simulation at runtime, allowing the users to further customise according to needs. In this section, the authors give the readers the methodology to create the controlling and monitoring panel using available HIL SCADA library widgets to test the distance protection relay.

The controlling and monitoring panel as shown in Fig. 13.21 consists of 6 main parts: (❶) One line diagram, (❷) Distance protection commands and measurements, (❸) Faults, (❹) Extras, (❺) Capture/Scope, (❻) Power Swing Control.

These main parts are created by using available widgets in the "Monitoring" and "Action" core of HIL SCADA library. These widgets are adjusted by performing a double-click to open the properties window. In the window settings, the users can do navigation commands that require widgets to perform requests of the users. These commands are structured in the Python programming language (What is Python Executive Summary 2022).

The first element (❶) is used to display the state of model values; frequency in the grid, contactor status, and also the presence and location of the transmission line fault.

The frequency in the grid can be visualised by using the *"Digital Display"* widget from *"Monitoring"* of the library. The users need to perform a double-click on the widget and selects the desired signal to observe. In this widget, the frequency of two meters, the element ➎of the *Test System* model (see Sect. 13.3.1), is selected (see Fig. 13.22).

The state of the model; contactor status, and location of the transmission line fault can be visualised by using the visual image. To create this element, the users need to select the *"Image"* widget in the *"Visual"* of the HIL SCADA widget library; then, the users perform a right-click to define the direct path of the desired image (see Fig. 13.23).

The visual image used in the HIL SCADA panel is in SVG (Scalable Vector Graphics) format to reconstruct the image under two cases, nominal operating and short-circuit fault. The illustrative detail of rebuilding this visual image can be performed via the creation of the element ➍of the panel.

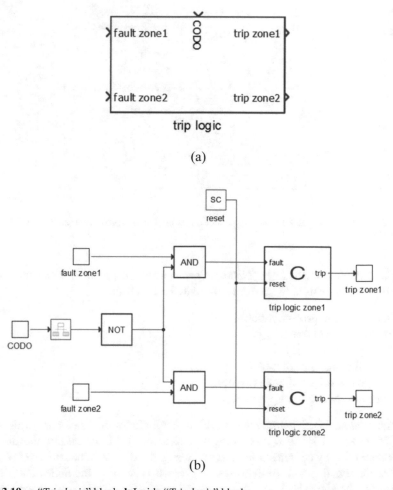

Fig. 13.19 a *"Trip logic"* block. b Inside *"Trip logic"* block

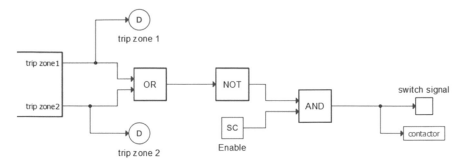

Fig. 13.20 Output of "*Trip logic*"

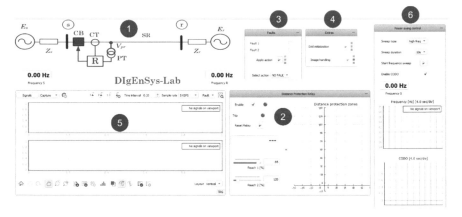

Fig. 13.21 Controlling and monitoring panel for testing distance protection relay created by using HIL SCADA

The second element (❷) "*Distance protection commands and measurements*" (see Fig. 13.24) is used to perform several actions as follows:

– Enable/disable protective relay.
– Observe the trip status.
– Reset protective relay.
– Check the presence of a fault.
– Change the values of zone reaches.
– Track the change in impedance.

The "*Enable/disable protective relay*" action includes the "*Tick box*" widget and "*LED*" widget. The "*Tick box*" widget is used to enable or disable the distance protection relay by defining "*Macro code*" inside the widget. When the box is ticked, the enable signal is sent to the distance protection relay in the Schematic Editor; otherwise, the signal will not be sent, which means the distance protection relay is in the disabled state. The "*LED*" widget is used to show this enable signal as "sent"

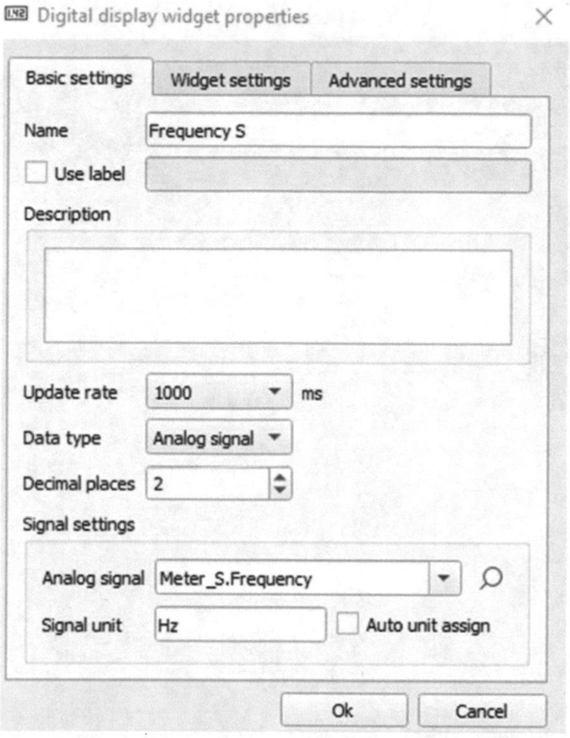

Fig. 13.22 Select the signal inside the *"Digital Display"* widget to observe the frequency of the *Test System*

or "unsent". It will be coloured in green if the enable signal is sent. The detail of the implementation is shown in Fig. 13.25.

The *"Observe the trip status"* action is used to display the trip status of the distance protection relay. By using *"LED"* widget, whether the trip signal comes from zone 1 protection or zone 2 protection, the *"LED"* will be highlighted in red colour (see Fig. 13.26).

The *"Reset protective relay"* action is performed to reset the distance protective relay when the protective relay is in a trip state. This action can be created by using the *"Button"* widget. The reset trip signal is defined in the *"Macro code"* inside this widget. When this button is clicked, the reset trip signal is sent to the *"Trip logic"* functional block of the distance protective relay (see Sect. 13.3.1), and then the trip signal is cleared (see Fig. 13.27).

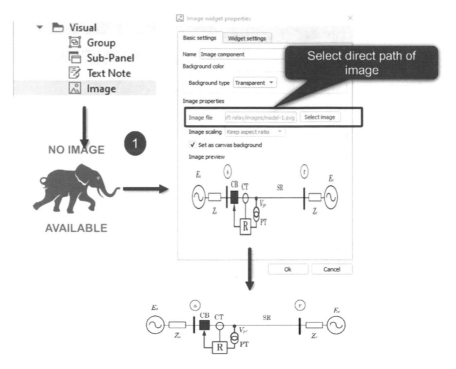

Fig. 13.23 Select the direct path of the image inside the "*Image*" widget

The "*Check the presence of fault*" action can be created by using the "*Digital Display*" widget. In "*Macro code*" inside this widget, the authors defined the visualised text based on the trip signal coming from zone 1 protection and zone 2 protection. In case of no trip signal coming from these zones, the text '*No fault*" is visualised; otherwise, it will indicate which protection zone is sending the trip signal (see Fig. 13.28).

The "*Change the values of zone reaches*" action includes two "Slider macro" widgets to pre-determine the reach zone values of two protection zones. These are the inputs named "*reach*" of two functional protection blocks, "*zone 1 protection*" and "*zone 2 protection*" of "*Zone Determination*" (see Sect. 13.3.1). The implementation is shown in Fig. 13.29.

The "*Track the change in impedance*" action can be performed using the "*XY graph*" widget to visualise the diagram zoomed graph with zones, transmission line and observed fault impedance. The users need to select the resistance (R) and reaction (X) signals for the X-axis and Y-axis of the graph. Additionally, the zones, transmission line and observed fault impedance are defined in the "*Advance setting*" section inside the widget. The detailed implementation is shown in Fig. 13.30.

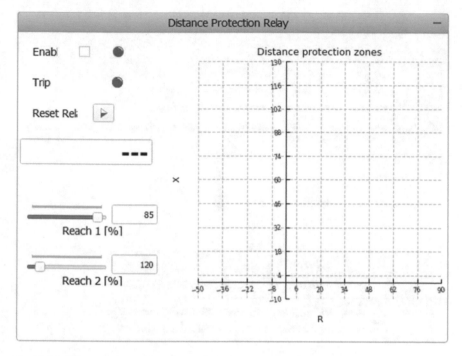

Fig. 13.24 *"Distance protection commands and measurements"*

The third element (❸) *"Fault"* (see Fig. 13.31) is indicated to allow the users to choose which types of faults users want to inject; short-circuit cases in the middle of the transmission line or at the end of the transmission line, or both short-circuit cases simultaneously.

Several actions can be performed in this element named *"Fault 1"*, *"Fault 2"*, *"Apply action"*, and *"Select action"*.

"Fault 1" represents the short-circuit cases in the middle of the transmission line, and "Fault 2" represents the short-circuit cases at the end of the transmission line. *"Fault 1"* and *"Fault 2″* is the *"Fault"* element in *Test System"* (see Sect. 13.3.1). They are created by using the *"Tick box"* widget; then, the commands to define them can be performed in *"Macro code"* inside the widget. If the box is ticked, the fault element is activated. The implementation can be shown in Fig. 13.32.

For the *"Apply Fault"*, the users select the *"Macro"* widget in *"Action"* of HIL SCADA library. This button is used to define the sample rate and time interval for captured signal in *"Capture/Scope"*. The users should right-click the *"Capture/Scope"*, then select *"Copy Widget ID"* and use this ID number for *"Macro widget settings"* of the *"Apply Fault"* button. Additionally, the types and starting times of short-circuit cases can be defined in these widgets via Python scripts. For the convenience of observing short-circuit events, the authors selected the starting time of the short-circuit at 0.1 s.

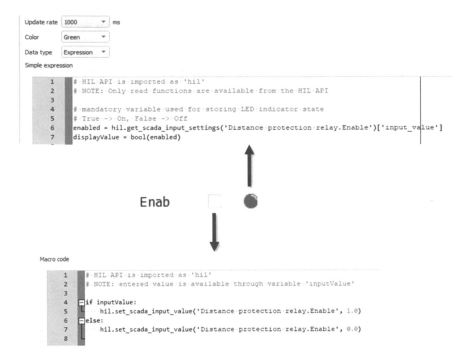

Fig. 13.25 *"Enable/disable protective relay"*

For the "*Fault selection*", the "*Combo Box*" widget in "*Action*" of the HIL SCADA library is used to perform this action (see Fig. 13.33). The users can define the name of short-circuit cases in the "*Available values*" of the widget properties, then, according to declared types of faults in the Fault element of the test system (see Sect. 13.3.1). The nominal operating case is named "*NO FAULT*". At the same time, short-circuit fault cases consist of many types of short-circuit faults (three-phase, three-phase-to-ground, single-phase-to-ground, phase-to-phase, and two-phase-to-ground) are explicitly named in each case according to the A, B, C phases and the grounded neutral N as shown in Table 13.3. Users must select one of these scenarios to conduct the appropriate tests.

The fourth element (❹) contains two "*Macro*" widgets for grid initialisation and for handling online diagram image.

For "*Grid initialisation*", the values of the external grid element (see Sect. 13.3.1) are set at the "*Macro*" code inside the widget (see Fig. 13.34).

For "*Image handling*", the users need to use a right-click to open the properties window and start to parse SVG images as XML (the Extensible Markup Language). The "contactor", "Fault 1", and "Fault 2" are the input data to execute the change in distance protection relay and fault icon in the visual image. The "*xml.etree.ElementTree*" module (xml.etree.ElementTree 2022) with associated commands are used in this section. The implementation can be shown in Fig. 13.35.

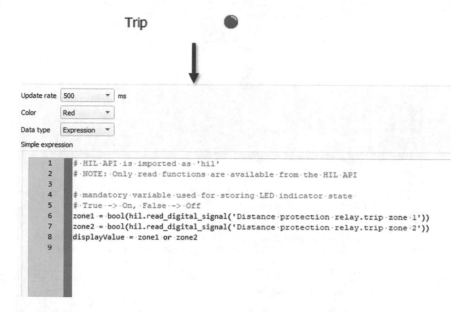

Fig. 13.26 *"Observe the trip status"*

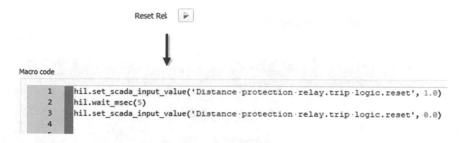

Fig. 13.27 *"Reset protective relay"*

The fifth element (**❺**) Capture/Scope, is created to allow the users to follow grid voltage values during frequency sweep, capture injecting of fault, or observe any other signal of interest. The users can right-click to select *"Switch to embedded mode"*, then select *"Signal"* to capture desired signals to observe during the real-time simulation (Fig. 13.36).

In the sixth element (**❻**) *"Power Swing Control"*, the users can select to do the following actions for power swing scenarios. This element consists of several actions, including *"Sweep type"*, *"Sweep duration"*, *"Start frequency sweep"*, and *"Enable CODO"*. The displays in this element include *"Frequency S"*, the graph of frequency and the graph of CODO.

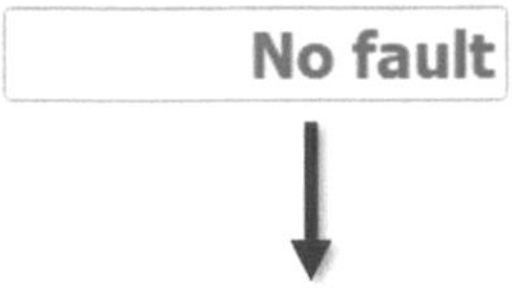

No fault

↓

```
1   no_fault_text = {
2       "text": "No fault",
3       "text_color": 'green',
4   }
5   zone1_text = {
6       "text": "Zone 1 fault",
7       "text_color": 'red',
8   }
9   zone2_text = {
10      "text": "Zone 2 fault",
11      "text_color": 'red',
12  }
13  zone1 = hil.read_digital_signal('Distance protection relay.trip zone 1')
14  zone2 = hil.read_digital_signal('Distance protection relay.trip zone 2')
15  if zone1:
16      textDisplayData = zone1_text
17  elif zone2:
18      textDisplayData = zone2_text
19  else:
20      textDisplayData = no_fault_text
21  displayValue = textDisplayData
22
```

Fig. 13.28 *"Check the presence of fault"*

```
hil.set_scada_input_value('Distance protection relay.Fault detection.zone 1 reach', inputValue)
```

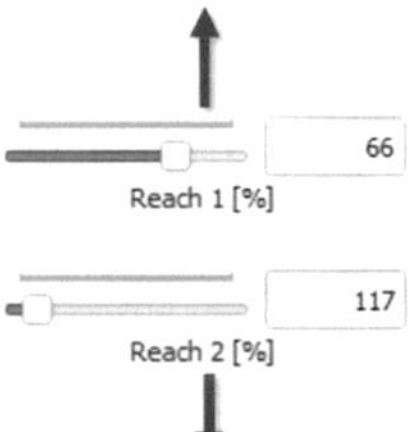

↑

```
                                    66
            Reach 1 [%]

                                    117
            Reach 2 [%]
```

↓

```
hil.set_scada_input_value('Distance protection relay.Fault detection.zone 2 reach', inputValue)
```

Fig. 13.29 *"Change the values of zone reaches"*

The "*Sweep type*" action is created by using the "*Combo box*" widget. There are two selections in this widget that are low frequency and high frequency (see Fig. 13.37).

The "*Sweep duration*" defines the duration time of power swing events using the "Combo box" widget. This widget has five selections according to five time periods, 2 s, 4 s, 6 s, 8 s and 10 s (see Fig. 13.38).

The "*Start frequency sweep*" action is created by using the "*Button*" widget. Two selections from "Sweep type", low frequency and high frequency, are defined inside

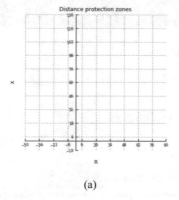

(a)

```
Z       = hil.get_scada_input_settings('Distance protection relay.Fault detection.magnitude')['input_value']
theta   = hil.get_scada_input_settings('Distance protection relay.Fault detection.angle')['input_value']/180*np.pi
reach1 = hil.get_scada_input_settings('Distance protection relay.Fault detection.zone 1 reach')['input_value']/100
reach2 = hil.get_scada_input_settings('Distance protection relay.Fault detection.zone 2 reach')['input_value']/100
R = Z*np.cos(theta)
X = Z*np.sin(theta)

# Zone impedance calculations
Z1 = reach1 * Z
mag1 = Z1/2
R1 = reach1 * R
X1 = reach1 * X
Z2 = reach2 * Z
mag2 = Z2/2
R2 = reach2 * R
X2 = reach2 * X

# reference curve data points
angles = np.linspace(0, 2*np.pi, 100)
zone1 = []
for angle in angles:
    x = R1/2 + mag1*np.cos(angle)
    y = X1/2 + mag1*np.sin(angle)
    zone1.append([x,y])
zone2 = []
for angle in angles:
    x = R2/2 + mag2*np.cos(angle)
    y = X2/2 + mag2*np.sin(angle)
    zone2.append([x,y])
line = [[0,0], [R,X]]
# specify curve options ('data' part is mandatory)
global zone1_curve
zone1_curve = {
    # list of x,y point pairs: [[x1, y1], [x2, y2],...,[xn, yn]]
    "data": zone1,
    # curve line style: 'solid', 'dashed', 'dashdot', 'dotted'
    "line_style": "dashed",
    # curve line color: 'red', 'green', 'blue', 'cyan', 'magenta', 'yellow'
    "line_color": 'blue',
    # curve line width: float number
    "line_width": 2.0,
    # reference curve title
    "title": "Zone 1",
    # show a title in the legend
    "show_in_legend": True
}
zone2_curve = dict(zone1_curve)
zone2_curve["data"] = zone2
zone2_curve["title"] = "Zone 2"
line_curve = {
    "data": line,
    "line_style": "solid",
    "line_color": 'red',
    "line_width": 2.0,
    "title": "Transmission line",
    "show_in_legend": True
}

# list of reference curves data: [ref1, ref2,...refN]
referenceCurves = [zone1_curve, zone2_curve, line_curve]
```

(b)

Fig. 13.30 a *"Track the change in impedance"*. b *"Advance setting"* section

Faults
Fault 1
Fault 2
Apply action
Select action
NO FAUL

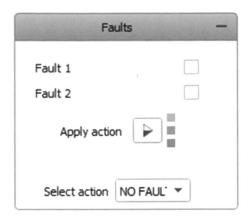

Fig. 13.31 "*Fault*" element

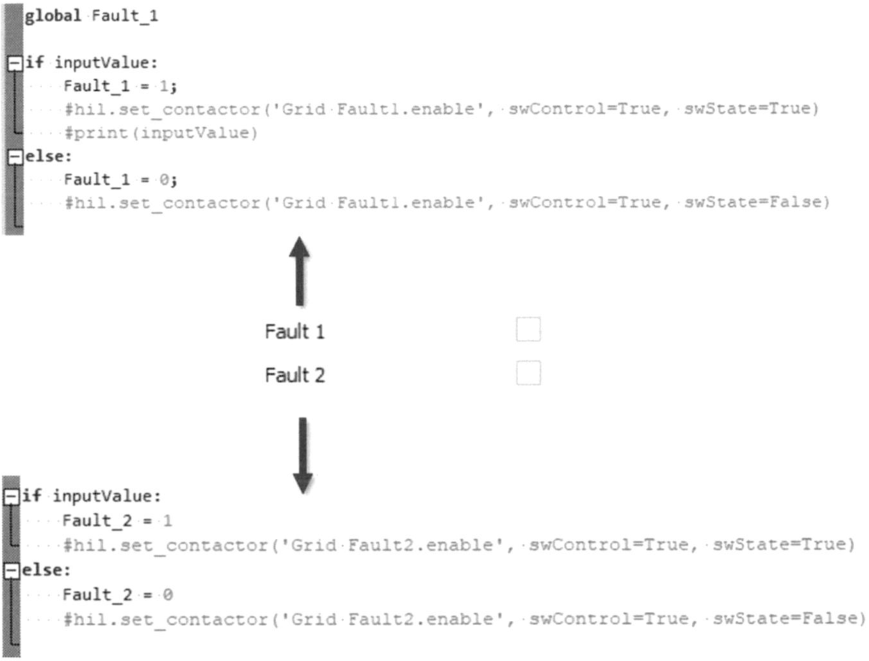

global Fault_1
if inputValue:
Fault_1 = 1;
#hil.set_contactor('Grid Fault1.enable', swControl=True, swState=True)
#print(inputValue)
else:
Fault_1 = 0;
#hil.set_contactor('Grid Fault1.enable', swControl=True, swState=False)
Fault 1
Fault 2
if inputValue:
Fault_2 = 1
#hil.set_contactor('Grid Fault2.enable', swControl=True, swState=True)
else:
Fault_2 = 0
#hil.set_contactor('Grid Fault2.enable', swControl=True, swState=False)

Fig. 13.32 "*Fault 1*" and "*Fault 2*"

this button. The low frequency is set as 55 Hz, while the high frequency is set as 65 Hz. The sine waveform from two sources is executed with the new frequency in the period of time selection in "*Sweep duration*" (see Fig. 13.39).

The "*Enable CODO*" action is created to send the enable CODO signal to the CODO algorithm by using the "*Tick box*" widget from the HIL SCADA library.

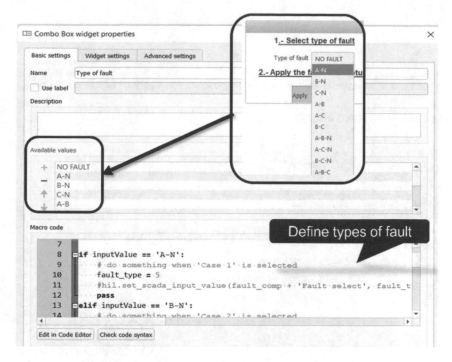

Fig. 13.33 *"Fault selection"*

Name	Short-circuit fault cases
A-N	Single-line-to-ground
B-N	
C-N	
A-B	Phase-to-phase
A-C	
B-C	
A-B-N	Two-phase-to-ground
A-C-N	
B-C-N	
A-B-C	Three-phase
A-B-C-N	Three-phase-to-ground

Table 13.3 Nameplates corresponding with short-circuit fault cases

When the box is ticked, the CODO algorithm is enabled; otherwise, it is disabled (see Fig. 13.40).

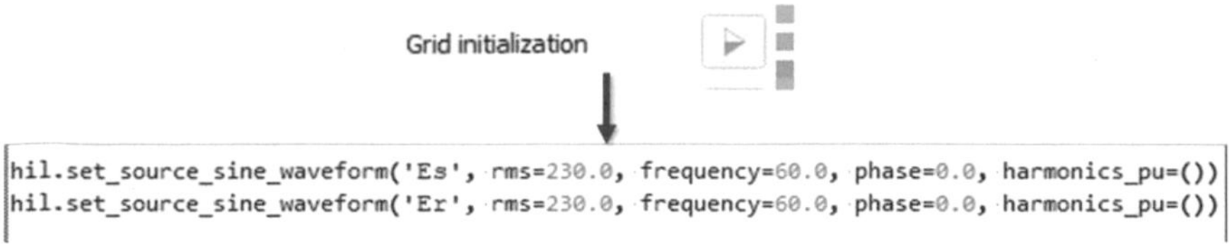

Grid initialization

```
hil.set_source_sine_waveform('Es', rms=230.0, frequency=60.0, phase=0.0, harmonics_pu=())
hil.set_source_sine_waveform('Er', rms=230.0, frequency=60.0, phase=0.0, harmonics_pu=())
```

Fig. 13.34 *"Grid initialisation"*

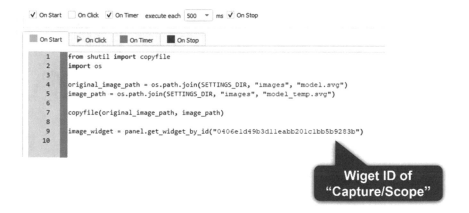

Fig. 13.35 *"Image Handling"*

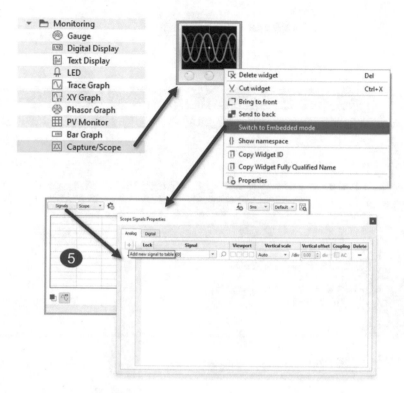

Fig. 13.36 *"Capture/Scope"*

13.4 DIgSILENT PowerFactory simulation with SEL-411L relay

The 230 kV transmission model with distance protection relay is built-in DIgSILENT PowerFactory to validate the results for the distance protection testing from Typhoon HIL. The distance protective relay used in this section is SEL-411L, a protective relay product of *"Scheweitzer Engineering Laboratories"* (SEL-411L 2022). The distance protection relay model of SEL- 411L is available in the library in DIgSILENT PowerFactory and makes it possible to implement distance protection testing (Fig. 13.41).

The system information implemented can be shown in Tables 13.4 and 13.5.

The SEL-411L relay model in the PowerFactory consists of protection functions including: differential, overcurrent, voltage, frequency control, distance and out-of-step. However, for the primary purpose of this chapter, distance protection is focused and another function is set out of service.

The settings for the protection relay model SEL-411L in PowerFactory for distance protection can be shown in Table 13.6.

Fig. 13.37 *"Sweep type"*

There are both phase distance elements (Z1P and Z2P) and ground distance elements (Z1MG and Z2MG). Phase distance elements protect phase-to-phase, phase-to-ground, and 3-phase faults, while ground distance elements are responsible for faults involving the ground. These elements have the same settings for zone 1 (Z1P = Z1MG) and zone 2 (Z2P = Z2MG) since they have equal requirements for the zone reaches.

13.5 Protection Zone Settings

This section shows the results of two testing cases: Testing zone 1 with a three-phase fault on the transmission line and Testing zone 2 with single-phase-to-ground on the transmission line.

For two cases, the authors used three different approach methods, theory-based calculations, Typhoon HIL simulation and DIgSILENT PowerFactory. To get easier in comparison of three methods, the same settings of the distance protection zone of the relay is used.

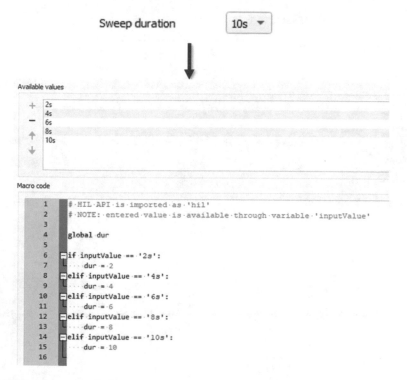

Fig. 13.38 *"Sweep duration"*

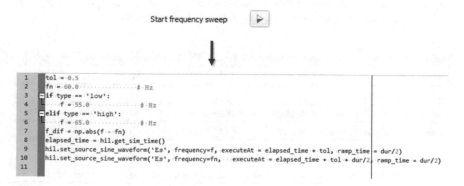

Fig. 13.39 *"Start frequency sweep"*

Zone 1 is the first protection zone of the protective relay and is the first o trip with no intentional time delay if a fault is detected in this zone. This zone is set at 80% of the transmission line length. Zone 2 must cover the total length of the transmission line and in addition, gives backup protection to t the adjacent line. The reach for zone 2 is set to 120%. The impedance values of the two protection zone

Fig. 13.40 *"Enable CODO"*

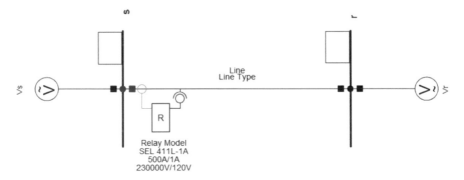

Fig. 13.41 Single line diagram of 230 kV transmission model with distance protection relay on the beginning of transmission line modelled in DIgSILENT PowerFactory

Table 13.4 AC voltage sources parameters for the 230 kV transmission model in PowerFactory

Parameter	Value	Unit
Rated system voltage	230	kV
Positive and negative sequence of sending side impedance	2.645 + j0.1847	Ω
Positive and negative sequence of receiving side impedance	10.68 + j0.3233	Ω

Table 13.5 Transmission line parameter for the 230 kV transmission model in PowerFactory

Parameter	Value	Unit
Length of transmission line	100	km
System frequency	50	Hz
Positive and negative sequence of line impedance per km	0.20631 + j0.74769	Ω/km
Zero sequence of line impedance per km	0.5157 + j1.91319	Ω/km

Table 13.6 Relay settings for SEL-411L in PowerFactory

Element	Define
CT	Current Transformer Ratio
VT	Voltage Transformer Ratio
Z1P	Zone 1 phase distance element
Z2P	Zone 2 phase distance element
Z1MG	Zone 1 ground distance element
Z2MG	Zone 2 ground distance element
Z1PD	Phase distance—Zone 1 time delay
Z2PD	Phase distance—Zone 2 time delay
Z1GD	Ground distance—Zone 1 time delay
Z2GD	Ground distance—Zone 1 time delay

Table 13.7 Protection zone

Protection zone	Value	Unit
Zone 1	$16.133\angle74.547\,°$	Ω
Zone 2	$24.2\angle74.547\,°$	Ω

can be calculated by multiplying the pre-determined protection percentage by the secondary positive sequence impedance of the transmission line. These values can be shown in Table 13.7.

Protection zone settings in Typhoon HIL simulation

The protection zone settings can be implemented for the virtual relay model by double-clicking to the distance protection relay component; then, the properties window will appear (see Fig. 13.42) to allow users to input distance protection relay settings.

The input values of the distance protection relay are the magnitude and angle impedance of the transmission line (primary side) and reached percentage values of the two zones. Users can click to button "*preview*" to preview the R-X diagram (see Fig. 13.43). The blue circle is the reach for protection zone 1, and the orange circle is the reach for zone 2.

Protection zone settings in DIgSILENT PowerFactory

The distance protection relay settings are based on parameters in Table 13.4. For two pre-determined protection zones, the parameters needed can be shown in Table 13.8. R-X diagram retrieved from plot page of PowerFactory in Fig. 13.44.

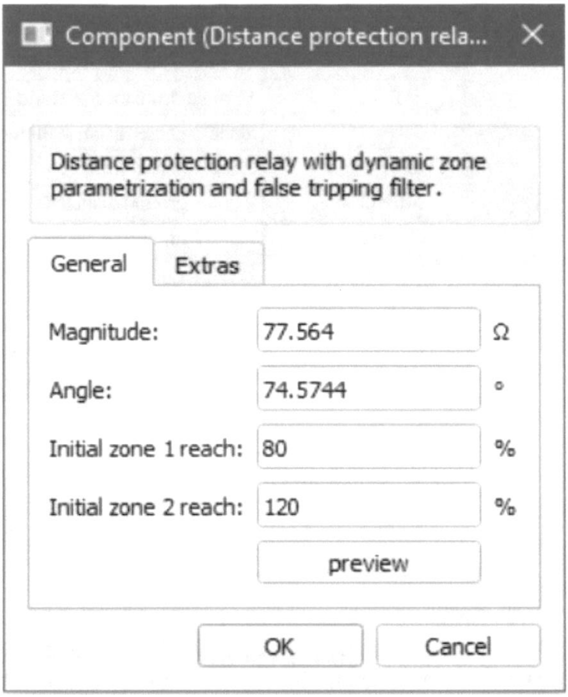

Fig. 13.42 Positive sequence line impedance in distance protective relay settings

13.6 Results

This section shows the fault impedance results based on three different methods in two testing cases; a three-phase short-circuit at the middle of the transmission line and a single-line-to-ground short-circuit at the end of the transmission line, according to the pre-determined reach zone 1 and zone 2. The results based on calculation correspond to Eqs. (13.1)–(13.5), while Typhoon HIL simulation and PowerFactory can measure the fault impedance results.

Table 13.9 shows the comparison between calculation and measured fault impedance.

For Typhoon HIL Simulation and DIgSILENT PowerFactory, an R-X diagram with line impedance characteristics can be shown for users to get a visual understanding of the testing.

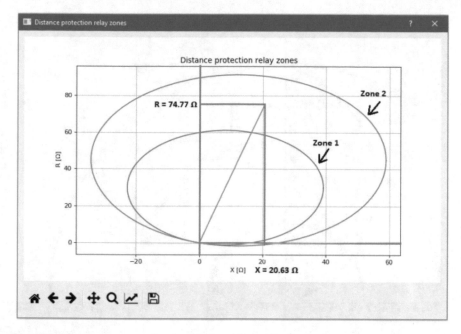

Fig. 13.43 Positive sequence line impedance with zone 1 and zone 2 reach represented in R-X diagram

Table 13.8 Relay settings for SEL411L in PowerFactory

Element	Setting values
CT	Primary: 500 A Secondary 1A
VT	Primary: 230,000 V Secondary 120 V
Z1P	Impedance = 16.15 Angle = 74.57
Z2P	Impedance = 24.20 Angle = 74.57
Z1MG	Impedance = 16.15 Angle = 74.57
Z2MG	Impedance = 24.20 Angle = 74.57
Z1PD	Time setting: 0 cycles
Z2PD	Time setting: 20 cycles
Z1GD	Time setting: 0 cycles
Z2GD	Time setting: 20 cycles

In Typhoon HIL Simulation, for testing zone 1 with three-phase short-circuit in the middle transmission line, button "*Fault 1*" is selected, while testing zone 2 with a

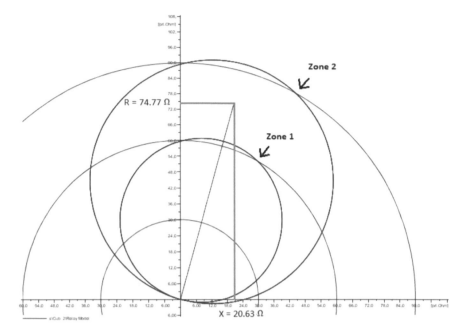

Fig. 13.44 R-X diagram with line impedance characteristics; reach zone 1 and zone 2 protection in PowerFactory

Table 13.9 Comparison between calculated and measured fault impedance in Typhoon HIL simulation and DIgSILENT PowerFactory

	Three-phase short-circuit	Single-line-to-ground short-circuit (phase A)
Theory-based calculation	10.4∠69.046 °	30.620∠74.741 °
Measured fault impedance in Typhoon HIL simulation	12.32∠78.106 °	34.98∠72.43 °
Measured fault impedance in DIgSILENT PowerFactory	10.969∠64.74 °	32.58∠71.27 °

single-line-to-ground short-circuit at the end of the transmission line, button "*Fault 2*" is selected. Figure 13.45 shows the results in testing zone 1, while Fig. 13.46 shows the results in testing zone 2. The R-X diagram retrieved from DIgSILENT PowerFactory can be shown in Figs. 13.47 and 13.48.

For three-phase short-circuit in the middle of a transmission line case, fault impedances retrieved from three methods are less than the pre-set zone 1 protection impedance. Therefore, the protective relay operates in this case. The trip signal of the protective relay is enabled in Typhoon HIL simulation from zone 1 protection. From the R-X diagram gained from the HIL simulation, the fault impedance, in this

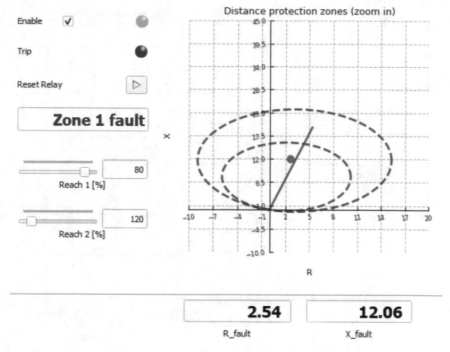

Fig. 13.45 Testing zone 1 protection with three-phase short-circuit in the middle of transmission line in Typhoon HIL

case, is in zone 1 protection of the protective relay, similar to the R-X diagram from DIgSILENT PowerFactory.

For a single-line-to-ground short-circuit at the end of the transmission line case, fault impedance, in this case, is higher than the pre-set zone 1 and zone 2 protection impedance. In this case, the fault is out of the protection zone of the distance protective relay; therefore, there is no operating trip signal. From the R-X diagram retrieved from Typhoon HIL simulation and DIgSILENT PowerFactory, the fault impedance, in this case, is out of zone 2 protection.

The measured fault impedances from DIgSILENT PowerFactory are a little higher than results retrieved from Typhoon HIL simulation, but the same results with fault detection in zone 1 with a three-phase short-circuit in the middle on the transmission line and failure in zone 2 protection of not detecting a single-line-to-ground short-circuit at the end of the transmission line.

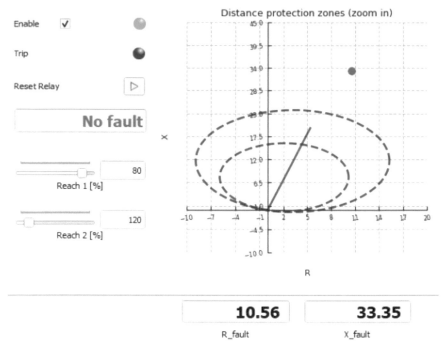

Fig. 13.46 Testing zone 2 protection with single-line-to-ground short-circuit at the end of the transmission line in Typhoon HIL

13.7 Conclusions

This chapter presented the proposed methodology for testing the distance protection relay using model-based engineering toolchains of Typhoon HIL. The virtual HIL simulation showed similar fault impedance of two other methods, theory-based calculation and DIgSILENT PowerFactory, in two short-circuit cases, three-phase short-circuits and single-line-to-ground short-circuits. However, this approach allows users to access the testing performance of a physical HIL device with all restrictions of the real-time environment. In the future, the virtual protection relay can be developed to conduct protection function testing in more complex power systems.

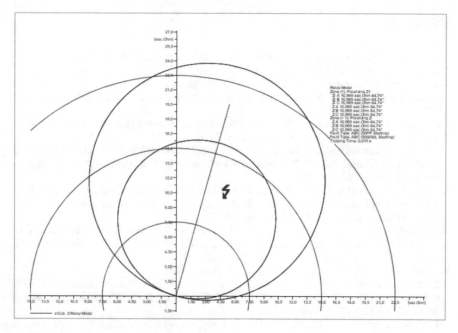

Fig. 13.47 Testing zone 1 protection with three-phase short-circuit in the middle of transmission line in DIgSILENT PowerFactory

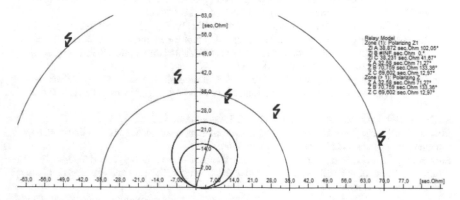

Fig. 13.48 Testing zone 2 protection with single-line-to-ground short-circuit at the end of transmission line in DIgSILENT PowerFactory

References

Anderson PM (1999) Power system protection. New York: McGraw-Hill : IEEE Press

Camarillo-Peñaranda JR, Aredes M, Ramos G (2020) Hardware-in-the-loop testing of virtual distance protection relay. In: 2020 IEEE/IAS 56th industrial and commercial power systems technical conference (I&CPS), Jun. 2020, pp 1–6. https://doi.org/10.1109/ICPS48389.2020.9176775

Component library. https://www.typhoon-hil.com/documentation/typhoon-hil-software-manual/topics/component_library.html. Accessed 12 Jul 2022

Distance protection relay with false tripping prevention. https://www.typhoon-hil.com/documentation/typhoon-hil-application-notes/References/distance_protection_relay.html#concept_dpr__figure_01. Accessed 18 Jul 2022

Dede A, Della Giustina D, Franzoni F, Pegoiani A (2014) IEC 61850-based logic selectivity scheme for the MV distribution network. In: 2014 IEEE international workshop on applied measurements for power systems proceedings (AMPS), pp 1–5

Erezzaghi ME, Crossley PA (2003) The effect of high resistance faults on a distance relay. In: 2003 IEEE power engineering society general meeting (IEEE Cat. No.03CH37491), Jul. 2003, vol 4, pp 2128–2133, https://doi.org/10.1109/PES.2003.1270943

Horowitz SH, Phadke AG (2014) Power system relaying. Wiley

IEEE Standard Electrical Power System Device Function Numbers, Acronyms, and Contact Designations. IEEE Std C372–2008 Revis. IEEE Std C372–1996, pp 1–48, Oct. 2008, https://doi.org/10.1109/IEEESTD.2008.4639522

Latif U, Moldesæther R, Dhakal P (2020) Development a cyber-physical testbed for relay-in-the-loo. University of South-Eastern Norway

Montaña DAM, Rodriguez DFC, Ivan Clavijo Rey D, Ramos G (2018) Hardware and software integration as a realist SCADA environment to test protective relaying control. IEEE Trans Ind Appl 54(2):1208–1217, https://doi.org/10.1109/TIA.2017.2780051

Pazdcrin AV, Samovlenko VO, Tashchilin VA, Chusovitin PV, Dymshakov AV, Ivanov YV (2018) Platform for testing Iec 61850 control systems using real-time simulator. In: 2018 international youth scientific and technical conference relay protection and automation (RPA), Sep. 2018, pp 1–14, https://doi.org/10.1109/RPA.2018.8537188

Rodriguez DFC, Osorio JDP, Ramos G (2018) Virtual relay design for feeder protection testing with online simulation. IEEE Trans Ind Appl 54(1):143–149, https://doi.org/10.1109/TIA.2017.2741918

Saadat H (1999) Power system analysis. United States of America

SEL-411L advanced line differential protection, automation, and control system. selinc.com. https://selinc.com/products/411L/. Accessed 21 Jul 2022

Sorrentino E (2014) Comparison of five methods of compensation for the ground distance function and assessment of their effect on the resistive reach in quadrilateral characteristics. Int J Electr Power Energy Syst 61:440–445

What is Python? Executive Summary. Python.org. https://www.python.org/doc/essays/blurb/. Accessed 16 Jan 2022

xml.etree.ElementTree—The ElementTree XML API—Python 3.10.5 documentation. https://docs.python.org/3/library/xml.etree.elementtree.html. Accessed 30 Jul 2022

Chapter 14
Cyber-Physical Co-simulation Framework Between Typhon HIL and OpenDSS for Real-Time Applications

Raju Wagle, Pawan Sharma, Mohammad Amin, and Francisco Gonzalez-Longatt

Abstract Cyber infrastructures have been extensively used for power system monitoring, control, and operation because of the development of new information and communications technology (ICT) in power systems. Deployment of cyber-physical co-simulation in the case of a realistic distribution network is still a big challenge. Hence, to solve the problem of modelling complex distribution networks, a cyber-physical co-simulation framework is proposed in this chapter. The proposed framework consists of a cybernetic layer, a physical layer, and a co-simulation framework between OpenDSS and Typhoon HIL. The cyber layer consists of software and tools to model the distribution system and communicate with the physical layer. The physical layer is the Typhoon HIL real-time simulator consisting of virtual or real controllable devices. A realistic framework to execute the real-time simulation using the Typhoon HIL SCADA system and Python-based co-simulation is created in this chapter. The real-time simulation demonstrates the proposed framework's effectiveness in observing the distribution network's voltage profile due to real-time variation in reactive power from the PV.

Keywords Cyber-physical testbed · Co-simulation · Typhoon HIL · OpenDSS · Python · Real-time simulation

R. Wagle (✉) · P. Sharma
Department of Electrical Engineering, UiT The Arctic University of Norway, Narvik, Norway
e-mail: raju.wagle@uit.no

P. Sharma
e-mail: pawan.sharma@uit.no

M. Amin
Department of Electric Power Engineering, NTNU, Trondheim, Norway
e-mail: mohammad.amin@ntnu.no

F. Gonzalez-Longatt
Centre for Smart Grids, University of Exeter, Exeter, UK
e-mail: fglongatt@fglongatt.org

© The Author(s), under exclusive license to Springer Nature Singapore Pte Ltd. 2023   425
S. M. Tripathi and F. M. Gonzalez-Longatt (eds.), *Real-Time Simulation and Hardware-in-the-Loop Testing Using Typhoon HIL*, Transactions on Computer Systems and Networks, https://doi.org/10.1007/978-981-99-0224-8_14

14.1 Introduction

With the growing integration of converters-based renewable energy sources with advanced ICT (information and communication technology) in the electrical distribution network, the distribution system operation demands more robust and adequate monitoring and control infrastructure. Cyber-physical co-simulation is one of the suitable strategies for managing this complexity (Schloegl et al. 2015). The significance of co-simulation in optimal reactive power control is highlighted in Acosta et al. (2021). Cyber-physical co-simulation frameworks integrate the physical and digital or cybernetic systems to raise the standard of monitoring and control of the physical system by harnessing the powers of the cyber system. In addition, co-simulation is the synchronised use of two or even more simulation models that have different operating conditions (Steinbrink et al. 2018). Numerous tasks such as sensing, calculation, communication, and actuation are included in many applications like industrial automation, aeronautical and automotive control, power grid management, and inaccessible and dangerous investigations using a cyber-physical co-simulation framework (Meer et al. 2017).

Research on cyber-physical co-simulation is currently continuing in many fields, including the automobile industry, the military, building science, and energy. As a result, several general co-simulation systems are currently available, each with unique features and levels of usefulness. Even though cyber-physical co-simulation is growing, several problems still need to be resolved, especially while developing a framework for modelling the distribution network. A real-time cyber-physical test bed for microgrid control is proposed in Venkataramanan et al. (2016), where the power system network is designed in RSCAD, and the real-time digital simulator (RTDS) is used for real-time co-simulation. Similarly, in Cao et al. (2019), a real-time simulation test bed that operates between Opal-RT and Matlab Simulink is proposed. Real-Time Laboratory (RT-LAB) and the communication network simulation developed with OPNET are observed in Tang et al. (2017). A cyber-physical testbed for co-simulation for a real-time testbed for reactive power control in the smart inverter is proposed in Wagle et al. (2022). A detailed review of the real-time co-simulation testbed is studied in Sun et al. (2016). In most of the previous studies, a cyber-physical co-simulation testbed is proposed for microgrids or small distribution networks. However, to analyse a real distribution network, the distribution network solver considered in the co-simulation testbed should be able to solve all types of distribution networks. Also, the real-time operation in the case of distribution networks needs to be robust and fast while solving the complex distribution network. OpenDSS is a very powerful tool specially designed to solve the distribution network. Hence, this work intends to develop a cyber-physical co-simulation framework to run the real-time simulation between the OpenDSS and Typhoon HIL.

The chapter presents a stronger emphasis on standards and common software tools, as well as the application of these tools to formulate a cyber-physical co-simulation framework between the Typhoon HIL and OpenDSS. A script to create communication is attached with a short explanation to achieve the main goal of the cyber-physical co-simulation.

14.2   Development of Cyber-Physical Co-simulation Framework Between Typhoon HIL and OpenDSS

The proposed cyber-physical co-simulation framework is depicted in Fig. 14.1, and it consists of two layers: (a) the Cybernetic layer and (b) the physical layer (top and bottom of Fig. 14.1, respectively). A classic PC, called workstation, uses the local CPU to run all the pieces of software of the cybernetic layer. It includes software to define and control de interactions between the layers and the distribution system solver. Finally, the physical layer includes the Typhoon HIL real-time simulator in this case.

Development of Typhoon HIL and OpenDSS co-simulation scenarios requires a significant amount of time, especially when dealing with multiple numbers of distributed resources.

To develop a co-simulation framework three main tasks should be achieved:

(i)   Create appropriate models of a distribution network and distributed resources in power system analysis software, in this case, OpenDSS,
(ii) Build an appropriate system to control and communicate with the real-time model and simulation using Typhoon HIL Schematic and SCADA.
(iii) Create a suitable python program to exchange information between OpenDSS and Typhoon HIL and vice versa.

The first task is creating a power network model, in this case, a distribution system; it includes the information of connected loads and PVs in OpenDSS, which is used as a base model for further studies. The size, location, and number of PVs interconnection depend on different factors; however, in this study, the PVs and load are connected as per the test study created by CIGRE. The second task involves designing

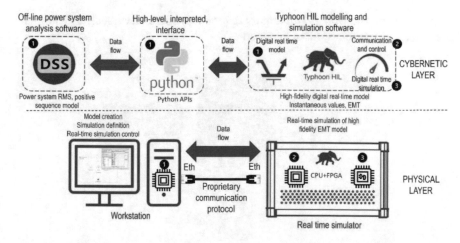

Fig. 14.1 Framework for cyber-physical co-simulation framework using OpenDSS and the Typhoon HIL real-time modelling and simulation framework

the typhoon HIL schematic editor and the typhoon HIL SCADA to exchange information between the Typhoon HIL real-time simulator and the Typhoon HIL SCADA. In the schematic editor, a subsystem is created to get the information from SCADA and to send the information to Typhoon HIL. The number of subsystems depends on the number of devices to be controlled from the SCADA terminal. The Typhoon HIL schematic editor has abundant libraries to create an interface dependent on the requirement. In the third task, a Python program must be created to initialise, execute the overall process, get the information from the real-time simulator, and send the signal to the real-time simulator. In this chapter, the considered distributed network is a European medium voltage network created by CIGRE Task Force C6.04 presented in the report *"Benchmark Systems for Network Integration of Renewable and Distributed Energy Resources"* (Cigre 2014). Therefore, in this chapter, only PVs are considered as distributed resources.

14.2.1 Creating Models of Distribution Networks and Distributed Resources in OpenDSS

The Open Distribution System Simulator (OpenDSS) is a simulation tool for modelling electrical distribution systems. It supports all types of steady-state analyses in distribution systems. The most popular application of OpenDSS is Distribution system analysis and planning, Multiphase power flow analysis, interconnection studies of Distribution generations, Harmonics, and inter-harmonics analysis. Different modes of operation like snapshot power flow, time series power flow, Harmonic study, Fault analysis, etc. can be achieved by using OpenDSS.

The overall structure of the OpenDSS (Sunderman et al. 2014) is shown in Fig. 14.2.

A distribution network might require several types of devices, such as distributed generators, transformers, lines, and loads.

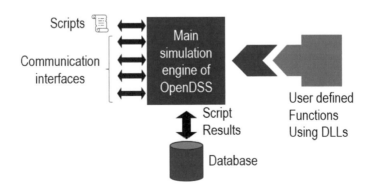

Fig. 14.2 OpenDSS database and communication interfaces and mechanisms

Complete detail of the network components is required to model a distribution network appropriately. Modelling details of power components models and the process of creating the power network models in OpenDSS can be found in the OpenDSS manual (Krishnamurthy 2017).

This subsection presents an explanation of the detailed modelling specific to the power network considered in this chapter. To model a network with thorough knowledge of the network components, the following steps, described in the subsequent subsection, are followed to create a distribution network.

14.2.1.1   Creating a Network as a Circuit

The first step in creating a model in OpenDSS is defining a new circuit. When a new circuit is created, it is installed as a three-phase voltage source named "Source" connected to a bus named "SourceBus" with a reasonable short circuit strength. The new circuit has many parameters that users can define or set to default values. To create the European MV CIGRE network (Krechel et al. 2019), the default values are replaced by the ones provided in CIGRE. Then, in the script of OpenDSS, the new circuit is defined, as shown in the code below.

```
// creating the slack bus and the main circuit
New Circuit.EuropeanMV
~ basekv=110 pu=1.0 phases=3 bus1=SourceBus
~ R0=1e-6 R1=1e-6
~ X0=1e-6 X1=1e-6
```

14.2.1.2   Creating Transformers

Transformers are one of the important components in an electric distribution network. To model a transformer in OpenDSS, the following script can be used.

```
//SUB TRANSFORMER DEFINITION
New Transformer.TR1 Phases=3 R=0.001 XHL=1.92
~ wdg=1 bus=SourceBus kV=110 kVA=25000 conn=Delta Numtaps=8
Mintap=0.9 Maxtap=1.1
~ wdg=2 bus=bus1 kV=20 kVA=25000 conn=Wye Numtaps=32 Mintap=0.9
Maxtap=1.1 Tap=1.03125 !1.1-11*0.00625

New Transformer.TR2 Phases=3 R=0.001 XHL=1.92
~ wdg=1 bus=SourceBus kV=110 kVA=25000 conn=Delta Numtaps=8
Mintap=0.9 Maxtap=1.1
~ wdg=2 bus=bus12 kV=20 kVA=25000 conn=Wye Numtaps=32 Mintap=0.9
Maxtap=1.1 Tap = 1.03125 !1.1-11*0.00625
```

In this model, there are two distribution transformers connected to the grid. It is considered that each transformer has $\pm10\%$ load changing taps, with each tap change corresponding to $\pm0.625\%$. However, in the transformer used here, the taps are fixed.

14.2.1.3 Creating Lines

The lines of the network are created as an object that connects two buses in the network. The line parameters are represented by the line codes, which can be defined separately in the script of OpenDSS. The code for creating line codes and lines in OpenDSS is given below. The data are modified as per our requirements.

```
// creating line codes for the lines
//considering single phase values with C
New LineCode.1 nphases=3 R1=0.501 X1=0.716 R0=0.817 X0=1.598
C1=151.24 C0=151.24 Units=km
New LineCode.2 nphases=3 R1=0.501 X1=0.366 R0=0.658 X0=1.611
C1=10.101 C0=4.0764 Units=km
// creating the lines for the network
New Line.LINE1 Bus1=bus1 Bus2=bus2 phases=3 LineCode=1 Length=2.82
Units=km
New Line.LINE2 Bus1=bus2 Bus2=bus3 phases=3 LineCode=1 Length=4.42
Units=km
New Line.LINE3 Bus1=bus3 Bus2=bus4 phases=3 LineCode=1 Length=0.61
Units=km
New Line.LINE4 Bus1=bus4 Bus2=bus5 phases=3 LineCode=1 Length=0.56
Units=km
New Line.LINE5 Bus1=bus5 Bus2=bus6 phases=3 LineCode=1 Length=1.54
Units=km
New Line.LINE6 Bus1=bus6 Bus2=bus7 phases=3 LineCode=1 Length=0.24
Units=km
New Line.LINE7 Bus1=bus7 Bus2=bus8 phases=3 LineCode=1 Length=1.67
Units=km
New Line.LINE8 Bus1=bus8 Bus2=bus9 phases=3 LineCode=1 Length=0.32
Units=km
New    Line.LINE9    Bus1=bus9    Bus2=bus10    phases=3    LineCode=1
Length=0.77 Units=km
New    Line.LINE10    Bus1=bus10    Bus2=bus11    phases=3    LineCode=1
Length=0.33 Units=km
New    Line.LINE11    Bus1=bus11    Bus2=bus4    phases=3    LineCode=1
Length=0.49 Units=km
New    Line.LINE12    Bus1=bus3    Bus2=bus8    phases=3    LineCode=1
Length=1.30 Units=km
New    Line.LINE13    Bus1=bus12    Bus2=bus13    phases=3    LineCode=2
Length=4.89 Units=km
New    Line.LINE14    Bus1=bus13    Bus2=bus14    phases=3    LineCode=2
Length=2.99 Units=km
New    Line.LINE15    Bus1=bus14    Bus2=bus8    phases=3    LineCode=2
Length=2.0 Units=km
```

14.2.1.4   Creating Loads

Loads can be defined by a specific piece of script in OpenDSS. The loads can be modelled as static loads or time-dependent loads. Also, they can also be defined as single-phase loads or three-phase loads. In the code below, the modelling of the load is done. In this study, the loads are modelled as static 3-phase loads.

```
//creating residential load for the networks
New Load.LOAD1 Phases=3 Bus1=bus1 kV=20 kva=15300 PF=0.98
New Load.LOAD2 Phases=3 Bus1=bus3 kV=20 kva=285 PF=0.97
New Load.LOAD3 Phases=3 Bus1=bus4 kV=20 kva=445 PF=0.97
New Load.LOAD4 Phases=3 Bus1=bus5 kV=20 kva=750 PF=0.97
New Load.LOAD5 Phases=3 Bus1=bus6 kV=20 kva=565 PF=0.97
New Load.LOAD6 Phases=3 Bus1=bus8 kV=20 kva=605 PF=0.97
New Load.LOAD7 Phases=3 Bus1=bus10 kV=20 kva=490 PF=0.97
New Load.LOAD8 Phases=3 Bus1=bus11 kV=20 kva=340 PF=0.97
New Load.LOAD9 Phases=3 Bus1=bus12 kV=20 kva=15300 PF=0.98
New Load.LOAD10 Phases=3 Bus1=bus14 kV=20 kva=215 PF=0.97

//creating commercial load for the networks
New Load.LOAD11 Phases=3 Bus1=bus1 kV=20 kva=5100 PF=0.95
New Load.LOAD12 Phases=3 Bus1=bus3 kV=20 kva=265 PF=0.85
New Load.LOAD13 Phases=3 Bus1=bus7 kV=20 kva=90 PF=0.85
New Load.LOAD14 Phases=3 Bus1=bus9 kV=20 kva=675 PF=0.85
New Load.LOAD15 Phases=3 Bus1=bus10 kV=20 kva=80 PF=0.85
New Load.LOAD16 Phases=3 Bus1=bus12 kV=20 kva=5280 PF=0.95
New Load.LOAD17 Phases=3 Bus1=bus13 kV=20 kva=40 PF=0.85
New Load.LOAD18 Phases=3 Bus1=bus14 kV=20 kva=390 PF=0.85
```

14.2.1.5   Creating PV Systems

PV systems in OpenDSS can be modelled in several ways. In some studies, a user-defined control application is needed. In such cases, the user may model the PV system as a synchronous generator. However, in this study, the PV systems are modelled as defined in the manual of OpenDSS. The code below includes the definition of the PV systems considered in this study. The required number of PVs is created as per the simulation studies considered.

```
// creating PV in the network
// P-T curve is per unit of rated Pmpp vs temperature
// This one is for a Pmpp stated at 25 deg
New XYCurve.MyPvsT npts=4 xarray=[0 25 75 100] yarray=[1.2 1.0 0.8
0.6]
// efficiency curve is per unit eff vs per unit power
New XYCurve.MyEff npts=4 xarray=[.1 .2 .4 1.0] yarray=[.86 .9 .93
.97]
// per unit irradiance curve (per unit if "irradiance"
property)
```

```
New Loadshape.MyIrrad npts=24 interval=1
~mult=[0 0 0 0 0 .1 .2 .3 .5 .8 .9 1.0 1.0 .99 .9 .7 .4 .1 0 0 0 0]
// 24-hr temp shape curve
New Tshape.MyTemp npts=24 interval=1
~temp=[25, 25, 25, 25, 25, 25, 25, 25, 35, 40, 45, 50 60 60 55 40 35 30
25 25 25 25 25 25]

// pv definition at bus 3 of size 25 kW
New PVSystem.PV1 phases=3 bus1=bus3 kV=20 kVA=25 irrad=1
~Pmpp=20 temperature=25 PF=1 %cutin=0.1 %cutout=0.1
~ effcurve=Myeff P-TCurve=MyPvsT Daily=MyIrrad TDaily=MyTemp
// pv definition at bus 4 of size 25 kW
New PVSystem.PV2 phases=3 bus1=bus4 kV=20 kVA=25 irrad=1
~Pmpp=20 temperature=25 PF=1 %cutin=0.1 %cutout=0.1
~ effcurve=Myeff P-TCurve=MyPvsT Daily=MyIrrad TDaily=MyTemp
// pv definition at bus 5 of size 30 kW
New PVSystem.PV3 phases=3 bus1=bus5 kV=20 kVA=30 irrad=1
~Pmpp=30 temperature=25 PF=1 %cutin=0.1 %cutout=0.1
~ effcurve=Myeff P-TCurve=MyPvsT Daily=MyIrrad TDaily=MyTemp
// pv definition at bus 6 of size 30 kW
New PVSystem.PV4 phases=3 bus1=bus6 kV=20 kVA=30 irrad=1
~Pmpp=30 temperature=25 PF=1 %cutin=0.1 %cutout=0.1
~ effcurve=Myeff P-TCurve=MyPvsT Daily=MyIrrad TDaily=MyTemp
// pv definition at bus 8 of size 30 kW
New PVSystem.PV5 phases=3 bus1=bus8 kV=20 kVA=30 irrad=1
~Pmpp=30 temperature=25 PF=1 %cutin=0.1 %cutout=0.1
~ effcurve=Myeff P-TCurve=MyPvsT Daily=MyIrrad TDaily=MyTemp
// pv definition at bus 9 of size 30 kW
New PVSystem.PV6 phases=3 bus1=bus9 kV=20 kVA=30 irrad=1
~Pmpp=30 temperature=25 PF=1 %cutin=0.1 %cutout=0.1
~ effcurve=Myeff P-TCurve=MyPvsT Daily=MyIrrad TDaily=MyTemp
// pv definition at bus 10 of size 40 kW
New PVSystem.PV7 phases=3 bus1=bus10 kV=20 kVA=40 irrad=1
~Pmpp=40 temperature=25 PF=1 %cutin-0.1 %cutout-0.1
~ effcurve=Myeff P-TCurve=MyPvsT Daily=MyIrrad TDaily=MyTemp
// pv definition at bus 11 of size 10 kW
New PVSystem.PV8 phases=3 bus1=bus11 kV=20 kVA=10 irrad=1
~Pmpp=10 temperature=25 PF=1 %cutin=0.1 %cutout=0.1
~ effcurve=Myeff P-TCurve=MyPvsT Daily=MyIrrad TDaily=MyTemp
// this PV is the one that replaces wind turbine of size 1500 KW in
original paper
// pv definition at bus 7 of size 1500 kW
New PVSystem.PV9 phases=3 bus1=bus7 kV=20 kVA=1500 irrad=1
~Pmpp=1500 temperature=25 PF=1 %cutin=0.1 %cutout=0.1
~ effcurve=Myeff P-TCurve=MyPvsT Daily=MyIrrad TDaily=MyTemp
```

14.2.2 Creating Typhoon HIL Schematic and SCADA

The important part of real-time co-simulation using OpenDSS and Typhoon HIL is the creation of the Typhoon HIL schematic and the SCADA. In this subsection, detailed steps to create a schematic and SCADA are presented.

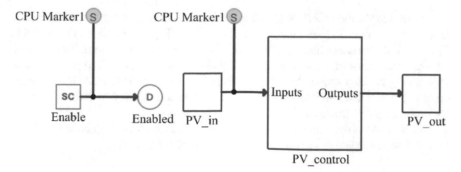

Fig. 14.3 Schematic diagram presenting the control of a PV distributed generator

14.2.2.1 Creation of Typhoon HIL Schematic

Typhoon HIL schematic editor is a graphical user interface-based modelling system. The model in the schematic works as a bridge between the Typhoon HIL and the OpenDSS simulator.

Variables are controlled in real-time in the Typhoon HIL SCADA. An appropriate definition of the controlled variables is performed in the Typhoon HIL Schematic. Finally, the controlled variables are passed to OpenDSS by using the Python API in the HIL SCADA. This process continues with the width of the simulation. Figure 14.3 shows a schematic to communicate between Typhoon HIL and OpenDSS. In the figure, the inputs and the outputs blocks are the signal processing block to receive and send the signal.

14.2.2.2 Creation of Typhoon HIL SCADA

Typhoon HIL SCADA is essential in realising the real-time co-simulation between Typhoon HIL and OpenDSS. SCADA contains different panels to initialise co-simulation, control, and monitor expected outputs. The beautiful part of designing SCADA is that the Python program can be utilised effectively. Three important factors to be considered while designing the Typhoon HIL SCADA for co-simulation are listed below.

- Initialising Co-Simulation framework between Typhoon HIL and OpenDSS
- Development of panel to execute real-time control
- Development of a panel for monitoring the output in a real-time framework.

Initialising Co-simulation Framework Between Typhoon HIL and OpenDSS

To initiate the co-simulation between Typhoon HIL and OpenDSS, first, the OpenDSS shared library is interfaced using the Python library. OpenDSSDirect.py is the python interpreter to interact with Python from Typhoon HIL. OpenDSSDirect.py is imported in Typhoon HIL SCADA to execute the desired work. More details on how to import Python-based modules to Typhoon HIL python core can be found in Typhoon HIL operating manual (Hil 2022). Second, a program to communicate with the OpenDSS on startup and during the real-time simulation needs to be scripted in Python. Figure 14.4 shows the panel developed in Typhoon HIL SCADA to initiate co-simulation. Different macro widgets can be placed in a panel to define programs for initialising co-simulation.

To initialise the real-time co-simulation framework, the following code is scripted to initialise the OpenDSS parameters in Typhoon HIL SCADA.

```
# NOTE: Variables and functions defined here will be
# Available for use in all Macro and Expression scripts.
# NOTE: This code is always executed prior simulation start.

# Variable 'SETTINGS_DIR' holds directory where loaded Panel
.cus file is located.
# Also you can call 'get_settings_dir_path()' function in any
# Macro and Expression scripts to get the same directory.
SETTINGS_DIR = get_settings_dir_path()

# The 'add_to_python_path(folder)' function can be used to add
custom folder
# with Python files and packages to the PYTHONPATH. After
folder is added, all Python
# files and Python packages from it can be imported into the
SCADA Namespace.

# HIL API is imported as 'hil'
```

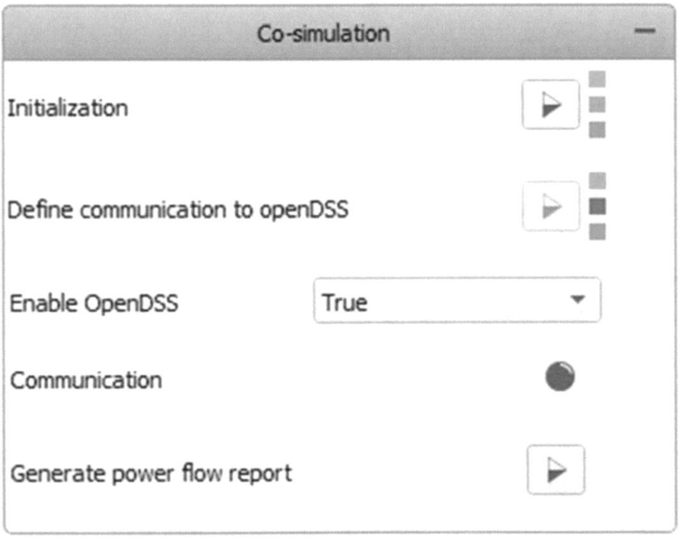

Fig. 14.4 Co-simulation setup panel

```
# SCADA API is imported as 'panel'
# SCADA API constants are imported as 'api_const'
# Numpy module is imported as 'np'
# Scipy module is imported as 'sp'
# Schematic Editor model namespace is imported as 'scm'
# Function for printing to HIL SCADA Message log is imported
as 'printf'.


import os
from os import path
import sys
import numpy as np
import pandas as pd
import random
import math

if not sys.platform.startswith("win"):
raise Exception('You are on Linux! In order to run this model you need
to manually instal opendssdirect package.')
SW_VERS = hil.get_sw_version()

#path for the Python library
sendto_dir = path.expandvars(r"%APPDATA%\typhoon\{}\python_
portables\python3_portable\Lib\site-packages".format(SW_VERS))
add_to_python_path(os.path.normpath(sendto_dir))

#import OpenDSS library
import opendssdirect as dss
dss.utils.run_command("clear")

#path for the OpenDSS model
path = SETTINGS_DIR +"\European_MV.dss"

#compile dss model
dss.utils.run_command("Compile \"{}\"".format(path))

#initialization parameters from OpenDSS
enable = 0
# NOTE: The code specified in this handler will be executed on
simulation start.

# NOTE: Variables specified here will be available in other
handlers.

# HIL API is imported as 'hil'

global powers1
global powers2
global powers3
global powers4
global powers5
global powers6
global powers7
global powers8
global powers9
```

```
global busV1
global busV2
global busV3
global busV4
global busV5
global busV6
global busV7
global busV8
global busV9
global Q_PV1
global Totalloss
global Activeloss
global Reactiveloss
# declaring initial values to the global variable
powers1=0
powers2=0
powers3=0
powers4=0
powers5=0
powers6=0
powers7=0
powers8=0
powers9=0
Reactiveloss=0
Activeloss=0
activepower=0
reactivepower=0

phases = ["A", "B", "C"]
Q_Power=["Q_PV"]
dss.Solution.SolveSnap()

# Assign voltages of buses where PV are installed to show in
SCADA panel
dss.Circuit.SetActiveBus("bus3")
busV1 = dss.Bus.VMagAngle()
dss.Circuit.SetActiveBus("bus4")
busV2 = dss.Bus.VMagAngle()
dss.Circuit.SetActiveBus("bus5")
busV3 = dss.Bus.VMagAngle()
dss.Circuit.SetActiveBus("bus6")
busV4 = dss.Bus.VMagAngle()
dss.Circuit.SetActiveBus("bus8")
busV5 = dss.Bus.VMagAngle()
dss.Circuit.SetActiveBus("bus9")
busV6 = dss.Bus.VMagAngle()
dss.Circuit.SetActiveBus("bus10")
busV8 = dss.Bus.VMagAngle()
dss.Circuit.SetActiveBus("bus11")
busV9 = dss.Bus.VMagAngle()
dss.Circuit.SetActiveBus("bus7")
busV7 = dss.Bus.VMagAngle()
```

```
# NOTE: The code specified in this handler will be executed on
timer event.
# HIL API is imported as 'hil'

x = bool(hil.read_digital_signal(name = "enabled"))
if x:
    # Sets new active and reactive power levels in OpenDSS for
all PVsystems in network
    powers1 = hil.read_analog_signal("PVcontrol.Q")
    powers2 = hil.read_analog_signal("PVcontrol1.Q")
    powers3 = hil.read_analog_signal("PVcontrol2.Q")
    powers4 = hil.read_analog_signal("PVcontrol3.Q")
    powers5 = hil.read_analog_signal("PVcontrol4.Q")
    powers6 = hil.read_analog_signal("PVcontrol5.Q")
    powers7 = hil.read_analog_signal("PVcontrol6.Q")
    powers8 = hil.read_analog_signal("PVcontrol7.Q")
    powers9 = hil.read_analog_signal("PVcontrol8.Q")

        #Edit the reactive powers of PVs in OpenDSS model in real
time
    dss.PVsystems.Name("PV1")
    dss.PVsystems.kvar(powers1)
    #dss.PVsystems.kvar(Q_PV1)
    dss.PVsystems.Name("PV2")
    dss.PVsystems.kvar(powers2)
    dss.PVsystems.Name("PV3")
    dss.PVsystems.kvar(powers3)
    dss.PVsystems.Name("PV4")
    dss.PVsystems.kvar(powers4)
    dss.PVsystems.Name("PV5")
    dss.PVsystems.kvar(powers5)
    dss.PVsystems.Name("PV6")
    dss.PVsystems.kvar(powers6)
    dss.PVsystems.Name("PV7")
    dss.PVsystems.kvar(powers7)
    dss.PVsystems.Name("PV8")
    dss.PVsystems.kvar(powers8)
    dss.PVsystems.Name("PV9")
    dss.PVsystems.kvar(powers9)

    #Solves a snapshot of the powerflow
    dss.Solution.SolveSnap()

    if dss.Solution.Converged():
            # Selects the bus where the inverter is installed and
collect voltages
        Totalloss=dss.Circuit.Losses()
        Activeloss=Totalloss[0]
        Reactiveloss=Totalloss[1]

        #dss.Circuit.SetActiveElement("Vsource.SOURCE")
        totalpower = dss.Circuit.TotalPower()
        activepower = totalpower[0]
        reactivepower = totalpower[1]
```

```
# Assign voltages of buses where PV are installed to show
in SCADA panel
    dss.Circuit.SetActiveBus("bus3")
    busV1 = dss.Bus.VMagAngle()
    dss.Circuit.SetActiveBus("bus4")
    busV2 = dss.Bus.VMagAngle()
    dss.Circuit.SetActiveBus("bus5")
    busV3 = dss.Bus.VMagAngle()
    dss.Circuit.SetActiveBus("bus6")
    busV4 = dss.Bus.VMagAngle()
    dss.Circuit.SetActiveBus("bus8")
    busV5 = dss.Bus.VMagAngle()
    dss.Circuit.SetActiveBus("bus9")
    busV6 = dss.Bus.VMagAngle()
    dss.Circuit.SetActiveBus("bus10")
    busV8 = dss.Bus.VMagAngle()
    dss.Circuit.SetActiveBus("bus11")
    busV9 = dss.Bus.VMagAngle()
    dss.Circuit.SetActiveBus("bus7")
    busV7 = dss.Bus.VMagAngle()
```

Development of Panel to Execute Real-Time Control

To run the co-simulation of Typhoon HIL and OpenDSS in real-time, real-time signals must be given from the SCADA to the OpenDSS. For this purpose, different panels with widgets are designed. Figure 14.5 shows the widget setting for the slider to change the reactive power of PV in real-time. The signal processing component available in the Typhoon HIL library is used to process the signal between the cyber layer and the physical layer,

Development of Panel to Execute Real-Time Control

To monitor the output variables in SCADA in real-time, a panel needs to be designed. An appropriate widget to show the desired signal is selected from the library. In the widget, a program is written to gather an appropriate signal to be monitored. Figure 14.6 shows the setting of the widget to monitor the reactive power injection from SCADA in real-time. The number of this widget in SCADA depends on the number of real-time controllable PVs in the network. Similarly, Fig. 14.7 shows the program written in the widget to monitor the voltage at the bus where PV is connected. The number of this widget depends on how many bus voltages we need to monitor.

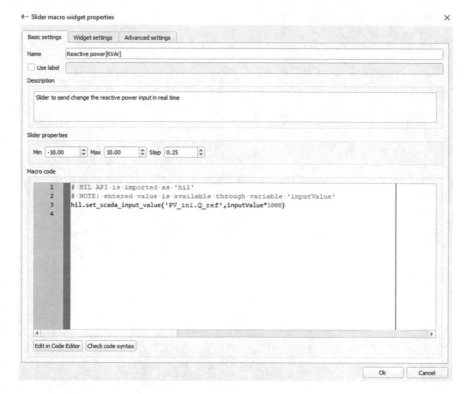

Fig. 14.5 Widget for changing reactive power of PV in real-time

14.3 Implementation of Typhoon HIL and OpenDSS Co-simulation Framework

The overall process of implementing the co-simulation between Typhoon HIL and OpenDSS is shown in Fig. 14.8. First, the test distribution system is modelled in OpenDSS. The PVs are placed on the distribution network in the OpenDSS model. The OpenDSS modules are executed through a Python interface. A Python program is written inside the SCADA of the Typhoon HIL to interact with the OpenDSS. In a schematic editor on Typhoon HIL, a communication interface between the SCADA and the Typhoon HIL real-time simulator is modelled. This model in the schematic editor can interact with the SCADA and the Typhoon HIL Real-Time Simulator. The SCADA of Typhoon HIL consists of a python program to get the signals from OpenDSS, process the signal, and display the real-time outputs. The SCADA also consists of different sliders for sending the real-time signal to the OpenDSS. At each change, the signal is fed to the OpenDSS, the load flow is executed inside the OpenDSS, and the outputs are fed into the Typhoon HIL.

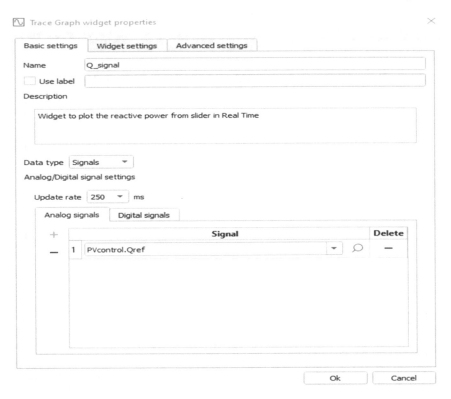

Fig. 14.6 Widget to display reactive power of PV in real-time

14.4 Simulation Results

The research was carried out in the Digital Energy Systems Laboratory (DIgEnSys-Lab). The DIgEnSys-Lab has physical equipment for real-time monitoring and control (see https://fglongattlab.fglongatt.org for further information). Each part of the simulation study is described in the following subsection. Typhoon HIL 604 is used to model the cybernetic and physical layers in this paper. The European medium voltage distribution network produced by CIGRE Task Force C6.04 in their publication "Benchmark Systems for Network Integration of Renewable and Distribution Energy Resources." It is assumed that the network is symmetrical and balanced. As illustrated in Fig. 14.9, the test system comprises two typical 20 kV, 50 Hz, three-phase feeders named feeder 1 and feeder 2. By turning on or off the switches S1, S2, and S3, the feeder can be operated in a radial or meshed topology. In this analysis, all the switches are assumed to be closed.

The wind source considered in the original study is replaced with a PV of the same size to test the effectiveness of real-time reactive power control with smart inverters of PVs in this situation. The load and other network information are kept the same as in the original study. Table 14.1 shows the ratings of PVs considered in this study.

Fig. 14.7 Widget to display the voltage of bus in real time

The screenshot of the SCADA of the proposed cyber-physical co-simulation framework is shown in Fig. 14.10. The SCADA consists of sliders to change the reactive power in real-time. The bus voltages in the network are continuously monitored during the real-time simulation. Digital and graphical displays can be inserted into the system as per the requirement of the observation.

To demonstrate the successful operation of the proposed cyber-physical co-simulation framework, Fig. 14.11 shows the reactive power profile of PVs that are changed in real-time from the SCADA. In this case, the observation is made with only a change in reactive power from the PV. However, a similar observation can also be made for other controllable variables. For the change of the reactive power of the PV, Fig. 14.12 shows the corresponding voltage profile of the network. The

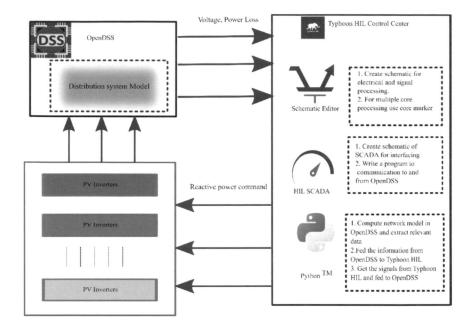

Fig. 14.8 Overall frameworks for co-simulation between Typhoon HIL and OpenDSS

voltage profile of the network changes in real-time for the real-time change in the
reactive power.

14.5 Discussion and Conclusion

This chapter proposed and demonstrated a cyber-physical co-simulation framework
between Typhoon HIL and OpenDSS.

It can be easily customised and enhanced by the reader; as a consequence, it opens
the door for new research approaches to real-time control and monitoring studies for
distribution networks.

This research can include real-time optimisation of the distribution network with
numerous PVs to determine the appropriate reactive power requirements for smart
inverters and compensate for voltage fluctuations caused by loads and PV generation
changes.

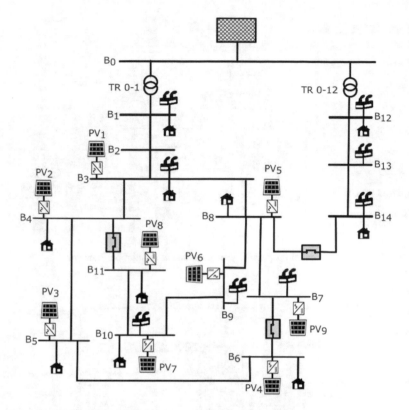

Fig. 14.9 Test system: Modified CIGRE medium voltage distribution system (Cigre 2014)

Table 14.1 MV distribution network benchmark application: parameters of PV units (Cigre 2014)	Bus number	Type of DER	$P_{max}(kW)$
	3	PV	20
	4	PV	20
	5	PV	30
	6	PV	30
	7	PV	1500
	8	PV	30
	9	PV	30
	10	PV	40
	11	PV	10

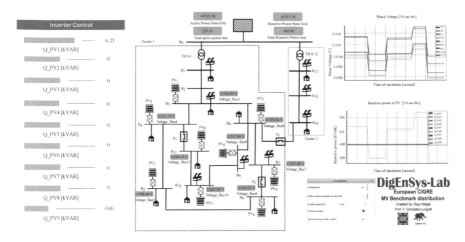

Fig. 14.10 Output window of proposed Cyber-physical co-simulation framework

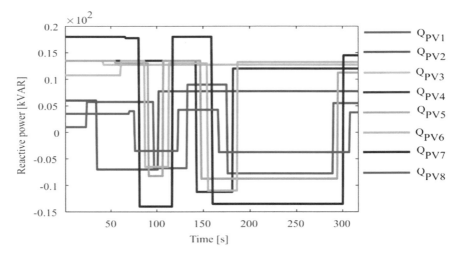

Fig. 14.11 Reactive power profile of PV obtained by dynamically changing the reactive power input

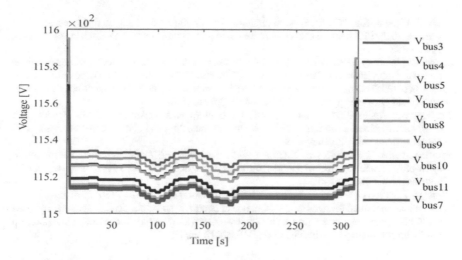

Fig. 14.12 Voltage profile of CIGRE network obtained from Co-simulation framework

Acknowledgements The authors are very grateful to the Arctic Centre for Sustainable Energy (ARC) (project number 740108), UiT The Arctic University of Norway, Norway, for providing an opportunity for Mr Raju to visit and work in DIgEnSys-Lab. Authors and especially Prof F. Gonzalez-Longatt, acknowledge the technical support provided by the teams of Typhoon HIL and EPRI.

References

Acosta MN, Gonzalez-Longatt F, Andrade MA, Torres JR (2021) Optimal reactive power control of smart inverters: vestfold and Telemark regional network. In: 2021 IEEE Madrid PowerTech, pp 1–6

Cao G et al (2019) Real-time cyber–physical system co-simulation testbed for microgrids control. IET Cyber-Phys Syst Theory Appl 4(1):38–45

Cigre (2014) Benchmark systems for network integration of renewable and distributed energy resources

HIL T (2022) Typhoon HIL manual. typhoon-hil, 2022. [Online]. Available https://www.typhoon-hil.com/documentation/

Krishnamurthy D (2017) OpenDSSDirect.py. dss-extensions.org, 2017. [Online]. Available https://dss-extensions.org/OpenDSSDirect.py/notebooks/Example-OpenDSSDirect.py.html

Krechel T, Sanchez F, Gonzalez-Longatt F, Chamorro HR, Rueda JL (2019) A transmission system friendly micro-grid: optimising active power losses. In: 2019 IEEE Milan PowerTech, pp 1–6

Schloegl F, Rohjans S, Lehnhoff S, Velasquez J, Steinbrink C, Palensky P (2015) Towards a classification scheme for co-simulation approaches in energy systems. Proceedings—2015 international symposium smart electrical distribution system technology EDST 2015, pp 516–521

Steinbrink C, Schlogl F, Babazadeh D, Lehnhoff S, Rohjans S, Narayan A (2018) Future perspectives of co-simulation in the smart grid domain. 2018 IEEE international energy conference energycon 2018, pp 1–6

Sun CC, Hong J, Liu CC (2015) A co-simulation environment for integrated cyber and power systems. 2015 IEEE international conference smart grid communication. SmartGridComm 2015, pp 133–138

Sunderman W, Dugan RC, Smith J (2014) Open source modeling of advanced inverter functions for solar photovoltaic installations. In: 2014 IEEE PES T&D conference and exposition, pp 1–5

Tang Y et al (2017) A hardware-in-the-loop based co-simulation platform of cyber-physical power systems for wide area protection applications. Appl Sci 7(12)

van der Meer AA et al (2017) Cyber-physical energy systems modeling, test specification, and co-simulation based testing. In: 2017 Workshop on modeling and simulation of cyber-physical energy systems (MSCPES), pp 1–9

Venkataramanan V, Srivastava A, Hahn A (2016) Real-time co-simulation testbed for microgrid cyber-physical analysis. In: 2016 Workshop on modeling and simulation of cyber-physical energy systems (MSCPES), pp 1–6

Wagle R, Tricarico G, Sharma P, Sharma C, Rueda JL, Gonzalez-Lonzatt F (2022) Cyber-physical co-simulation testbed for real-time reactive power control in smart distribution network. In: 2022 IEEE PES innovative smart grid technologies—Asia (ISGT Asia), Singapore, Singapore, pp 11–15. https://doi.org/10.1109/ISGTAsia54193.2022.10003553

Printed in the United States
by Baker & Taylor Publisher Services